AF291231

Lightweight Concrete

Lightweight Concrete
Properties, Design And Applications

Edited by

Ahmad Mousa

and

John W. Bull

First published in 2026 by Porto Professional

All rights reserved. No part of this publication may be reproduced, stored in a retrieval system, or transmitted in any form or by any means, electronic, mechanical, photocopying, recording or otherwise, without prior permission in writing from the Publisher.

The Editor has made every effort to ensure the accuracy of information contained in this publication, but assume no responsibility for any errors, inaccuracies, inconsistencies and omissions. Likewise, every effort has been made to contact copyright holders. If any copyright material has been reproduced unwittingly and without permission the Publisher will gladly receive information enabling them to rectify any error or omission in subsequent editions.

Copyright © 2026 Ahmed Mousa, Mohab Hussein, Jose Alexandre Bogas, Kazi M. A. Sohel, Pengpeng Ni, Kaiwen Liu, Qian Su, Payam Shafigh, Muhammad Aslam, Lina María Chica Osorio and Albert Leonard Alzate Ramírez

British Library Cataloguing in Publication Data
A CIP record for this book is available from the British Library

Print ISBN: 9781849955676
PDF ISBN: 9781849956741
EPUB ISBN: 9781849956765

The rights of Ahmed Mousa, Mohab Hussein, Jose Alexandre Bogas, Kazi M. A. Sohel, Pengpeng Ni, Kaiwen Liu, Qian Su, Payam Shafigh, Muhammad Aslam, Lina María Chica Osorio and Albert Leonard Alzate Ramírezto be identified as the authors of this work has been asserted by them in accordance with the Copyright, Design and Patents Act 1988.

Cover design by Out of House
Typesetting by Riverside Publishing Solutions
Printed and bound in Great Britain by CPI

To order please go to our website www.portopress.com or contact our distributor, BookSource, 50 Cambuslang Road, Clydesmill Industrial Estate, Glasgow G32 8NB. Telephone 0141 642 9192

Porto Press Ltd
3 Connaught Road
St Albans
AL3 5RX

www.portopress.com

Paper from responsible sources

The publisher and authors have used their best efforts in preparing this book, but assume no responsibility for any injury and/or damage to persons or property from the use or implementation of any methods, instructions, ideas or materials contained within this book. All operations should be undertaken in accordance with existing legislation, recognized codes and standards and trade practice. Whilst the information and advice in this book is believed to be true and accurate at the time of going to press, the authors and publisher accept no legal responsibility or liability for errors or omissions that may have been made.

Contents

List of Contributors

Editors

Dr Ahmad Mousa: Assistant Professor in the Civil Engineering Department at the University of Nottingham Ningbo China. He holds a Ph.D. in Civil Engineering from the Georgia Institute of Technology and an M.Sc. from Purdue University. He is a licensed Professional Engineer in the State of California. Dr. Mousa's research interests are broad, with significant contributions to construction materials and sustainable building practices. He has published extensively in high-impact journals. His work on lightweight concrete and sustainable building materials is widely cited. Prior to his current roles, he served as a Senior Lecturer at Monash University Malaysia from 2014 to 2022.

Professor John Bull: Head of Civil Engineering at Northumbria University.

Contributors

Muhammad Aslam: Department of Civil Engineering, School of Engineering & Technology, Institute of Southern Punjab, 60000 Multan, Pakistan.

José Alexandre Bogas: Associate Professor with Habilitation at the DECivil, Instituto Superior Técnico (IST), Universidade de Lisboa (UL). His track-record involves 28 years of experience in teaching, structural design, research and consultancy in civil engineering. As a researcher at the CERIS research Unit, hosted by IST/UL, his expertise lies in the behaviour of construction materials, particularly in the following domains: cement-based materials; special concretes (high-performance, lightweight, low-carbon, recycled, self-compacting, fibre-reinforced, and nanoreinforced concretes); durability; and service life prediction. He is the author/co-author of over 80 papers in international journals (>80% Q1 WoS/ISI, >4500 citations, h-index 38), 5 book chapters, 1 international patent, and 83 conference communications; supervisor of 5 concluded PhD and over 50 MSc; principal coordinator of 3 competitive R&D projects funded with 0.7 M€ and member of 12 research projects funded through competitive calls with around 130 M€.

Mohab A. Hussein: Section Chief, Division of Bridge Engineering & Infrastructure Management, New Jersey Department of Transportation.

Kaiwen Liu: Key Laboratory of High-Speed Railway Engineering of the Ministry of Education, Southwest Jiaotong University, Chengdu, 610031, China.

Mohamed Mahgoub: Associate Professor, New Jersey Institute of Technology, New Jersey, USA.

Pengpeng Ni: Obtained his BEng degree in Civil Engineering from Northeastern University, China, in 2010, his MSc degree in Earthquake Engineering from Université Joseph Fourier – Grenoble 1, France, and Istituto Universitario di Studi Superiori di Pavia, Italy, through the Erasmus Mundus Master's Programme in 2012, and his PhD degree in Geotechncial Engineering from Queen's University, Canada, in 2016. He worked at Nanyang Technological University, Singapore, as a Research Fellow from 2017 to 2019. He returned back to China in 2019, and now is a Professor at Sun Yat-sen University. He has co-authored over 100 publications, and received the Tso Kung Hsieh Award from the Institution of Civil Engineers, UK (Best Paper on Structural and Soil Vibration in *Géotechnique*). He serves as an Associate Editor for *Geotechnical and Geological Engineering* and *International Journal of Geotechnical Engineering*, and an Editorial board Member for Geotextiles and Geomembranes. His research interests include soil-structure interaction, pipeline and trenchless technology, tunnelling, and foundation engineering, etc.

Lina María Chica Osorio: (Medellín, Antioquia, 1984) is a Colombian Mining and Metallurgical engineer with a Ph.D. on Engineering - Materials Science and Technology from Universidad Nacional de Colombia. Associated professor at Universidad de Medellin, Colombia, with 12 years of academic experience. Main topics of research are related to construction and building materials design and performance.

Albert Leonard Alzate Ramírez: (Rionegro, Antioquia, 1976) is a Colombian civil engineer with a Ph.D. in Advanced Structural Design from the Polytechnic University of Madrid. His areas of expertise include structural analysis and design, structural pathology and strengthening with fiber-reinforced polymers (FRP). Currently, he is a full-time professor at the Surcolombiana University in Colombia.

Payam Shafig: Department of Architecture, Faculty of Built Environment, University of Malaya, 50603 Kuala Lumpur, Malaysia.

Dr. Kazi Md Abu Sohel: Associate Professor in the Department of Civil and Architectural Engineering at Sultan Qaboos University, Oman. Dr. Sohel has been engaged in teaching and research for over 20 years at different universities in Singapore, Bangladesh, Bahrain and Oman.

Qian Su: Key Laboratory of High-Speed Railway Engineering of the Ministry of Education, Southwest Jiaotong University, Chengdu, 610031, China.

Preface

Lightweight concrete has transitioned from a specialized application to a standard material in modern sustainable construction. Its unique combination of reduced density, improved thermal insulation, and structural efficiency has made it an increasingly attractive choice for a wide range of applications—from high-rise buildings and long-span bridges to insulation systems and urban infill. The growing global emphasis on energy efficiency, carbon reduction, and material optimization has further propelled lightweight concrete to the forefront of innovative building material research and practice.

This book, *Lightweight Concrete: Properties, Design and Applications,* aims to present a comprehensive and contemporary overview of selected aspects of this special concrete. It brings together contributions from researchers and practitioners across the globe, offering a blend of theoretical insight, practical guidance, and perspectives on the development and use of lightweight concrete.

The book covers a diverse spectrum of topics, beginning with an evaluation of construction market drivers and the future potential of lightweight concrete. Subsequent chapters delve into specific technical areas, including mix design for structural and non-structural applications, the behavior of concrete containing novel lightweight aggregates such as expanded polystyrene, and the performance of specialized types, such as self-compacting and cellular lightweight concrete. Several contributions focus on advanced mechanical and dynamic properties—including flexural fatigue and response under high-speed loading—highlighting the material's suitability for demanding infrastructure applications.

The international scope of the work reflects the global nature of both the challenges and the innovations in lightweight concrete technology. We hope that this book will serve as a valuable resource for civil engineers, architects, contractors, and advanced students who seek to understand, design with, and advance the use of lightweight concrete.

We extend our sincere gratitude to all the contributing authors for sharing their expertise and for their commitment to this project. Our thanks also go to the reviewers for their constructive feedback, and to the team at Porto Press for their professional support and guidance throughout the publication process. We are indebted to the late Professor John W. Bull, who initiated this project and whose vision guided its early development.

Finally, we acknowledge that the field continues to evolve rapidly. While every effort has been made to ensure the accuracy and relevance of the content, we welcome dialogue and corrections from our readers as part of the ongoing collective effort to refine and expand the knowledge about this versatile building material.

Ahmad Mousa and John W. Bull

Chapter 1
The merits and future of lightweight concrete: Insights from the US construction market

Ahmad Mousa, Mohab Hussein and Mohamed Mahgoub

1. Introduction

The requirements and prerequisites of global sustainability have been shaping modern construction in the last two decades (Ries and Holm, 2004; Ries *et al.*, 2010; Van Vliet *et al.*, 2012). The need to use more effective construction materials and repairs has been subsequently triggered by an increasing demand for upgrades to the aging infrastructure worldwide. The adaptability and sustainability of lightweight concrete (LWC) provide a range of technical qualities that allows it to outperform many conventional construction materials (Ries *et al.*, 2010; Aslam *et al.*, 2016; Roberz *et al.*, 2017). LWC is commonly used for precast and cast-in-place panels, bridge decks, slabs-on-grade, elevated building decks, and prestressed or post-tensioned columns and facade elements (Holm and Ries, 2007). Nonetheless, the presence of LWC in construction does not seem to measure up to its potential. To this end, the question arises: How well do we appreciate the merits of LWC in construction? This chapter will aim to answer this critical question by studying the presence of LWC in the ready mix concrete market – with emphasis of the USA.

LWC was used for the first time in the USA during the First World War to build the warship SS *Selma* (Holm *et al.*, 1984; Clarke, 1993). The key historical developments of LWC in the USA are illustrated in Figure 1.1. LWC is still considered unpopular in the USA compared to normal weight concrete (NWC). This is so, despite its irrefutable esthetic merits, structural versatility and sustainable benefits (NRMCA, 2003). There is a need to identify the root causes and key challenges behind this seemingly stagnant or even declining interest in LWC in the USA. A structured survey was undertaken to estimate the receptiveness and perception of the construction market to LWC. Based on industry indicators, a situational analysis was conducted to assess further the current state-of-the-practice and to provide a realistic outlook for the future use of LWC in construction. In this chapter a value engineering analysis on LWC is presented in an attempt to provide owners and end-users with a systematic decision-making tool to estimate the cost of alternatives. Based on the input from the market and from selected experts, the authors finally propose the use of the Kotter model to seek a paradigm change in the current standing of LWC in the American construction market.

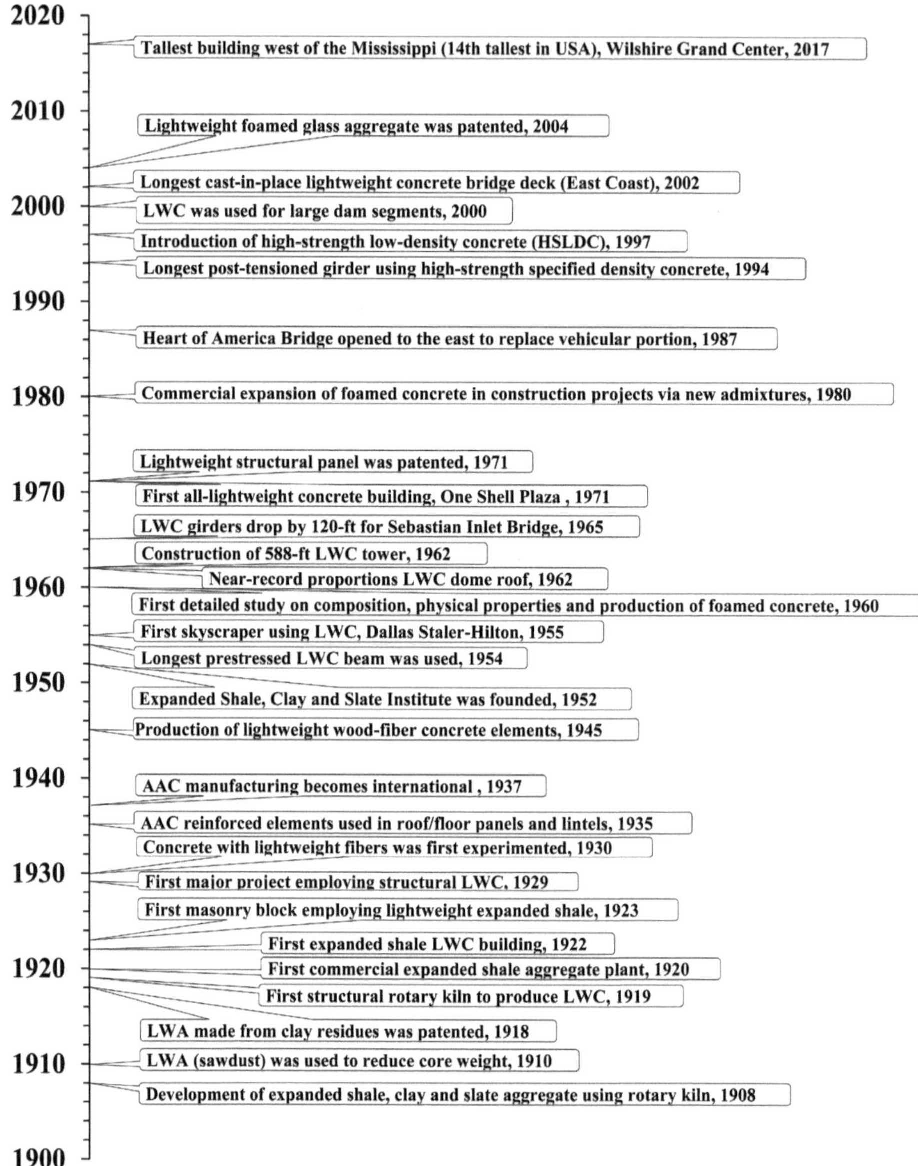

Figure 1.1 *The historical evolution of the lightweight concrete industry in the USA.*

2. Characteristics and merits

Unlike NWC, LWC provides a wide array of structural characteristics and architectural solutions, that can be previewed from the sustainability perspective (World Commission on Environment and Development, 1989; Alhaddi, 2015). Indirect benefits of using LWC include transportation, labor, time, energy, environmental impact, economical building, and durability. It is our intention in this chapter to provide an overview of the major sustainability gains of LWC that have been repeatedly reported in the literature (e.g. Jin *et al.*, 2015). It should be noted that the figures for savings, properties and sustainability measures presented herein could be source-dependent. Additionally, there is a wide range of LWA that come from different sources and have various characteristics. Therefore, proper utilization of the native properties of the LWA, combined with adequate concrete mix design, is necessary to maximize the benefits of LWC. In view of this, the research findings may not apply across the board for all LWC types in all situations, especially when poor judgement in design or application is exercised, inappropriate LWA is used, or if lack of practical experience is evident. As such, pertinent references should be visited for specifics and further details.

2.1 Material savings

The in-place density of LWC can vary widely between 800 kg/m^3 and 1750 kg/m^3 (50 pcf and 110 pcf), which obviously reduces the dead load. *ACI 213* defines structural-grade lightweight concrete as that having a dry density of between 1450 kg/m^3 and 1850 kg/m^3 (90 pcf and 115 pcf) (Akers *et al.*, 2014). The lighter weight of structural LWC provides an economical structural solution that does not compromise strength, i.e. smaller columns and foundations with reduced rebar requirements (Holm and Ries, 2007; Kim *et al.*, 2012). Despite the average weight reduction of 25–35%, the structural capacity of LWC is still comparable to NWC (Harmon, 2007). In fact, LWC can offer design compressive strengths of up to 83 MPa (12,000 psi) (Kahn *et al.*, 2004). Figure 1.2 shows typical densities for different types of LWC as compared to NWC (Holm and Ries, 2007). The following definitions can be used for the different types of concrete:

Normal weight concrete – concrete having a density of between 2250 kg/m^3 and 2500 kg/m^3 (140 pcf and 155 pcf) made with ordinary aggregates (sand, gravel, crushed stone).

Specified density concrete (SDC) – structural concrete having a specified equilibrium density of between 800 kg/m^3 and 2250 kg/m^3 (between 50 pcf and 140 pcf) or greater than 12,500 kg/m^3 (55 pcf). SDC may consist of one type of aggregate or a combination of lightweight and normal weight aggregates. This concrete is project specific and should include a detailed mixture testing program and aggregate supplier involvement before design.

Semi-lightweight concrete – concrete made with a combination of lightweight aggregates (expanded, clay, shale, slag, slate or sintered fly ash) and normal weight aggregates and having an equilibrium density of 1675–1925 kg/m^3 (105–120 pcf) (ACI, 2007).

Structural lightweight concrete – concrete made with structural lightweight aggregate as defined in ASTM C330/C330M-17a (2017). The concrete has a minimum 28-day compressive strength of 17 MPa (2,500 psi), an equilibrium density between 1125 kg/m³ and 1925 kg/m³ (70 pcf and 120 pcf) and consists entirely of lightweight aggregate or a combination of lightweight and normal-density aggregate.

Insulating concrete – insulating concretes are very light non-structural concretes, employed primarily for high thermal resistance, incorporating low-density, low-strength aggregates such as vermiculite and perlite. With low densities – seldom exceeding 800 kg/m³ (50 pcf) – thermal resistance is high. These concretes are not intended to be exposed to the weather and generally have a very low compressive strength, ranging from about 0.75 MPa to 7.00 MPa (100–500 psi).

Material savings in construction could be achieved by increasing the efficiency and functionality of the materials used or by replacing financially or environmentally expensive materials with a waste or by-product material. The use of recycled materials and by-products in lieu of normal aggregates (e.g. gravel, crushed stone) decreases pollution and preserves natural resources (Lo and Cui, 2004; Dulsang *et al.*, 2016). LWC can host a range of waste materials as aggregate replacement, which enhances its sustainable gains without compromising performance. For example, the compressive strength of LWC at 91 days can be as high as 56 MPa by using a 5% waste glass sand replacement (Hunag *et al.*, 2015). Such material savings could reduce the overall

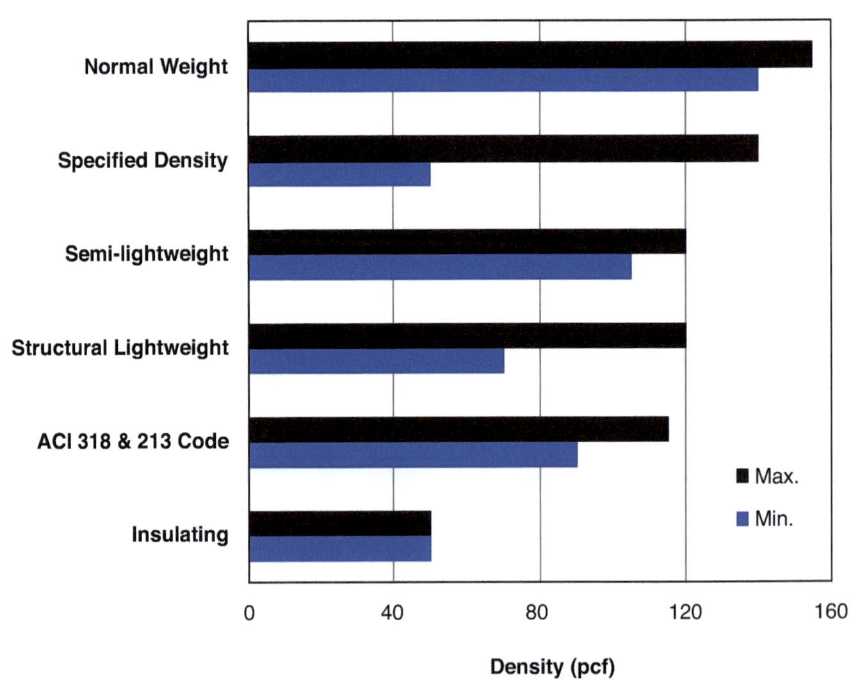

Figure 1.2 *Typical density of LWC compared to NWC (Holm and Ries, 2007).*

cost by more than 10% after offsetting the higher initial cost of LWA. The growing utilization of lightweight by-products and processed materials is indicative of sustainability changes in construction. In addition, the lighter weight of LWCs reduces transportation and handling cost. Obviously, further use of alternative aggregates is an environmentally attractive solution to the depleting finite resources of natural/regular aggregates (e.g. crushed stone and gravel). The use of LWC could result in an average weight saving of 25–30% on a bridge superstructure, not to mention the subsequent reduction in reinforcing and prestressing steel in piers and foundations (Raithby and Lydon, 1981).

2.2 Structural efficiency

Structural efficiency is defined as the strength-to-density (S/D) ratio of the concrete. The entire hull structure of the warship SS *Selma* (and of the more than 100 subsequent ships) was constructed with high-strength low-density concrete (HSLDC) in a shipyard in Mobile, AL, and launched in 1919. The structural efficiency achieved in these warships was higher than 50 (Figure 1.3). This S/D value was truly a breakthrough, unheard of for that time. The upward trend in the 1950s with the introduction of prestressed concrete, followed by production of high-strength normal-density concrete (HSNDC) for columns of very tall cast-in-place concrete-frame commercial buildings, constituted the second wave of improvement in the structural efficiency of concrete (Figure 1.3). This extremely high S/D ratio (raising it to above 60) came approximately 40 years after the warship SS *Selma*, and was the result of improvements in the cementitious matrix induced by modern admixtures (e.g. high-range water reducers) and the high-quality pozzolans (e.g. silica fume, metakaolin, and fly ash). The United States Army Corps of Engineers (USACE) defines concrete types according to their strength-to-density as follows (Holm and Bremner, 2000):

LDC (LWC):	Low-density concrete – compressive strength 20–35 MPa (3000–5000 psi)
NDC (NWC):	Normal-density structural concrete – compressive strength 20–35 MPa (3000–5000 psi)
HSLDC (HSLWC):	High-strength low-density concrete – compressive strength >35 MPa (5000 psi)
HSNDC (HSNWC):	High-strength normal-density concrete – compressive strength >35 MPa (5000 psi)

Obviously, USACE indiscriminately specifies the same strength requirements for both NWC (NDC) and LWC (LDC). It is important to note that the compressive strength of the HSLDC cores taken from the warship SS *Selma* and tested in 1980 was twice the 28-day specified strengths (Holm and Bremner, 2000). From a structural efficiency point of view, these figures are not noticeably different from that of modern HSNDC. Likewise, HSLDC of First and Second World War vessels, as well as numerous recent bridges, was found to demonstrate a successful 80-year long-term performance (Bremner *et al.*, 1994).

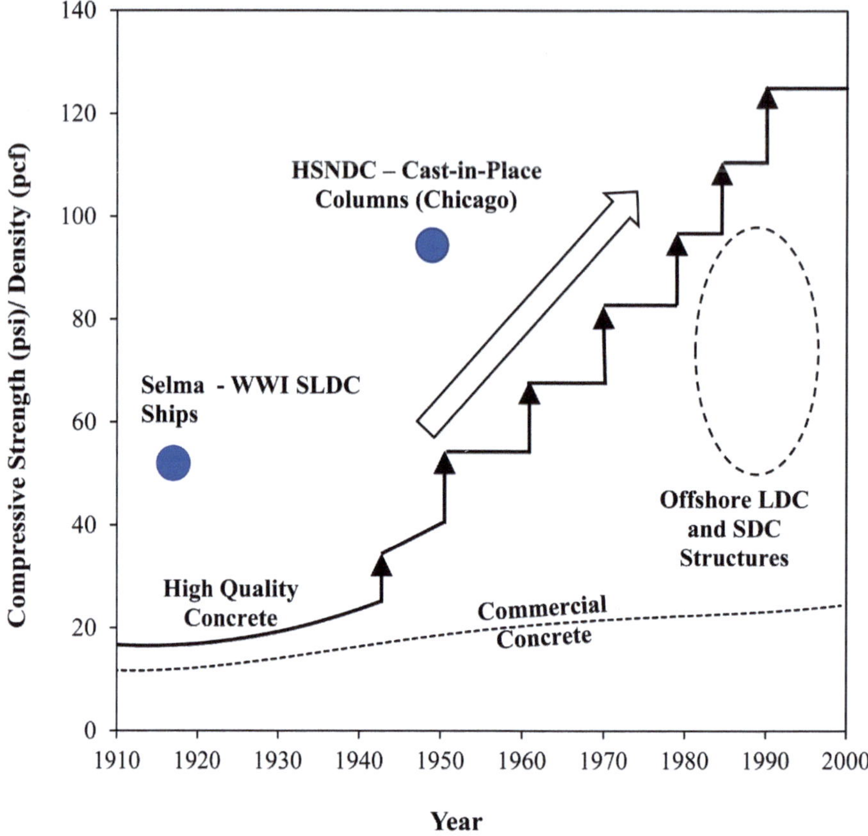

Figure 1.3 *Structural efficiency of concrete in terms of the compressive strength-to-density ratio (after Holm and Bremner, 1994). LDC: low-density concrete (compressive strength 20–35 MPa (3000–5080 psi)); SDC: specified-density concrete with partial or total replacement of normal-density aggregates with low-density structural grade aggregates.*

2.3 Carbon emission

The effects of global warming and climate change have become manifest through an increase in the average global temperature (more than 1 °C) over the past 170 years (NASA, 2019). This rise in the average global temperature is now growing at a faster pace of 0.2 °C per decade. Much of this is directly attributed to an increase in greenhouse gases (GHGs), which trap heat in the atmosphere and slow the rate at which the energy escapes to space (US EPAgency, 2017). The primary GHG emissions are carbon dioxide (CO_2), methane (CH_4), nitrous oxide (N_2O), and fluorinated gases. As shown in Figure 1.4, for the breakdown of GHGs for the USA CO_2 makes an 82% contribution and is the largest primary source of GHGs (Hockstad and Hanel, 2018). It is worth noting that in the 2008 recession that hit many business sectors in the USA the construction sector was also affected, and was associated with a sharp decline in the CO_2 emission levels (Figure 1.4).

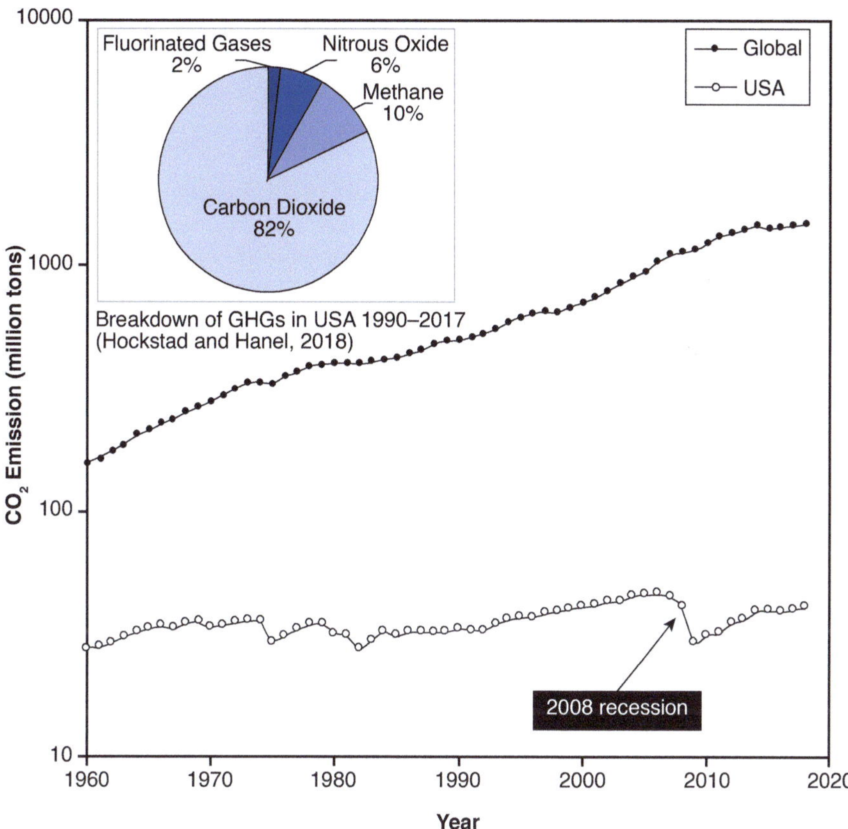

Figure 1.4 *CO₂ emission levels for the cement sector over the past six decades (USA v. Global).*

Cement constitutes up to 15% of concrete by weight. Approximately 96% of the CO_2 emissions embodied in concrete are attributed to the cement content, which is significantly impacted by the cement manufacturing process, fuel type, the raw ingredients used and the energy efficiency of the cement plant (US EPA, 1995; Nisbet *et al.*, 2002). It is estimated that 77–227 kg (170–500 lbs) of CO_2 is embodied for every 0.765 m³ (one cubic yard) of concrete. Although there have been noticeable improvements in energy reduction in cement production and concrete-making in the last few decades, CO_2 emissions have been reduced by only 7% (Hockstad and Hanel, 2018). In any event, the slimmer LWC structural elements require less cement and aggregate, and thus the generated CO_2 footprint is assumed to be reduced.

2.4 Energy

Cement production accounts for 0.33% of the total energy consumption in the USA, as compared to 1.8% for steel (NRMCA, 2008). The energy needed to manufacture cement amounts to approximately 85% of the total energy used to make concrete (Medgar, 2007). The energy consumed in mining and transporting virgin aggregates, mixing at a batch plant and transporting concrete to the construction site is relatively small. It is estimated to be approximately 0.02 MBtu/ton for sand and gravel, 0.03 MBtu/ton for crushed stone, 0.03 MBtu/ton for concrete plant operations, and 0.06 MBtu/yd3 for transportation (Marceau *et al.*, 2006). Such levels of emitted CO_2 are fairly low compared to that associated with cement production.

The industrial sector – which encompasses manufacturing, mining, agriculture and construction – accounted for almost a third of total US energy use in 2012. And energy-intensive manufacturing accounts for a little more than half the total energy use for this sector. While the cement industry consumes a very small portion of the total US energy, it is by far the most energy-intensive of all manufacturing industries. According to the US Energy Information Administration (US-EIA, 2019), although cement contributes little to the U.S. economy, it consumes nearly ten times that share in energy. For comparison, on other energy-intensive industries consume on average twice that share in energy. Cement is also unique in its heavy reliance on coal and petroleum coke. Over the long term, the projections of the US-EIA (US-EIA, 2019) show increasing energy consumption from the cement industry, which is in line with the increasing share of the total gross output of goods and services (Figure 1.5). Cement output is directly related to various types of construction. Obviously, the increase in consumption by the energy-demanding industries is manifest in the gradual decline in the projected non-industrial energy uses (red dashed line in Figure 1.5).

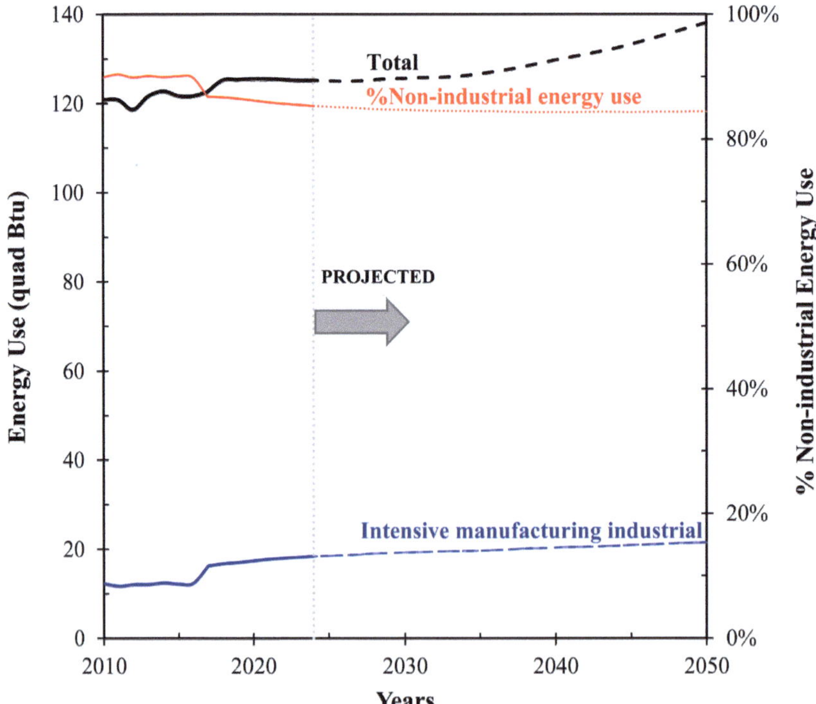

Figure 1.5 *Intensive manufacturing industrial compared to total energy use in USA. Dashed portions represent future projections (US-EIA, 2019).*

In view of this, it is instrumental to note that any attempt to shift our reliance on NWC toward LWC can gradually reduce the energy share of the intensive manufacturing industries. The reduced cement and aggregate usage in LWC – due to the smaller structural elements – is the key factor here. Energy consumption in the concrete industry can also be reduced further by replacing cement with supplementary cementitious materials (SCMs). Every 1% replacement of cement with SCM can result in approximately a 1% reduction in energy consumption per unit volume of concrete (Sabnis, 2015). In a comparative study on the environmental impact of construction using steel versus concrete, the emission levels and associated energy were calculated for a 5-story building made of structural steel frames and reinforced concrete frames, considering the life cycle analysis (Guggemos and Horvath, 2005). The CO_2 emission and embodied energy for the concrete frame were estimated to be 550 kg/m^2 and 8300 kJ/m^2, respectively, as opposed to 620 kg/m^2 and 9500 kJ/m^2, respectively, for the steel frame. Although the study considered NWC frames, it is possible to foresee the much-reduced energy and environmental impact of using slimmer structural sections (10–25% volume reduction), should LWC is used.

2.5 Thermal conductivity

Efficient use of energy in buildings is another area where global sustainability dictates that changes be made in construction. This is especially critical in view of the growing environmental restrictions and the rising cost of energy (Kumar *et al.*, 2017). As such, the thermal insulation properties of LWC are deemed very attractive. The thermal conductivity of a material (λ) is the quantity of heat transmitted through unit thickness in a direction perpendicular to a surface of unit area under a unit temperature gradient.

The use of LWA significantly reduces the thermal conductivity of concrete (Ünal *et al.*, 2007; Benazzouk *et al.*, 2008; Sengul *et al.*, 2011; Kim *et al.*, 2012). The pores in LWA increase the porosity of the concrete, which is the prime factor affecting the thermal conductivity of concrete. Demirboğa and Gul (2003) have shown that replacing NWA with the expanded perlite increases the total porosity of concrete. The enclosed pores subsequently reduce the thermal conductivity because of the very low thermal conductivity of air entrapped in the voids. For example, replacing NWA with LWA reduces the thermal conductivity of concrete from more than 0.9 W/m.K to around 0.65 W/m.K (Zhang and Poon, 2015). As such, LWC provides excellent thermal insulation for exterior walls (Holm and Ries, 2007) as well as for the storage of hot water and petroleum products (Akers *et al.*, 2014).

There is a range of thermal insulation levels that LWC can offer. For example, the thermal conductivity of LWC containing furnace bottom ash (FBA) decreases with increasing ash content. This can be attributed to the lower density of FBA compared to natural crushed fine stone. When 100% FBA was used to replace natural fine aggregates in LWC, the thermal conductivity was reduced to less than 0.5 W/m.K (Zhang and Poon, 2015). Similarly, expanded polystyrene (EPS) foamed concrete with a strength of 363 psi (2.5 MPa) has a thermal conductivity of merely 0.206 W/m.K (Bouvard *et al.*, 2007). Values for the thermal conductivity of EPS concrete are generally much lower than those of earlier LWCs of similar density or strength (Kan and Demirboğa, 2009). In fact, the combination of low thermal conductivity, low weight,

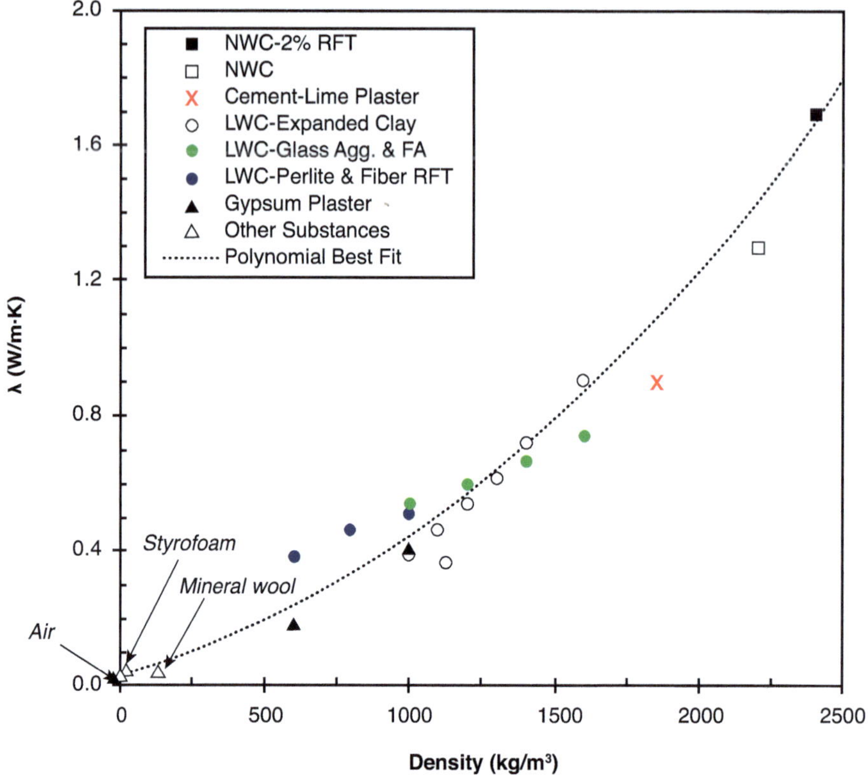

Figure 1.6 *Density v. thermal conductivity (λ) for selected NWC and other materials and different types of LWC.*

relatively high strength, and the flexibility of castings, enables EPS foamed concrete to be considered as a very sustainable material for use as building blocks or partition walls (Chen and Liu, 2013).

Kurpinska *et al.* have compiled density-thermal conductivity findings for LWC with granulated foam glass and granulated sintered fly ash together with other materials, reported in BS EN 12524: 2000 (Kurpinska *et al.*, 2019). These values appear to adequately fit a second-order polynomial (Figure 1.6). As shown, the perlite and fiber-reinforced LWC (average $l \cong 0.5$) seems to offer a thermal conductivity that is approximately one-third that for NWC (average $l \cong 1.5$).

2.6 Fire rating

The fire resistivity of concrete components in buildings is crucial consideration for structural integrity during fire events: the modulus of elasticity decreases with an increase in temperature. LWC has a superb rating for fire resistance and provides a higher R-value because of its lower coefficient of thermal expansion and fire stability (ACI, 2007). As with thermal insulation discussed earlier, the fire endurance of reinforced or prestressed concrete elements has been found to be highly dependent on the type of aggregate used (Bilow and Kamara, 2008). Carbonate aggregates include limestone and dolomite; siliceous aggregate include silica-based materials, including granite and sandstone. Apart from natural LWA (e.g. pumice), LWAs are usually manufactured by heating shale, slate, or clay.

The fire resistance of concrete elements (e.g. floor and roof slabs) is measured either by their heat transmission or by the end point of structural failure. Building codes normally have two requirements for fire rating: the minimum slab thickness to limit heat transmission and the minimum size of concrete cover to limit steel temperatures. ACI 216.1-07/TMS-0216-07 (ACI, 2007) states that plain and reinforced concrete bearing or nonbearing walls and floor and roof slabs are required to provide fire-resistance ratings of from 1 to 4 hours and shall comply with specified minimum equivalent thickness values (Figure 1.7). Equivalent thickness is the thickness obtained

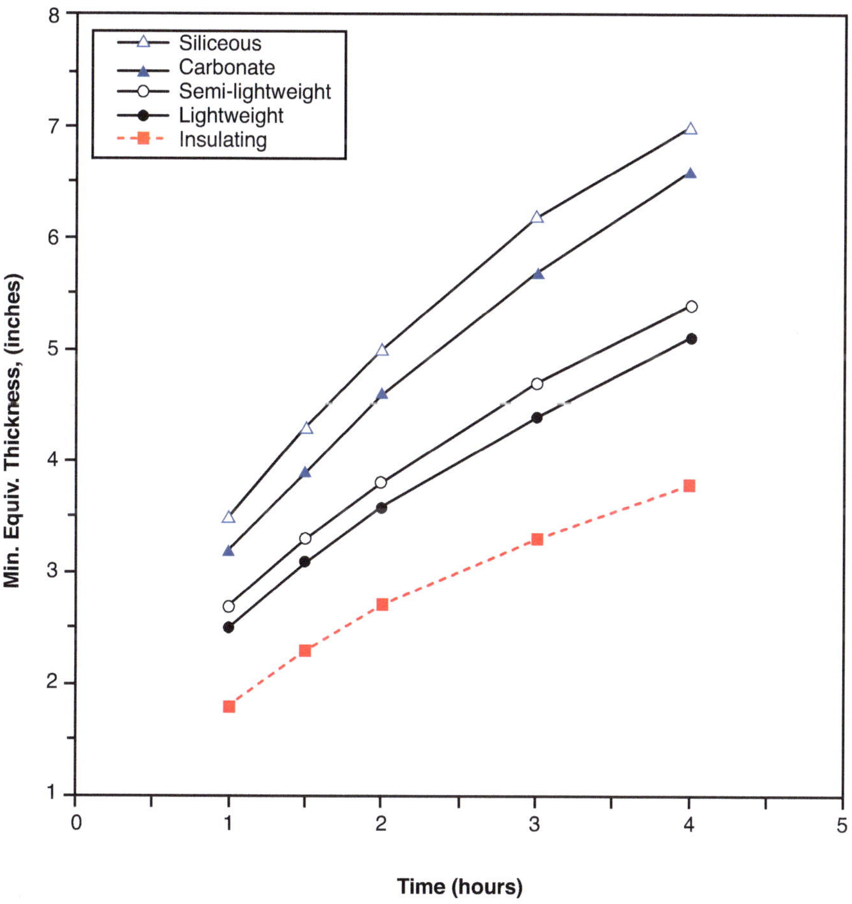

Figure 1.7 *Minimum equivalent thickness (inches) v. nominal fire rating (hours) for single-layer concrete walls, floors and roofs (Holm and Ries, 2007).*

by considering the gross cross-sectional area of a wall less the area of voids or undulations in hollow or ribbed sections, all divided by the width of the member. Clearly, for solid sections the equivalent thickness is the same as the actual thickness .The use of LWA in lieu of NWA to make a concrete element obviously results in a considerable reduction in the equivalent thickness and still provides the same required level of fire rating. At only 117 mm (4.625 in), the LWC floor system of the Bank of America in Charlotte (the tallest building in the southeast USA) achieved the required 3-hour fire rating (Holm and Bremner, 1994). By comparison, NWC sections utilizing silica aggregate would need an additional 38 mm (1.5 in) approximately to achieve the same fire endurance level (Figure 1.7).

2.7 Acoustic insulation

The sound insulation of solid walls and floors is dependent on their mass. However, the acoustic insulation for LWC units containing closely textured surface aggregates is better than predicted by their mass (Clarke, 1993). For example, an LWC wall of the same thickness offers the same acoustic insulation but has the advantage of lighter weight. Similarly, A 175-mm thick LWC wall made using sintered pulverized-fuel ash (density 1600 kg/m^3) provided the same acoustic insulation as a NWC wall (density 2400 kg/m^3) of the same thickness (Clarke, 1993). The velocity of an ultrasonic pulse travelling through LWC (3300–4200 m/s) is by default lower than that for NWC (3500–4400 m/s) (Hunag *et al.*, 2015). These values can vary depending on the curing duration.

The first increase in wall mass per unit area (conventionally expressed in weight per unit area) (0.72 kPa or 15 psf) reduced sound transmission by about 40 decibels, but adding the next 0.72 kPa (15 psf) only reduced it by 5 decibels. Tests on cast-in-place lightweight concrete (LWC) panels confirmed this pattern, as described by the Harris and Knudsen curve. The tested LWC contained 4.5% air, had a fresh density of 18.22 kN/m^3 (116 pcf), and a compressive strength of about 20.68 MPa (3000 psi).

Sound Transmission Class (STC) is a widely used measure for rating the sound isolation of building materials. The sound isolation is directly proportional to the STC rating. Figure 1.9 presents typical STC values for different building assemblies against standard levels of speech audibility through a barrier. LWC is clearly well-positioned on this scale too (Figure 1.9). For example, the sound insulation of a 12-cm (4.7-in.) LWC panel outperforms acoustically a 20-cm (8-in.) thick NWC slab floor.

Collectively, the inherent acoustic insulation characteristics of LWC panels represent a direct cost saving in building, particularly given that they can simultaneously offer attractive thermal insulation, a good fire rating and high structural efficiency. It should be remembered that ad-on acoustic solutions come with extra cost and that, typically, competitive sound-insulation panels do not offer the structural and thermal benefits that LWC intrinsically provides.

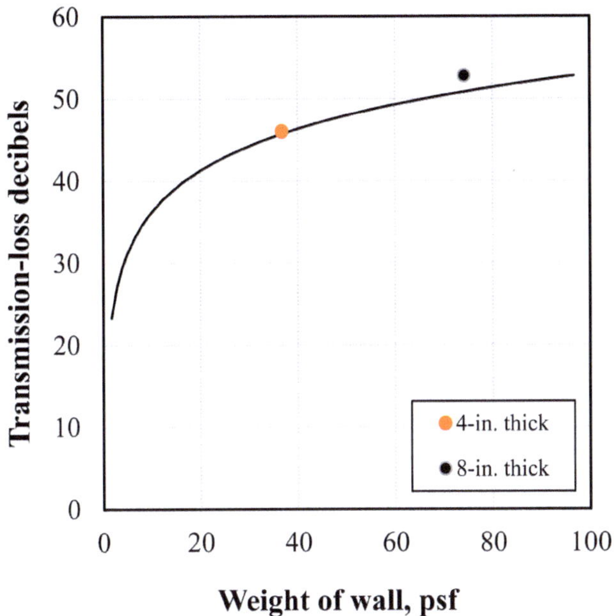

Figure 1.8 *Sound transmission loss through a LWC panel as a function of the wall weight. The results of 10-cm (4-in.) and 20-cm (8-in.) structural LWC are superimposed. The LWC had a compressive strength of 20.68 MPa (3000 psi), was 4.5% air and had a fresh density of 18.22 kN/m³ (116 pcf) (Harris and Knudsen, 1950).*
Note: CMU is concrete masonry unit; *volcanic pumice.

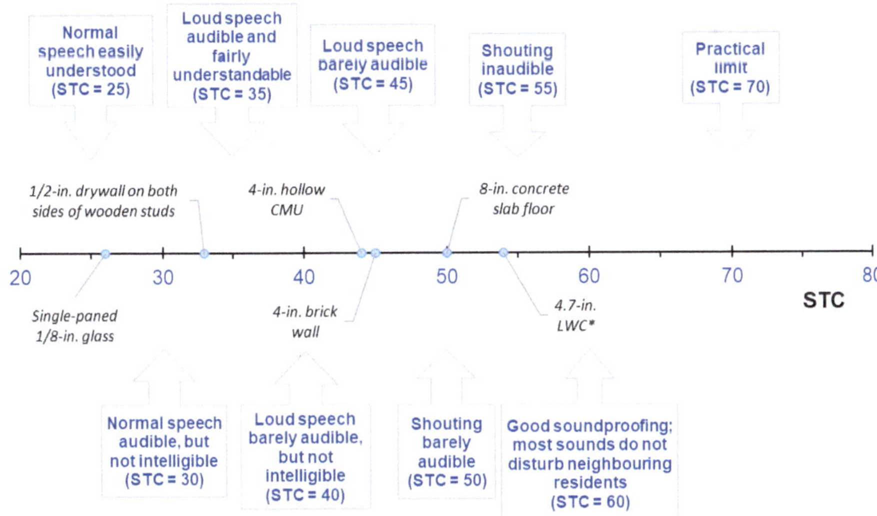

Figure 1.9 *STC levels and acoustic performance of typical wall sections. Presented STC limits and typical wall sections were collected from SBCA (2002), Acoustics.com (2004), Hossain and Lachemi (2005), Cowan (2014) and CSRI (2016).*

2.8 Durability

Durability is by far one of the most important and desirable characteristics in building components. Regrettably, LWC has been perceived (wrongly, as will be discussed) as a subclass under regular concrete from the durability standpoint. This misconception is probably based on its light weight, with the consequence that it is associated with mediocre or poor predictions of long-term performance. To demonstrate the durability of LWC, some of its most critical manifestations are discussed herein, including absorption, permeability, freezing and thawing resistance, alkali-aggregate reaction and chemical resistance, carbonation and corrosion resistance, and abrasion resistance (Figure 1.10). Given the extensive research conducted in this regard, the authors have elected to simply highlight the key findings that show the durability aspects of LWC in a global sense. LWC to NWC have been compared, where appropriate, for the selected sustainability factors and indicators. The interrelationships between these aspects and other sustainability benefits should not be downplayed.

2.8.1 Aggregate absorption

The water absorption of LWC is not directly related to its durability (Clarke 1993). Characteristics of pore size, continuity and distribution, particularly for those pores close to the surface, largely control the rate of absorption of LWA. The cellular nature of structural-grade low-density aggregates enables a 24-hr absorption value ranging from 5% to more than 25% of moisture by mass of dry aggregate (Holm and Bremner, 2000). By comparison, the moisture absorption of normal weight/ density aggregates (NWA) is typically less than 2%. LWA is capable of retaining

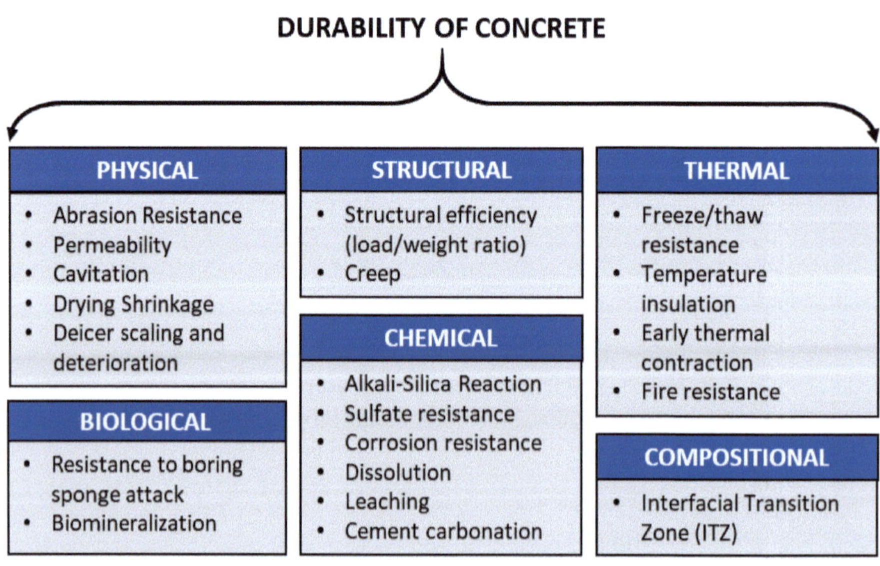

Figure 1.10 *Key durability factors of concrete.*

the absorbed moisture largely within the interior of the particles. By comparison, water primarily absorbed by NWA is surface moisture (ASTM, 2020). The internally entrapped water within LWA does not count towards its water-to-cement (w/c) ratio and thus can exist for extended periods of hydration. This enhances the aggregate/matrix interface transition zone (ITZ) as internal curing continues to reduce permeability by extending the hydration process, forming additional hydrated products in the pores and capillaries of the binder. As such, potential pozzolanic reaction in LWC is effective over a longer time. To this end, standard laboratory testing that is usually completed in less than a few months may not adequately take this reaction into account (Akers *et al.*, 2014).

2.8.2 Permeability

Permeability is a key property that controls the long-term durability of concrete. It governs the rate of entry of aggressive liquids as well as the movement of water during wetting or drying. Permeability reduces with concrete age (Teo *et al.*, 2010). The mechanical and physical properties of LWA directly affect the strength and permeability of LWC (Russell, 2008). Intuitively, one expects the permeability of LWC to be higher than that of its NWC counterpart because of the high porosity of LWA (Ozyildirim, 2008) Several studies, however, reported LWC as having the same or lower permeability than NWC (Akers *et al.*, 2014). For example, Kahn *et al.* observed low permeability values for high-strength lightweight concrete mixtures (Kahn *et al.*, 2004). Yazdani and Goucher (2015) attributed this observation to the superior ITZ between the aggregate and the cement binder in LWC. This has been observed using non-destructive means. For example, using scanning electron microscopy, Zhang and Gjorv (1991) found that LWA with a porous external layer produced a denser and more homogeneous ITZ.

The improved aggregate–cement binder interface in LWC is a result of mechanical interlocking between the LWA and the cement binder, combined with a chemical interaction (Chandra and Berntsson, 2002). The latter involves a pozzolanic reaction between the cement matrix and the LWA. This, combined with the similar elastic properties of the LWA and the cement matrix, contributes to the quality of the ITZ. Additionally, the high porosity of LWA enhances its absorption capacity, which subsequently creates a compatibility in moisture content between the LWA and the surrounding water-rich cement binder (hygral equilibrium). Similar results and conclusions by Russian, Japanese and English investigators have confirmed these findings (Akers *et al.*, 2014). All pertinent research has emphasized that the high-integrity ITZ is responsible for the low permeability of LWC. This does not occur in NWA because of its very low absorption (typically around 2%) and permeability. In fact, the high disparity in water content (if not accounted for in the mix design) on the opposite sides of ITZ may cause bleeding and thus increase the permeability of NWC.

2.8.3 Freezing and thawing resistance

As with NWC, countering freezing and thawing has necessitated using air-entrained mix designs for LWC without compromising strength (Wagner, 2015). Air-entrained

LWC has shown that it can perform comparably if not better than NWC with respect to freezing and thawing (Ozyildirim, 2008). This is attributed to the fact that the permeability of LWC has been found to be equal to or less than that of NWC. LWC enjoys elastic compatibility of its constituents as well as enhanced bond between the LWA and the cement paste (Akers *et al.*, 2014). But there is another inherent reason for the better freezing and thawing resistance of LWC: the pores within the LWA can act as pressure relief chambers when the hydraulic pressure develops as the chemically uncombined water freezes. This has been shown to occur even in the absence of air-entraining agents (Holm and Bremner, 2000). LWC made of crushed vesicular brick has also exhibited good freeze-thaw protection.

2.8.4 Alkali-aggregate reaction and chemical resistance

Replacement of reactive or non-reactive NWC with LWC appears to significantly reduce the disruptive expansion associated with alkali–aggregate reaction (Boyd, 1998; Bremner *et al.*, 1998). In fact, LWC made of either natural and artificial LWA was found not to be negatively affected by the long-term interaction between silica-rich aggregates and the alkalies in the cement, or from the ingress of alkalies from natural sources such as seawater (Holm, 1980). The porous nature of LWA provides the necessary space to minimize the deleterious expansion of reactive materials to precipitate in a non-harmful manner (Holm and Bremner, 2000). These characteristics combined with the low permeability at the ITZ are advantageous in chemically harsh environments.

2.8.5 Carbonation

Carbonation of concrete is the reaction of carbon dioxide from the air with the calcium hydroxide liberated during the hydration process. While the rate at which the carbonation progresses through concrete has been noted, most studies addressed relatively short-term effects (Holm and Bremner, 2000). Carbonation is associated with a reduction in pH values from approximately 13 to 9, which in turn neutralizes the protective layer over the reinforcing steel, rendering it vulnerable to corrosion. The carbonation of concrete surfaces over a 100-year lifecycle can absorb from 33% to 57% of the CO_2 produced during the calcination process of cement (Pade *et al.*, 2007).

The carbonation depth in LWC depends highly on the aggregate type combined with w/c ratios. Thus, the greater carbonation depth generally observed in LWC compared to NWC is not necessarily true for all LWAs and w/c ratios (Bogas and Gomes, 2015). For example, Gündüz and Ugur reported a 15–55% increase in the carbonation depth of LWC made with pumice aggregate, at w/c ratios varying from 0.14 to 0.59 (Gündüz and Uğur, 2005). This highlights the effect of the paste in the carbonation resistance. A lower carbonation resistance was reported for LWC in which normal sand was replaced by lightweight fine aggregates (Al-Khaiat and Haque, 1999), which was attributed to the higher porosity of the LWA used. However, Bremner *et al.* (1994) reported lower carbonation levels in LWC compared with NWC of the same strength class, justifying the result on the basis of the lower w/c ratio required compared to NWC. In a comparative study, Lo *et al.* have shown that most of their LWC mixtures yielded lower carbonation than did NWC – despite their same strength rank (Lo *et al.*, 2008). They explained the superiority of LWC in terms of carbonation level on the basis of the lower w/c needed to achieve the required strength, and the lower 'carbonatable' constituents (less cement) in NWC mixes, which increases their carbonation potential compared to LWC.

2.8.6 Corrosion

Intuitively, high permeability allows easy penetration of deleterious agents into concrete which accelerate rebar corrosion. Based on the work by Keeton, the US Navy reported the lowest permeability with high-strength LWC (Keeton, 1970; Holm and Bremner, 2000). Bamforth (1987) incorporated LWC as one of four concretes tested for permeability to nitrogen gas at a pressure of 1 MPa (145 psi). Compared to NWC, sanded LWC at 50 MPa (7,250 psi), 6.4% air with a density of 2.00 kg/m³ (124 pcf) demonstrated the lowest water and air permeability of all mixtures tested. A combination of two primary factors controlled rebar corrosion in concrete elements: an adequate cover thickness, and the quality of concrete cover. Reduced permeability and the higher resistance to freezing and thawing collectively led to a more durable LWC that can resist steel corrosion more effectively.

2.8.7 Abrasion resistance

The abrasion resistance of concrete is determined by the quality of several mechanical properties. It depends on the strength, hardness, and toughness of both the cement matrix (the hardened cement paste) and the aggregates, as well as on how well these two components are bonded together (Akers *et al.*, 2014). Most LWA suitable for structural concrete is composed of solidified glassy material with a hardness comparable to quartz. The net resistance of LWA to abrasive forces is expected to be lower than that of a solid particle of most natural normal aggregates, due to the former's porous nature. However, LWC bridge decks that have been subjected to more than 100 million vehicle crossings, including truck traffic, show abrasive performance similar to that of NWC. Hoff reported that the measured ice abrasion of LWC exposed to arctic conditions demonstrated essentially similar performance to that of NWC (Hoff, 1992; Holm and Ries, 2007).

It is important to note that using well-selected LWA and proper mixing with cement will protect the aggregate and thus increase abrasion resistance. High compressive strength generally increases the abrasion resistance of LWC. Should the matrix of LWC be abraded and the aggregate particles exposed, it will deteriorate relatively faster than NWA because of the vulnerability of the LWA. Surface treatment together with combining relatively soft-coarse aggregate with a hard-fine aggregate enhances abrasion resistance (Clarke, 1993).

3. Supply and demand

There are increasing concerns about the future of LWC as a common building material, particularly amongst those involved in the construction industry. Production volume and sale trends of LWC are not readily available, however, partly because the records of concrete producers are considered confidential. As such, it is rather challenging to estimate the demand for LWC across the country with a high level of confidence. In the absence of definitive data from concrete suppliers, data from a related study will be used to depict the current market changes for the production of LWA and the existing geographical interest in the use of LWC (Mousa *et al.*, 2018). (Mousa *et al.*, 2018) also include their interpretation of LWC supply and demand – based on typical mix designs – using the available production levels of cement, LWA and ready-mix concrete (RMC) in the last few decades. This analysis could help obtain a clearer picture of the future status of LWC.

3.1 Production of LWA

Aggregates account for 70–85% of the weight of concrete. Aginam *et al.* estimate that the cost of aggregates will fall to within 5–10% of the total cost of modern transportation projects (ASCE, 2009; Aginam *et al.*, 2013). As a key constituent, the availability of suitable LWA controls the properties as well as the price of LWC. Significant research has been conducted in replacing traditional aggregates with LWA. These attempts span naturally occurring materials (e.g. scoria, diatomite, pumice and pumicite, slag stone, etc.) and materials from artificial sources. The latter category includes manufactured aggregates (e.g. common clay, perlite, vermiculite, slate, shale), industrial by-products (e.g. recycled glass, cinders, fly ash) and lightweight synthetic particles (LSP). The production and unit price of common LWAs in the USA are summarized in Table 1.1. It is important to note that price varies geographically in the USA.

Table 1.1 *Estimated production and prices of common LWA in the USA (by annual production).*

Type	Annual production ('000 metric tons)	Price ($/ton)	Availability (no. of US states)	Source
Common clays and shale	12,000	36[f]	38	Manufactured
Crushed glass aggregates	5,533[a]	2[b]	50[a]	By-product
Scoria	2,700[c]	7[d]	12	Natural
Diatomite	920	10	4	Natural
Perlite	670	72	28	Manufactured
Pumice and pumicite	610	33	5	Natural
Vermiculite	220	916[f]	11	Manufactured
Slate	43[e]	12[g]	6	Manufactured

[a] US EPA, 2011.
[b] Meyer *et al.*, 2001.
[c] Mortensen, 2006.
[d] Kogel *et al.*, 2006.
[e] USGS, 2014.
[f] USGS, 2016.
[g] USGS, 2018.
All other data from US Geological Survey (USGS, 2020).

Figure 1.11(a) depicts the US national production of ready mix concrete (RMC) (unpublished data), Portland cement and LWA, during the last four decades. As expected, the trends for cement and concrete production are highly correlated. Consideration should be given to the fact that the concrete volumes shown do not include site-mixed concrete and precast concrete. LWA, cement and concrete production dropped noticeably following the 2008 economic recession. The combined production of sand and gravel (S&G) had dropped in the early 1980s followed by a steady increase until 2008 (Figure 1.11(b)). It is important to note that natural aggregates in construction include sand, gravel and crushed stone. LWA had dropped concurrently but was stagnated until 2008 when both S&G and LWA plunged appreciably. Subsequently, the S&G production recovered somewhat but LWA remained at about the same level. As shown in Figure 1.11(b), the ratio of S&G to LWA productions has gone through three phases over the last 40 years. Between 1980 and 1995 the ratio was 15–21, increasing to 20–30 between 1995 and 2005. In the last 15 years the ratio increased again, to 30–37 in favor of S&G, signifying the stagnation of LWA production.

Disregarding minor fluctuations, the ratio of S&G to LWA production has significantly increased during the last few decades (Figure 1.11(c)). This comparatively declining use of LWA (shown by the high S&G/LWA ratio) between 1980 until 2020 is indicative of the potentially decreasing interest in LWC. Such observation is further supported by the increase in the ratio of RMC to LWA over the same period. With the recovery of cement and concrete production after 2011, the stagnant production of LWA (captured in the increasing S&G/LWA and concrete/LWA ratios) indirectly suggests a loss of some of the LWC market share to NWC in the last few years (Figure 1.11(c)). It is worth noting that the total LWA consumption shown includes those aggregates used for landscaping and other related applications.

3.2 Geographical demand for LWC

The available information on geographical interest in LWC is very scattered and limited. Based on the authors' personal communication with officials at the expanded shale, clay and slate institute (ESCSI), approximately 25% of LWAs go into the LWC market. The institute estimates that the yearly LWC production has fluctuated between a low of 1.4 million m^3 and a high of 2.14 million m^3 (1.8 and 2.8 million cubic yards) during the past two decades. On average, this amounts to less than 1% of the national concrete production. A major RMC company in the northeast region of the USA said that their LWC production has been relatively flat at approximately 5% of the concrete market during the last 30 years. A producer of this size has, surprisingly, supplied LWC to only 5–10 projects per year on average. The northeast region of the USA is relatively active in using LWC, particularly in the states of Pennsylvania and New Jersey, compared to other parts of the USA. In contrast, the northern and some southern states seem to exhibit a diminishing interest in LWC.

A survey was sent to bridge engineers in all 52 US states to determine the level of usage and satisfaction with LWC bridges and to identify problems and perceived LWC benefits (Cousins *et al.*, 2013). The survey gauged the LWC usage in bridge girders and decks, typical LWC mix designs, special design requirements, and numbers of

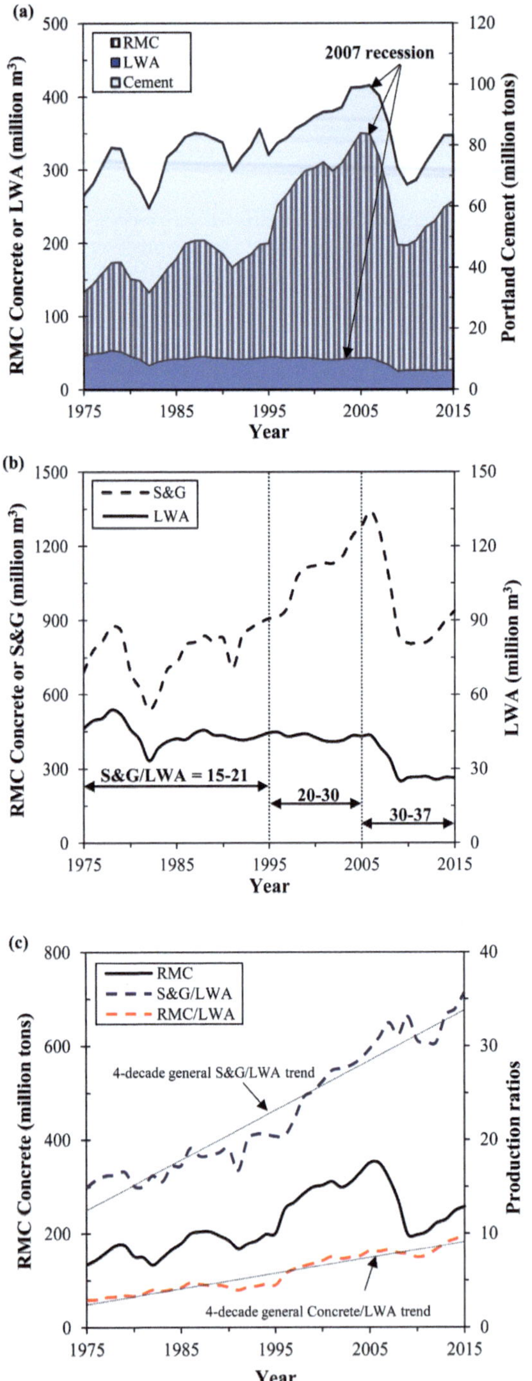

Figure 1.11 *National annual production of LWA over a four-decade period: (a) cement and concrete; (b) sand and gravel (S&G); (c) general trends. The figures for 2020 are projected.*

20

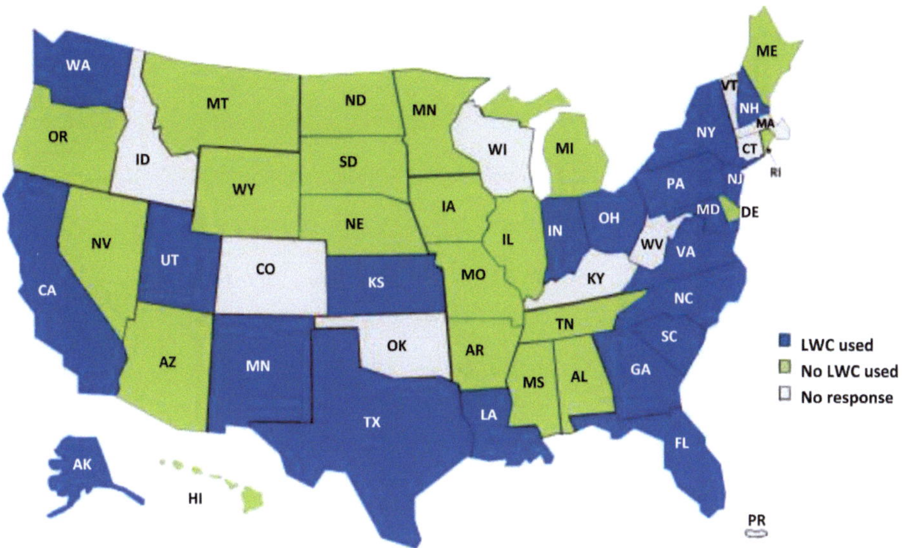

Figure 1.12 *LWC bridges in the USA (from Mousa et al., 2018; after Cousins et al., 2013).*

pertinent projects. The distribution of LWC bridges shown in Figure 1.12 is indicative of the geographical interest in the transportation sector. The responses received from 38 states (14 states abstained) included 228 bridges that were constructed with LWC decks. Two states (Virginia and Alaska) had used LWC prestressed girders and 15 states had used LWC in bridge decks (mostly in New York and Pennsylvania). Respondents to the survey reported 33 bridges (13 in Alaska and 20 in Virginia) built with LWC girders.

3.3 'Guesstimation' of LWC production

In order to estimate the production of LWC, we relied on the reported values of natural and manufactured LWA together with some typical LWC mix designs. The average annual consumption of vermiculite, perlite, pumice and common clays in concrete is estimated to be 46%, 29%, 18% and 2%, respectively (Bernhardt and Reilly, 2019). No usable data were found for slate, diatomite, and scoria or aggregates made of by-products. LWA used to make LWC can vary from 18% to 75% (by weight) of the total mixture (Abdullahi *et al.*, 2008; Dawood and Ramli, 2008; Simons, 2010; Yu *et al.*, 2015; Shafigh *et al.*, 2016). It is understood that this wide range reflects the targeted LWC properties. LWA contents of 67% and 26% (by weight) of the total mixture were reported for insulating and structural LWC, respectively (Holm and Ries, 2007; Norlite Lightweight Aggregate, 2013; Stalite Lightweight Aggregate, 2019). Based on these ranges, and for the sake of our 'guesstimation' of LWC production, the total weight of LWA used in concrete was converted to volume of LWC using the unit weight of two standard LWC mixture designs. This approach is believed to yield reasonably representative estimates of LWC production. Using this LWA content together with the aforementioned utilization figures for vermiculite, perlite, pumice and common clay,

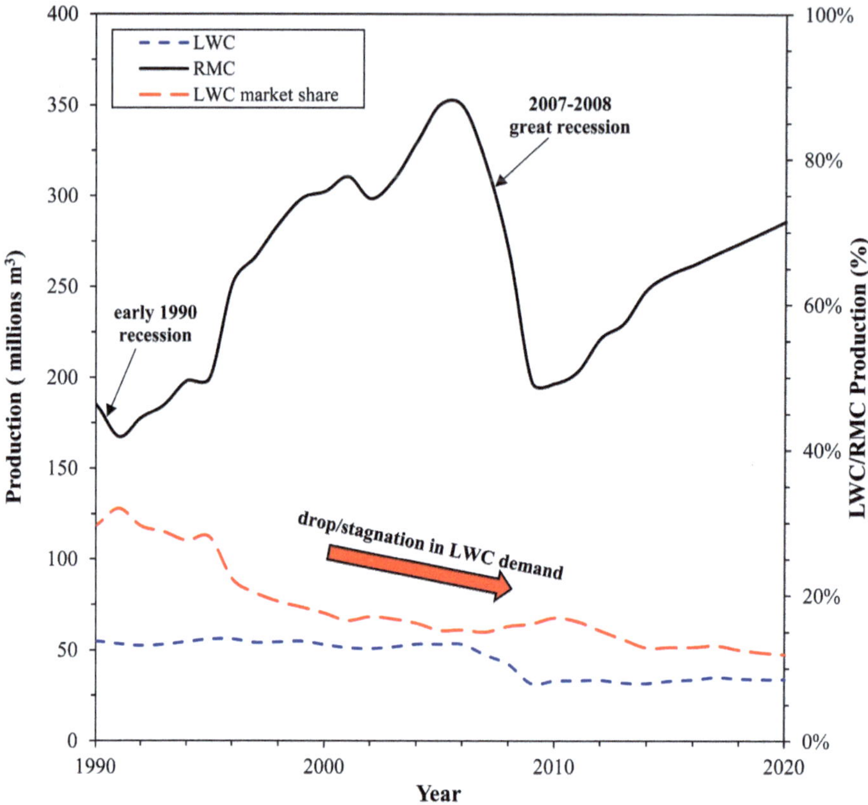

Figure 1.13 *Estimated LWC production and market trends compared to RMC (2020 RMC production projected).*

Figure 1.13 shows LWC production against that of RMC for the last three decades. The calculated LWC market share (using the total RMC production as a baseline) has dropped on average from approximately 30% in 1990 to a mere 12% in 2019. The true market share is believed to be much smaller – considering the unavailable data on precast concrete and site-mixed concrete.

Such a substantial drop in LWC usage in construction over the last three decades is quite alarming. Looking closely, LWC/RMC trends seem to go down incrementally, through phases of drop followed by stagnation of market demand. Despite the assumptions made, the relative production of LWC with respect to that of RMC is probably representative of the market trends. These 'guesstimated' LWC production rates are in line with S&G/LWA and RMC/LWA trends presented earlier in Figure 1.11I. The absence of published data on the production or sales of LWC can be chiefly attributed to business reasons. Apart from the unjustified – yet common – confidentiality excuse, the authors strongly believe that market rivalry is at the heart of this issue. It is suspected that RMC suppliers have an interest in promoting NWC due to the cheaper unit price and high sales volume. Expectations of sales volumes for LWC needed for the same project are noticeably lower. This seems to give grounds to the lack of collaboration from RMC suppliers, who were reluctant to provide us some representative LWC production rates.

4. Market indicators

Supported by the estimated trends in LWC production shown earlier, its presence and demand in construction seems to be at stake. In view of this, the concrete industry management (CIM) program was tasked in 2017 to investigate closely the receptiveness of the construction market in the USA to this novel and important concrete type. The study also attempted to determine future indicators. A survey was conducted by several teams working under the program's supervision. The details of this study can be found in Mousa *et al.* (2018) and the details presented in this section are extracted from their findings.

The survey was designed to address several characteristics of the LWC industry and to encompass a range of sectors in the construction market, including residential, commercial, institutional, governmental, retrofit and green building. In so doing, a broad spectrum of stakeholders was approached to participate in the survey, including concrete suppliers, construction material manufacturers, structural engineers, architects, concrete contractors, and LWA suppliers. In all, 6,203 participants from different regions in the USA were contacted via phone interviews and emails. However, of these only 517 responded, amounting to 8% of the targeted audience.

As with many surveys, this study sought to examine and test the validity of a number of hypotheses. These took the form of the following concerning the declining interest in LWC in the USA:

A1. Sustainability values for LWC are not well comprehended by stakeholders.
A2. Price is the prime variable that affects the demand for LWC.
A3. There seem to be technical concerns about LWC that affect demand.
A4. The low interest in LWC is directly related to the culture of the construction market.

Three key variables were selected in this survey: price, knowledge, and customer perception. The survey was designed to validate (or otherwise) these key hypotheses and to establish a correlation between these variables and the low interest in LWC. The survey also wanted to gauge the potential growth of LWC in the construction industry.

The survey outline is shown in Figure 1.14. The survey consisted of 36 questions: 26 questions were multiple-choice questions (objective/quantitative) and the remaining

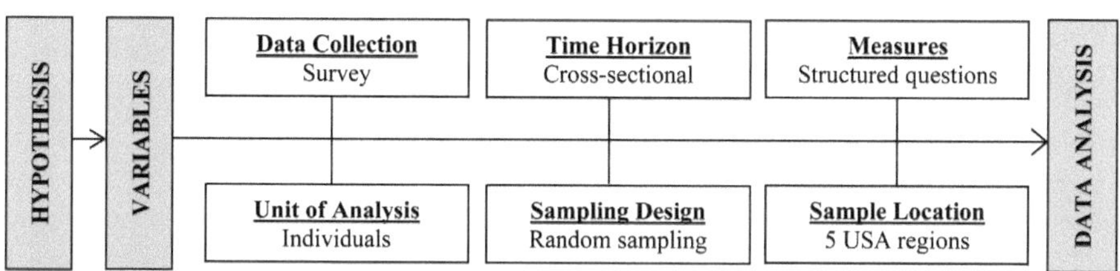

Figure 1.14 *Design and survey design and outline (questionnaire) (Mousa et al., 2018).*

questions were narrative open-ended (qualitative). It should be noted that some questions allowed multiple answers. For these questions we populated all answers as normalized responses rather than single answers, to better present the frequency of the selected choices. The study of the market should realistically portray a clearer picture of the participants' engagement, knowledge, inclinations and experience with LWC. This information can be used to project the future position of LWC in the concrete market. The 37 questions fall under four main categories:

1. *Level of engagement*: 9 questions were listed to survey the background of the audience, their entities, and their level of involvement in the construction market.
2. *Key obstacles*: 8 questions were listed to map the major obstacles facing LWC.
3. *Market perception*: 13 questions were listed to demonstrate a clear evaluation of the participants' perception on LWC.
4. *Perceived future*: 7 questions were listed on inclinations and future expectations for using LWC.

To this end, only selected questions and aspects of the survey are presented herein. An attempt is made to correlate the collected answers with the arguments made earlier. The compiled answers to the questionnaire are available in Appendix A. It should be noted that the participants' background and demographics were populated and analyzed for each question (not shown for reasons of conciseness).

4.1 Awareness and engagement

The professional demographics of the participants fall into five main groups: architects, manufacturers, contractors, engineers and suppliers. Unsurprisingly, architects constituted more than a quarter of the respondents, followed by concrete manufacturers (Figure 1.15(a)). Architects are naturally intrigued by the esthetics and nonstructural aspects of LWC. Approximately 75% of the respondents described themselves as executives and decision makers. Participants engaged in commercial, residential, and governmental sectors made up approximately 25%, 19%, and 17%, respectively (Figure 1.15(b)). This is indicative of the business diversity of their companies. At 13% involvement in green buildings projects, the respondents' background and familiarity with the sustainable construction principles is apparent. This level of involvement would also affect the responses to the questions addressing the exposure and appreciation to the sustainability dimension in construction.

With regard to the geographical extent of the bulk of the projects, the populated answers amount to 25%, 19%, 17%, 16% and 9%, in the Midwest, Southwest, Southeast, Northeast and Northwest regions, respectively (Figure 1.16). It is alarming that only 8% and 6% of all the participants had construction businesses across the entire USA and throughout North America, respectively. Approximately 80% of the respondents had been involved in at least one project that they have supplied, designed or constructed, using structural LWC. Probably, this involvement justifies the motivation behind their participation in the survey.

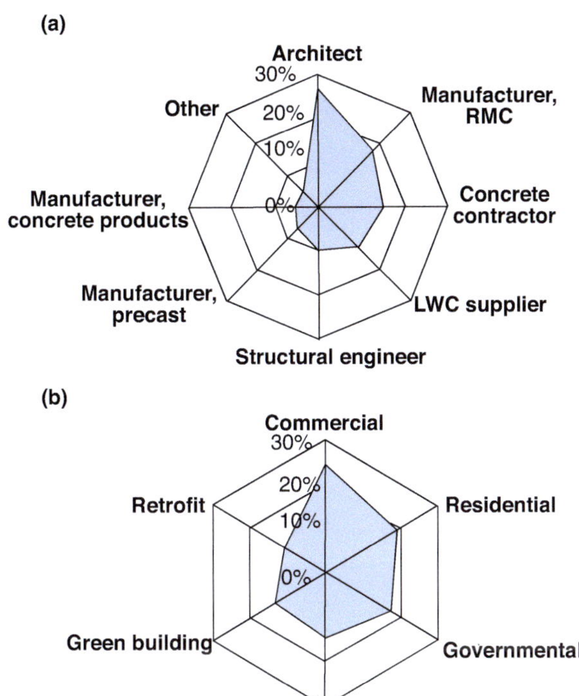

Figure 1.15 *Survey background: (a) classification of participants by occupation; (b) involvement by project type (Mousa et al., 2018).*

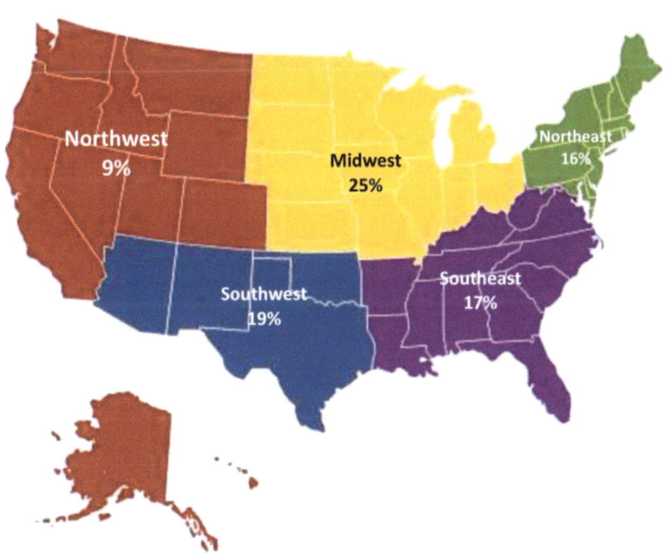

Figure 1.16 *Regional distribution of projects in which the survey participants were involved (Mousa et al., 2018).*

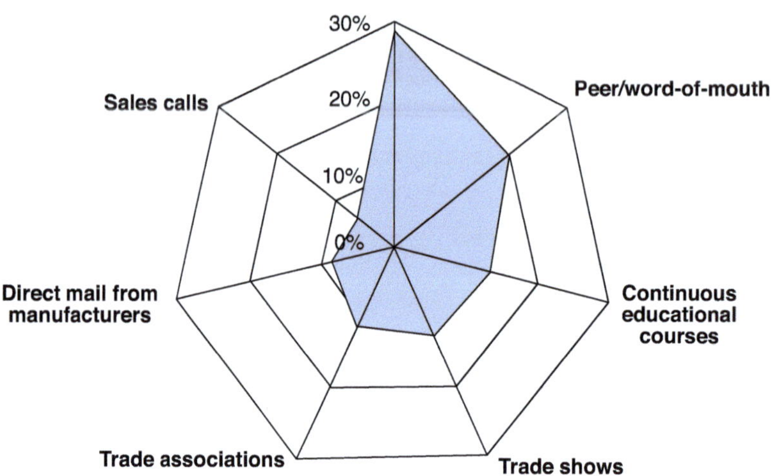

Figure 1.17 *Source of information on LWC (Mousa et al., 2018).*

Most of those asked about their source of knowledge on the latest developments in LWC construction replied that they were informed through trade magazines and websites (Figure 1.17). Peers/co-workers and word of mouth come in second place at 20% of the total responses. In response to the utilization of LWC in their projects, only 10% of the stakeholders indicated usage of 25% or higher, while the majority (68%) of the respondents reported usage of less than 10%, (Figure 1.18). These responses seem to support the presumed declining interest/demand in LWA shown in Figure 1.11.

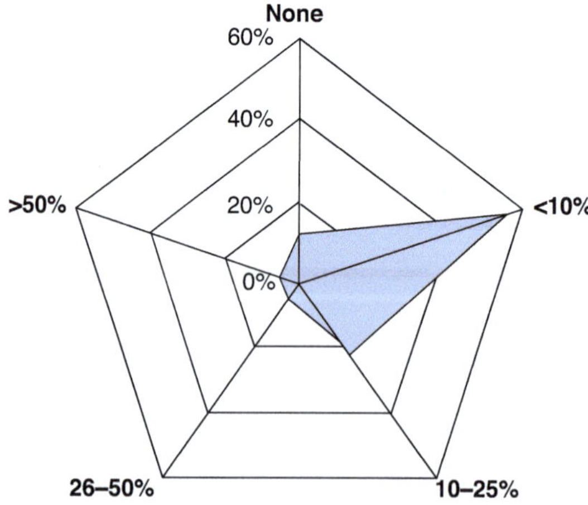

Figure 1.18 *Level of engagement with LWC (Mousa et al., 2018).*

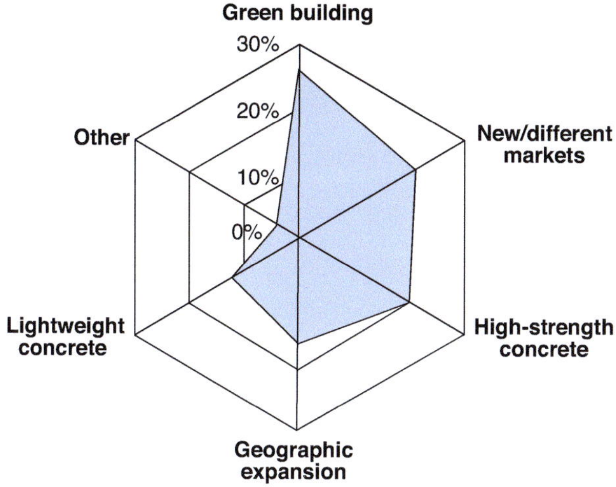

Figure 1.19 *Expected growth of participants' entities (Mousa et al., 2018).*

The participants were asked to rate the potential growth of their companies (Figure 1.19). Approximately 27% and 20% of the responses respectively believed that their company was growing in the field of green buildings and that this would open new markets. Only 12% believed their entities would expand in the field of LWC. Such a low rating is alarming.

With regard to the current and future availability of LWA, the populated responses indicated some balance between the available types of LWA (in the range of 11–23%) (Figure 1.20). The reluctance to provide answers (approximately 25%) demonstrates

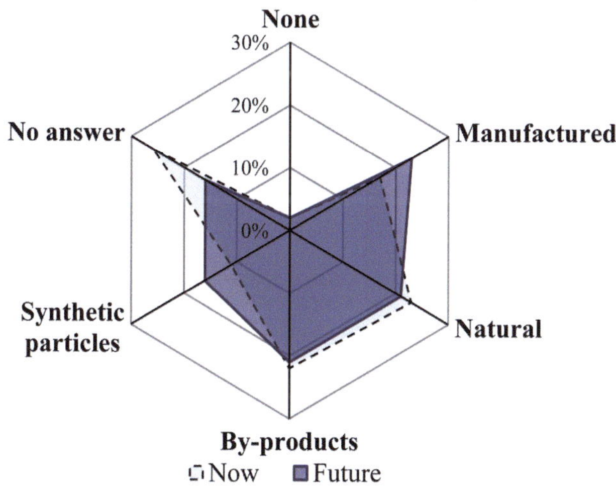

Figure 1.20 *Local availability of LWA (Mousa et al., 2018).*

that many participants are somewhat uncertain about the availability of LWA. Such sentiment probably stands behind the noted stagnant interest in LWC. Approximately 49% of the participants stated that at least one type is available in their vicinity. However, 48% were unaware of any available LWA types. Only 25% of participants believed that at least one type of LWA would be available in the vicinity of their work/residence in the future.

4.2 Current barriers

It is crucial to scrutinize the chief obstacles that could plausibly be behind the stagnant/declining use of LWC nationwide. For this purpose, the participants were asked to rate the top five hurdles limiting their use of LWC (a score of 1 being the least significant and a score of 5 being the most critical). At least one hurdle was flagged by approximately 74% of the respondents. Nearly one quarter of the rated responses considered LWC price as the primary turnoff. However, only 4% of the participants indicated that there are many other hurdles and challenges. These obstacles fall in a long list of claimed shortcomings: poor construction owing to the lack of technical knowledge, increased tension cracks, inadequate performance in cold weather, long setting times for flooring materials, absence of water tightness, unnecessary conditioning of aggregate, uncontrolled changes in bulk density, and the noticeable variations in materials. The results depicted in Figure 1.21(a) portray the relative impact level (RIL) of the main hurdles, calculated as the weighted average outcome of the impact. Approximately 64% (Q19) of the participants stated that they would likely use or recommend LWC should at least one hurdle be eliminated, including: low price, ability to adhere to codes, adequate placement of concrete, good pumpability, easy finishing process, estimated lifecycle cost and satisfactory long-term performance (Figure 1.21(b)). Yet, price remains the top-rated concern as shown by RIL values.

4.3 Consumer perception

To portray a realistic image of the collective perception of LWC in the American market—particularly with respect to price and quality—survey participants were asked to share their views on its value and application. With regard to the added value of employing structural LWC, 73% of respondents indicated at least one benefit. The key benefits by respondents are summarized in descending order (clockwise) in Figure 1.22(a). The responses show that fire resistance, followed by economic benefits such as cost savings, reduced column sizes, and smaller story heights are the main key benefits of LWC. The participants were asked to gauge the sustainable impact of LWC in order to evaluate the perception of the concrete industry in this department. Green or sustainable positive impact of LWC were not perceived by approximately 47% of participants. Material savings and the utilization of recycled materials were reported as other benefits. However, nearly 96% of the respondents stated that they were aware of at least one LWC technology.

(a)

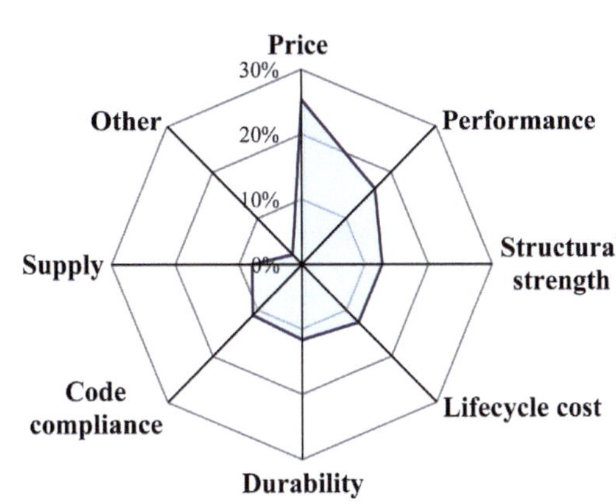

(b)

Figure 1.21 *LWC hurdles: (a) ranking of current obstacles using relative level of impact (RIL); (b) worst obstacle limiting future use. RIL = $(n_i \times IL) / (\sum_i n_i \times IL)$ where n is the frequency of responses for any given obstacle and IL is the respective impact level ranked on a scale of 1–5 (Mousa et al., 2018).*

Figure 1.22(b) depicts the quality and price perceptions of LWC compared to NWC. Approximately 33% of the respondents believed that LWC had a lower overall quality, while 37% of the respondents believed that LWC has the same overall quality. Only 11% of the participants thought that LWC has a greater quality than NWC – obviously under the impression that it provides superior compressive strength. Approximately 69% of the respondents reported LWC as more expensive than traditional concrete (91% of concrete suppliers versus 42% of architects). The fact that fewer architects think that LWC is expensive reflects that a majority of this group exercises good judgement and perceives the benefits as an added value that could offset the higher unit price of LWC. LWC was considered cheaper than traditional concrete by only 7%

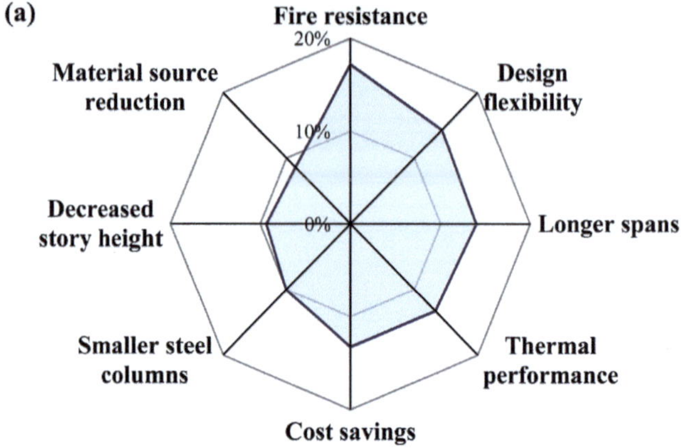

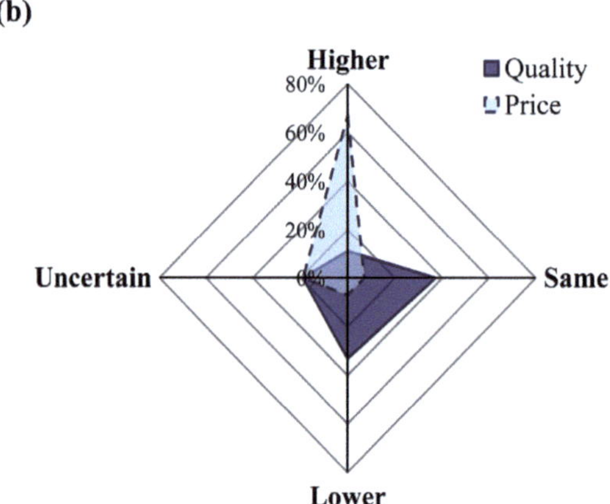

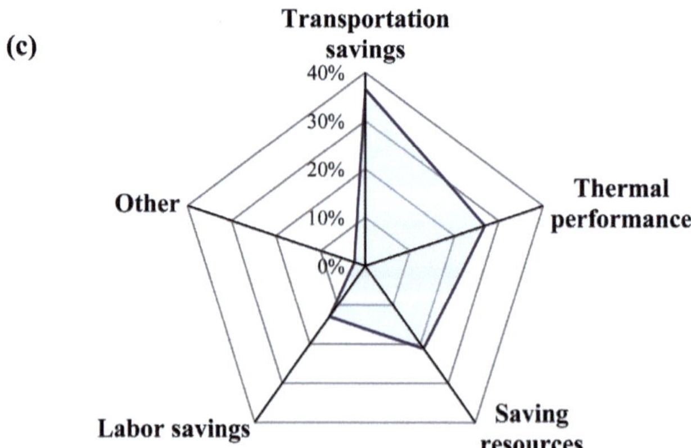

Figure 1.22 *LWC perception: (a) key benefits; (b) price and quality compared to regular concrete; and (c) rating of sustainability gains (Mousa et al., 2018).*

of the responses, while an equal number of participants believed that the cost was the same. Uncertainty about the prices was voiced by approximately 16% of respondents. Figure 1.22(c) displays the responses with regard to the green or sustainable impact of LWC. Transportation saving was among the top choices, followed by thermal performance, resource savings and labor savings. This may reflect some bias in appreciating the key sustainable benefits of LWC, even among those who opted for the many green qualities associated with using LWC.

4.4 Perceived future

In an attempt to gauge the future use of LWC in the market, participants were asked about their long-term anticipated use of LWC (Figure 1.23). Approximately 66% of respondents opted for the same current level of use. Approximately 25% of the respondents conveyed that they expect a growth in LWC usage, while a mere 3% of the respondents foresaw a dormant use. In response to potential legislative changes that would have an impact on LWC, approximately 91% of the respondents did not expect future improvements.

4.5 Synopsis

There are several concerns about LWC that most of the participants in the survey raised. The low response rate is indicative of low interest in LWC. Collectively, these observations signify a negative market impression with regard to LWC. It is likely that such market trends are driven by a predetermined false perception that LWC is of a lesser overall quality compared to NWC. At the same time, this highlights the limited

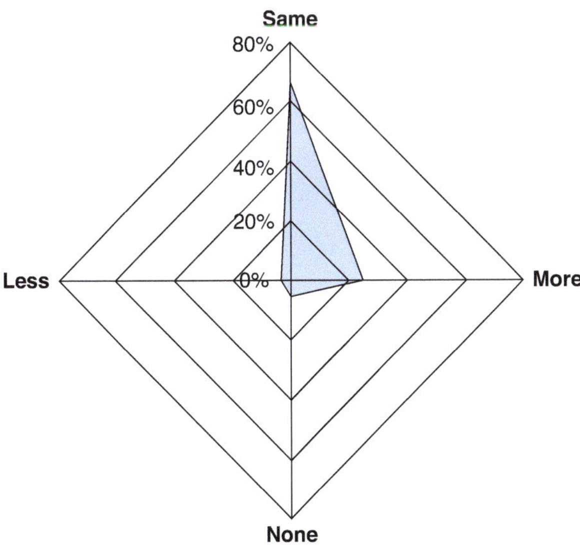

Figure 1.23 *Future use of LWC (Mousa et al., 2018).*

understanding and/or appreciation of the technical merits of LWC that we discussed earlier (see hypothesis A1). Likewise, LWC is wrongly perceived as a much more expensive alternative compared to NWC (hypothesis A2). The logical explanation for this perception is probably the inability of most market players to differentiate between unit price and total cost and a lack of appreciation for sustainability and long-term gains (Zidan *et al.*, 2013).

Such a misleading – yet prevailing – notion has shaped the consumption pattern in the construction market, apparently disregarding a proper total cost analysis and neglecting the inherent sustainability benefits of LWC. The responses relayed technical concerns – mainly the pumpability and strength (hypothesis A3). These concerns have likely stemmed from limited knowledge about manufacturing and construction using LWC. Stagnation of construction culture can be generally inferred from the collective responses (hypothesis A4). The survey revealed that a small population of construction stakeholders truly valued the beneficial traits that LWC is able to offer. Taken together, the results of the survey confirm our four hypotheses.

5. Situational assessment

The vague position of LWC in the construction market dictates the need for a keen investigation into the reasons behind what appears to be stagnant demand driven by declining or nebulous interest. Common technical misconceptions that are likely to limit the use of LWC should be closely assessed. Views of selected experts on the practical and business issues associated with LWC are being presented to provide further insights. Porter's analysis (Porter, 1979) was also employed to provide the authors of this chapter with a business perspective that can explain the current market mechanics of LWC supply and demand, which can guide the search for a potential resolution.

5.1 Technical misconceptions

The characteristics and performance of LWC are often a matter of debate among market players in construction. A number of technical concerns about LWC are usually – but mistakenly – brought up as excuses to limit or abandon its use. These concerns mostly lack complete scientific proof. The acceptance of such prevalent misconceptions could certainly be a cause of the global under-utilization of LWC as a sustainable building material. Most of the claimed issues could be minimized or circumvented given an adequate understanding of the material's behavior combined with practice and experience.

5.1.1 Pumpability

Pumping LWC is often considered problematic by contractors due to a range of factors, including inconsistent mixtures, slump loss, segregation and the need for excessive pump pressures. Commonly-used LWAs have high water absorption levels: insufficient saturation (presoaking) prior to mixing, therefore, unsurprisingly reduces the slump of the LWC and cause pumpability issues. The use of unsaturated LWA is likely to trigger uncontrollable workability during pumping.

On the other hand, the use of excessive mixing water to enhance pumpability can cause segregation (NRMCA, 2003). It is important to note that the total amount of water needed to saturate the pores in LWA (typically 15–35%) is substantially greater than the 'as received' natural moisture at delivery (typically 2–10%). Bringing the LWA to its saturated surface dry condition (SSD) prior to mixing typically requires a minimum of 24 hours of presoaking. Adequate presoaking of aggregate is deemed to minimize most pumping issues associated with LWC. In fact, pumping structural LWC could be as efficient as that of NWC in any project. A pumping record – a height of 315 m (1030 feet) – was achieved for both LWC and NWC (Goeb, 1985). Much to our surprise, a contractor, who was tasked to use LWC for rehab of bridge slabs for one of the departments of transportation (DOT) in the tristate region, made a request to replace LWC with NWC. We were informed (informal communications) that this change-order was based on a claim that a suitable LWC mix that can be pumpable to a maximum height of 30.5 m (100 ft) was not available. Such a claim is contrary to the testimony of Colin Chiluski, Prairie's QC Technician, who conducted testing at the batch plant and site of 150 North Riverside skyscraper in Chicago: 'The pre-wetted lightweight aggregate concrete mix was very easy to work with; the concrete maintained its slump throughout the entire job, it pumped well even with the addition of steel fibers. It made my job easy' (Morris, 2017).

5.1.2 Cold weather concreting

The presoaking of aggregate is clearly a major concern in the winter season, which is a bigger issue for the northern US states. Experience indicates that freshly mixed LWC loses heat more slowly compared to NWC. However, when exposed to cold temperatures, some lightweight concretes are still susceptible to damage from surface freezing (NRMCA, 2003). While manageable at a cost, these technical concerns are often used to justify the limited use of LWC in the northern states. A testimony by Trin Vega, Concrete Contractor of the 150 North Riverside skyscraper indicated no issues during construction using LWC (Morris, 2017): 'Through it all, we have actually had a very positive and productive experience. The lightweight concrete pumped, placed and finished very well; we finished ahead of schedule and we are completely satisfied with the overall experience.' Such feedback defies claims of LWC being unsuitable for cold weathers because of LWA presoaking requirements. The lack of experience of many contractors was found to be the most common reason for such claims.

5.1.3 Strength

Many practitioners are skeptical as to whether the strength of LWC is adequate. Compared with NWC, the same LWC mix design could yield a 22% reduction in compressive strength (Cousins *et al.*, 2013). Nonetheless, design strengths can be met economically with the use of appropriate LWA. Cousins *et al.* (2013) reported compressive strength values for LWC well above 55.0 mPa (8000 psi), and Kahn *et al.* (2004) reported compressive strength values for LWC of up to 83 mPa (12,000 psi).

'Strength ceiling' is a common concept and is used to indicate the maximum compressive strength of LWC. A mixture reaches its strength ceiling when it possesses a slightly higher strength with a higher cement content. Reducing the maximum size of coarse aggregate can increase the strength ceiling. A major RMC supplier in the Tristate region (New York, Pennsylvania and New Jersey) emphasized that structural LWC is about as strong as NWC, which together with the reduced weight allows builders to place smaller columns and footings.

5.1.4 Permeability and durability

As discussed earlier, the permeability of LWC is highly dependent on a number of competing factors. Typically, LWAs have large pores, which is likely to increase the vulnerability of LWC to adverse penetration by or exposure to deleterious chemicals. The stronger aggregate/cement paste bond induced by the porous nature of LWA generally reduces the flow paths at the ITZ, and thus reduces the permeability of LWC. Prolonged cement hydration driven by the increased moisture absorbed in LWA (internal curing) produces lower permeability and improves durability over time. The latter is also controlled by a range of factors. The increased free water/cement ratios required to presoak LWA can allow greater capillary action, thus resulting in water/humidity penetration. Thermal and shrinkage cracking can also allow further ingress of moisture. Conversely, the lower coefficient of expansion of LWC reduces cracks caused by temperature changes. LWC has an increased tensile strain capacity and reduced early drying shrinkage, both of which minimize cracking compared to NWC.

The technical issues commonly cited to portray LWC as a lower-grade concrete can be mitigated by adopting good practices: material selection and evaluation, adequate mix design, and high-quality workmanship. Such concerns are fully controlled for precast LWC. Many of the technical and practical misconceptions about LWC stem from the partial understanding of a few aspects of this material. Consistency of batched LWA is difficult to control, particularly if aggregate is obtained from different sources. As such, routine characterizations and QC on the LWA at site is of great importance to account for absorption variation and to control the mechanical properties of the produced concrete.

5.2 The experts' view

Qualitative unstructured interviews were conducted with a total of 15 practitioners and experts for their extensive industrial experience in construction. These interviews complemented our findings from the anonymous quantitative questionnaire. The participants included LWA and LWC suppliers, designers, owners, state officials, and executives. The opinions from the confirmatory interviews were used to validate the observed trends and key obstacles identified in the questionnaires. Their input also provided a deeper insight into the demand and potential of LWC in the construction market.

5.2.1 Technicalities and logistics

Aggregate presoaking, stockpile management and lack of skilled contractors were flagged by most of the interviewees as the most commonly encountered issues facing the presence of LWC. Experts from major concrete manufacturers signified pumpability control on site as a major practical concern that could discourage end-users.

An executive from a major LWA producer in the northeast region implied that mediocre technical understanding and awareness amongst engineers and end-users was the primary issue. An experienced structural engineer emphasized that consultants should be well versed in LWC design methodologies and should realize the different types of LWA, their respective characteristics, and common constructability issues. Some experts flagged the separation of LWA in dedicated bins at RMC plants as an operational limitation that concrete producers often exaggerate. The RMC suppliers we interviewed in the tristate region pointed out that it is a common practice to presoak LWA in their facilities in year-round temperature-controlled wetting areas to ensure a consistent product.

5.2.2 Economic dimension

An expert from a major RMC provider indicated that the main problems facing the acceptance and popularity of LWC are shortsighted concerns with costs, resistance to change, and a lack of understanding of the importance of LWC as a feasible solution that offers an array of sustainability benefits. The difference in unit price between NWC and LWC is mainly driven by the cost of LWA, disregarding savings in terms of quantities and other benefits. This seems to drive the preferences of owners and engineers for NWC as the mainstream building material. In fact, the cost of LWC elements – irrespective of long-term benefits – can be measurably less than that of NWC. For example, Skoyles *et al.* indicated the following in a comparative study on concrete floor slabs:

> The revised report updates the demonstration of a cost comparison between various types of lightweight floors each having the same foundation specifications, showing the ranking of lightweight structural concrete can range from approximately 20% less expensive to 9% more expensive than equivalent structural elements under the worst-case scenario. Moreover, in the case of solid slabs (which are not strictly comparable because of the stronger foundations required) lightweight concrete is about 7% cheaper than hollow tiles and 14% cheaper than dense concrete floors considering the best case. (Skoyles *et al.*, 1979).

This highlights that this market generally adopts cost comparisons between LWC and NWC that are flawed or based on incomplete data.

The RMC producers we communicated with indicated that, despite the low demand, LWC is fairly profitable for sellers and buyers. While the unit price of LWC is 20–30% higher than that of NWC on average, they emphasized that the saving of 1.5 inch (3.80 cm) in floor thickness combined with sparing the use of fire protection agents easily offsets this price difference. This is true even if the huge reduction in foundation size is discounted. To our surprise, one of the interviewed structural engineers employed by the DOT in a northeast state implied that the shift toward LWC would result in noticeable volume reduction of NWA and NWC sales. This was expected to alter the common business scheme of the aggregate and RMC suppliers even with the compensated LWC price, he added. Such opinion flags the potential rivalry between NWA and LWA suppliers. Another engineer at the same

DOT indicated that changing the mechanics of the supply and demand of aggregate and concrete could cause a price drop in NWC, which obviously could affect the profitability of RMC suppliers.

5.2.3 Common applications

Many experts highlighted the higher acceptance of LWC in certain projects, specifically those that require higher fire resistance (3–4 hour ratings). Such projects include hospitals, nursing and assisted-living facilities, dormitories, prisons, high-rise steel structures, long-span bridge decks, and structures built on poor soils. LWC is also well received by DOTs in many states across the USA for the rehabilitation and upgrade of infrastructures, and specifically for bridge decks. Bridges are often widened, and shoulders added without a significant increase in the dead load owing to the 25–30% weight reduction that structural LWC can provide. Honoring the generally higher strength of NWC, a combination of LWC slabs and high-strength NWC/LWC columns, or more conventional steel columns, can provide an optimized solution for multistory buildings. This approach utilizes the pertinent sustainable benefits of LWC in terms of thermal insulation and reduced weight.

5.2.4 Market mechanics

Approximately half the interviewees thought that there is an undeclared rivalry between the two camps in the market: the old school NWC advocates and the more modern builder supporting LWC. This is evident both in their concept and perception. The experts believe that the bias toward NWC is successfully fueled by the dominance of old school 'gurus' in many organizations and businesses related to construction. However, they do not believe that promoting LWC would negatively affect the market position of NWC. This belief is due to the long-term presence of NWC and the distinct differences in the properties and applications of NWC and LWC. Most experts emphasized the urgent need to counter the false perceptions of LWC in the construction industry. Among their suggestions are promoting awareness through conventions and professional events on the benefits of LWC, publishing articles on past experience, presenting service records and credible lab reports, and pertinent research. The experts are of the opinion that designers, followed by end-users, are the key influencers who can encourage the use of LWC. They indicated that certain cosmopolitan areas are more likely to host the desired change – given the need. This is geographically in good agreement with survey findings.

5.2.5 Practice and culture

NWC – being the mainstream concrete type – is perceived as a major marketing obstacle to LWC, particularly when many stakeholders are suspicious or unaware of the value of the substitute, and hence resist change (Zidan *et al.*, 2013). The experts interviewed advocated the need to include LWC into building codes and green building certifications, particularly compliance and acceptance criteria. However, a majority does not foresee that these changes will affect construction practices in the near future. The resistant culture was partly attributed to the technical myths as discussed earlier. This places research institutions and government agencies at the forefront of advocating sustainable construction. Other solutions could include public private partnerships (3P) as they pertain to public infrastructure. These groups are very interested in long-term performance since they must maintain a certain facility for a long period of time and hand it back to the public sector at an agreed time and condition.

5.2.6 Current and prospective trends

Most of the interviewees agreed that the trend of LWC demand is currently flat at a negligible market share of 5–7% of all the concrete produced in the nation. Rural suppliers are less likely to produce LWC unless a specialty project is involved (e.g. hospital, or bridge deck nearby). Aggregates and concrete suppliers as well as contractors indicated that LWC is rarely requested by designers and owners for structural work. A few interviewees noted some recent activity in the LWC works, primarily floor slabs in commercial and residential buildings. There is a concern that LWA production has not returned to the pre-recession levels that occurred in 2008 in many US states. Some experts emphasized that the key obstacles, such as unit price and availability of LWA, can be circumvented by improving production technologies and marketing strategies. Likewise, adequate geographical stockpiling of LWA is needed, as are new regulatory QC/QA procedures. Generally, LWA is increasingly supplying new and emerging markets besides the traditional concrete and concrete block industries, as indicated by an aggregate producer. These nonstructural applications include geofill, water filtration, green roofs, and soil blends for horticultural purposes.

5.3 Business reflections

The nature of competition in any industry is generally controlled by five forces or powers, referred to as the 5Ps (Figure 1.24(a)): threat of new entrants, threat of existing substitutes/competitors, bargaining power of consumers, bargaining power of suppliers, and the rivalry power among the existing competitors (Porter, 1979). This model can be used as a strategic business tool to analyze the attractiveness and potential of LWC in construction. Based on the survey results and the views of the experts, the authors of this chapter tried to provide their own assessment of the market forces controlling the presence of LWC in construction (Table 1.2). The influence levels associated with each factor were assigned a Likert score from 1 to 5 as follows: 1 – very low; 2 – low; 3 – moderate; 4 – high; and 5 – very high.

5.3.1 Forces of new entrants and competitors (threats)

LWC provides a unique and versatile blend of architectural and structural characteristics. To this end, the emergence of new entrants that can pose a threat to the status of LWC in the market is highly unlikely. Although NWC cannot substitute for LWC, particularly when weight reduction is targeted, it is perceived to be the prime market

(a)

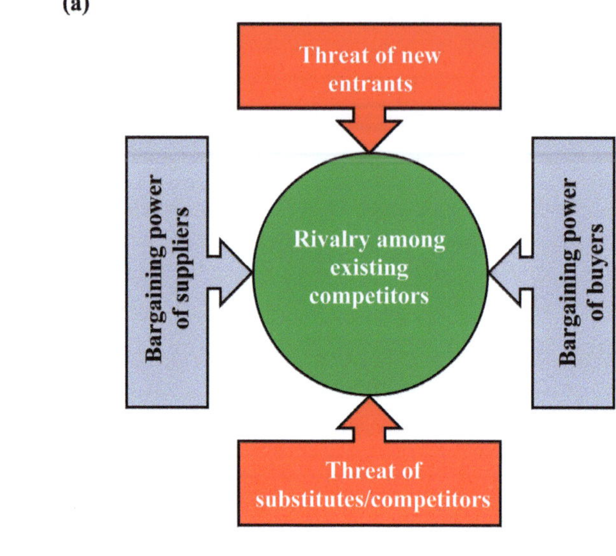

(b)

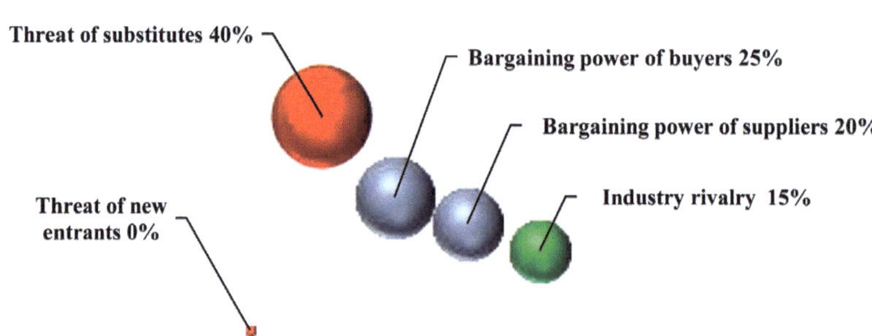

Figure 1.24 *Evaluation of the LWC industry: (a) Porter's 5Ps; (b) relative influence of forces (Mousa et al., 2018).*

competitor. LWC is obviously less competitive than NWC from a unit price perspective (very high influence). RMC suppliers and concrete manufacturers could be burdened with additional costs and lower efficiency by switching between different types of concrete to accommodate LWC during daily operations (increased influence). The availability of LWC is relatively limited compared to NWC, mainly because of the low demand. In this context, the strong attachment of buyers to NWC has a very high influence.

5.3.2 Bargaining powers (opportunities)

NWC will likely remain in high demand because end-users and designers appreciate immediate/short-term cost savings (increased influence). The choice of LWAs is technically sound for buyers seeking the sustainable benefits of LWC. This class

Table 1.2 *Competitive analysis for the LWC market using Porter's 5Ps (Mousa et al., 2018).*

Market Force	Influence
Threat of New Entrants	–
Threat of Substitutes/Competitors	
O and D: Relative price of NWC	5
RS and CM: Switching cost from NWC to LWC	4
RS and CM: Availability of LWC and LWA compared to NWC	2
O and D: Buyers' propensity to substitute to NWC	5
	Average = 4.0
Bargaining Power of Buyers	
O, D and CC: Loyalty to regular RMC suppliers	5
O, D and CC: Quality of LWA and LWC	1
O, D and CC: Size of LWC orders	3
O, D and CC: Ability to substitute LWA	1
	Average = 2.5
Bargaining Power of Suppliers	
LS: Geographical availability of LWA	1
LS: Increase production of LWA to provide attractive cost for the industry	1
LS, RS and CM: Consistent and steady supply of LWC	4
RS and CM: Loyalty to regular aggregate	2
	Average = 2.0
Industry Rivalry	
Lack of strategic diversity among competitors	1
Prevalent brand (NWC) loyalties	2
Highly differentiated types of concrete	1
Small number of LWC producers	2
	Average = 1.5

O:	Owners/End-users	1: Very low
D:	Designers (architects and structural engineers)	2: Low
RM:	RMC suppliers	3: Moderate
LS:	LWA suppliers	4: High
CM:	Concrete products manufacturers	5: Very high
CC:	Concrete contractors	

of buyers, however, can have some influence on mainstream LWC consumption. Owing to the small market share of LWC, customers for LWC have very little say on replacing LWA with a cheaper constituent. Suppliers, on the other hand, need to adopt effective and swift LWA stockpiling policies to make it more appealing. If the demand for LWA grows, mass production will create true competition among aggregate suppliers, who would then take the risk of investing to mass produce LWA.

With the presence of quarries around the USA, the geographical availability and production of LWC are probably of little concern. The consistency and steady supply of LWC depends largely on LWA. This seems to be highly controlled by aggregate producers, which could give them the upper hand in this aspect. RMC suppliers and concrete manufacturers have to follow the market mechanics. Therefore, their long experience and affinity to regular aggregate should be of no influence if the demand for LWC increases.

5.3.3 Industry rivalry

The absence of strategic diversity among concrete producers diminishes market competition, thereby positioning LWC as a nonviable construction alternative. In this environment, the strong attachment of buyers to NWC is high, which further slows competition. the word "act" interrupts the idea ..please simplify to: "The increased differentiation between LWC and NWC has established static consuming patterns as if the customers target each product act in isolation. Likewise, the small number of LWC producers does not enhance competition in the market, which in turn results in unappealing prices.

5.3.4 Possible market mechanics

This analysis reveals that the dominant presence of NWC seems to impede the market competitiveness of LWC. The relative average ratings of the five forces controlling the market are shown in Figure 1. 24(b). The threat of economically more appealing substitutes/competitors (in this case resorting to NWC at the expense of LWC) seems to be the major force driving the market. The strong preference for NWC exhibited by the end-users and designers, and their resistance to change current practices, equally pose a measurable barrier against increasing LWC demand. Mousa emphasized the low-market rivalry when a well-established construction material is replaced with a more sustainable but more expensive substitute (Mousa, 2015a). Apparently, there is a good balance between the bargaining power of the concrete suppliers and the buyers. With relatively low-market rivalry, both buyers and suppliers have a reasonable and equal chance to reshape the competition in the construction industry favorably towards LWC.

5.3.5 Hunting for incentives

Energy savings together with the tolerable levels of environmental impact enable designers and end-users of LWC to claim leadership in energy and environmental design (LEED) credits (Ries *et al.*, 2010). The LEED system was developed by the United States Green Building Council (USGBC) to encourage energy reduction, health, environmental and financial benefits, and social justice to the communities. The system has nine credit categories with corresponding points available up to a maximum number (Table 1.3), and awards buildings four levels of certification: certified (40–49 points), silver (50–59 points), gold (60–79 points), and platinum (80+ points) (https://www.usgbc.org/leed). The points are often calculated on a project-by-project basis.

Table 1.3 *Potential LEED points that LWC can receive.*

Concentration	Maximum Points[1]	LWC Contribution Points
Integrative Project Planning and Design	1	–
Location and Transportation	16	–
Sustainable Sites	10	1[2]
Water Efficiency	11	1[3]
Energy and Atmosphere	33	1[2]
Materials and Resources	13	7[2]
Indoor Environmental Quality	16	1[2]
Innovation	6	1[2]
Regional Priority	4	–

[1]USGBC, 2017.
[2]USGBC, 2020.
[3]Norlite Lightweight Aggregate, 2013.

Figure 1.25 summarizes the most common benefits of LWC (USGBC, 2017). From a business perspective, the LEED strategies can help property owners to reduce their operating costs, obtain tax incentives, enhance occupant efficiency (by improving productivity, comfort, and indoor environmental quality) and improve lifecycle performance over the building service life (US EPA, 2016). While such financial and social incentives can change market mechanics and give LWC some 'oomph' to acquire a higher market share, most owners seem to be oblivious to the benefits that LEED certification could bring.

6. Value engineering analysis

Many customers, and probably most average owners, are believed to lack the experience and/or the motivation needed for a holistic evaluation of value. In response to customer pressure to reduce cost, value engineering (VE) continues to identify capital improvements, pursue quality and increased profitability for market enhancement (Abidin and Pasquire, 2005). VE is a systematic and organized decision-making tool carried out on a product, project or service so that the required functions are rendered most effectively at the minimum possible cost. In so doing, consideration is given to viable alternatives without compromising the basic requirements. The use of VE at the initial phase of the construction projects has proven to reduce capital costs by 10–25% (Ellis *et al.*, 2005). Experience has also shown that the methodology has resulted in savings estimated at around 5–10% for a significant number of construction projects (Norton and McElligott, 1995). The method is widely accepted as an efficient tool for successful management in construction (Ellis *et al.*, 2005).

It is the authors' opinion that the controversial situation surrounding LWC in the construction market can be better evaluated using VE. This is of particular importance for novice owners and end-users who may not be aware of the merits of LWC and – equally important – its edge over NWC in terms of sustainability. To this

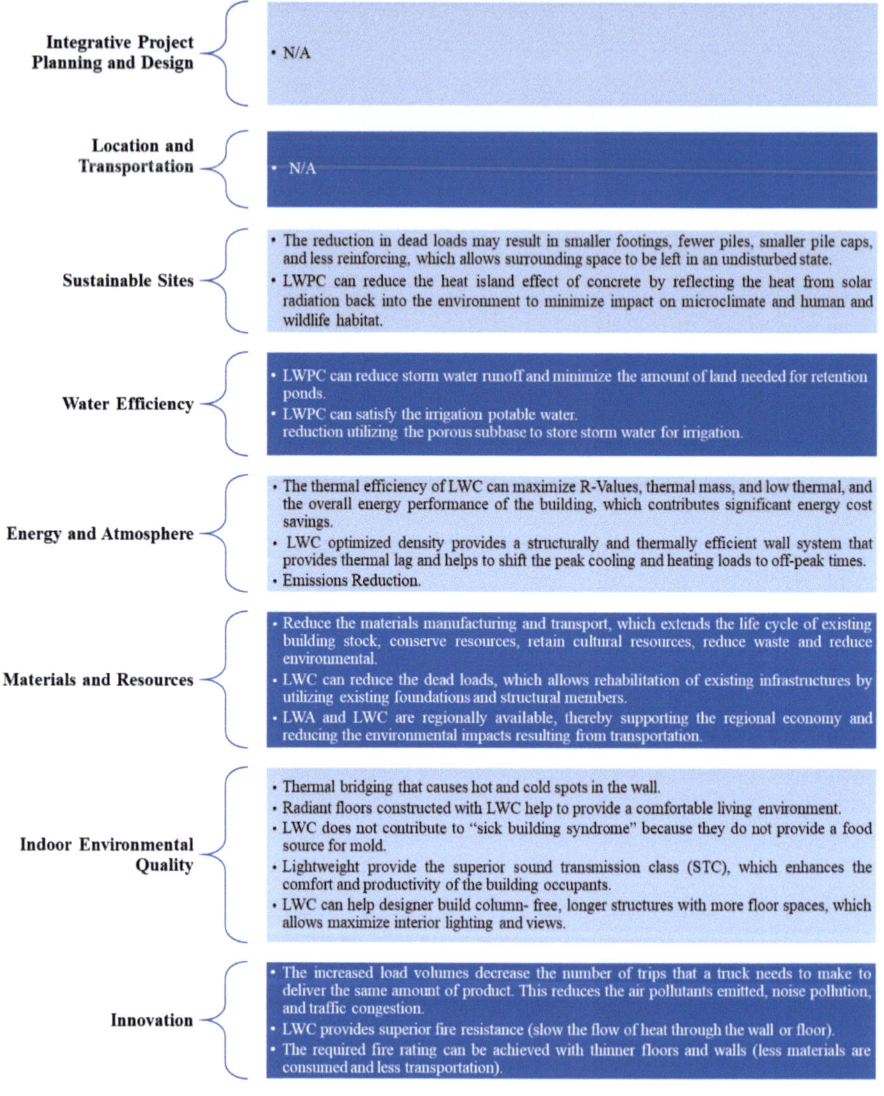

Figure 1.25 *The LEED system credit categories. LWPC = lightweight pervious concrete.*

end, VE can provide them with a simple numeric guide for sound decision making as regards choosing LWC over other types of concrete. For this purpose, the weighted evaluation matrix (WEM) procedure by Dell'Isola is a good candidate (Figure 1.26). The following section examines the appropriate steps in the decision-making process (Dell'Isola, 1982). The authors chose to evaluate seven types of concrete: non-structural NWC, SDC, semi LWC, structural LWC, high-strength LWC, high-strength NWC, and structural NWC.

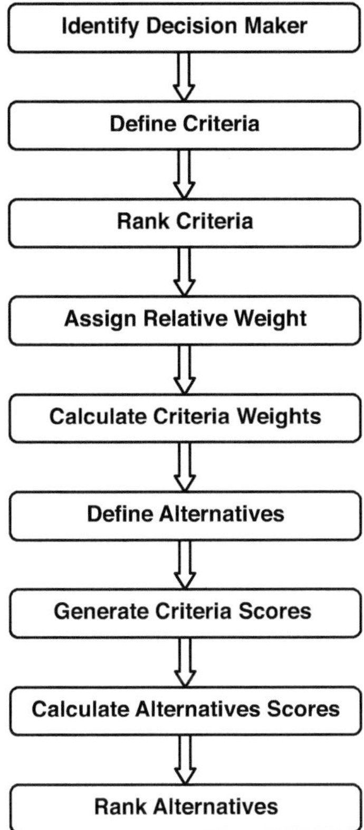

Figure 1.26 *Basic steps of the WEM procedure (after Dell'Isola, 1982).*

6.1 Evaluation of performance

In construction there are always alternatives (e.g. materials) that can be considered for the purpose of delivering targeted building functions. The selection of the optimum alternative requires evaluation of both cost (monetary) and performance/quality (non-monetary). The WEM technique was selected to assess the value of LWC against alternative concrete types. The WEM encompasses several factors for the performance criteria. The weightage of the factors of the performance criteria is set a priori. The performance criteria comprise the major value factors reported in the literature, which will be used subsequently to evaluate the proposed alternatives. Those factors are assigned relative weights, established by a paired-comparison technique (Dell'Isola, 1982). The evaluator then assigns a criterion score for each alternative.

In this technique, the weightage assignment is governed by the relative importance of any pair of factors. For example, the factor A is assigned a higher weightage compared to the factor B if the expert's judgement suggests that A is more important than B. This should be supported by experience and adequate understanding of the role of each factor. The method also allows adjustment to be made to the level of importance for any factor in the criteria. In view of this, the weightage can be set as follows: A1 for minor difference of importance – A is more important than B); A2 – the importance of A is twice that of B (or higher for major difference); and AB for equal importance. Upon completing the comparisons between all possible pairs of factors, the number of repetitions for each factor (including the importance multiplier and factors interaction) is reported in the matrix as a raw count (weight points). The normalized weight (%) is calculated as the ratio between the total number of repetitions (appearances) of any factor to the total number of all factors with their importance multiplier in the matrix (32 in this analysis).

In Figure 1.27, the alternative concrete types considered are listed below the criteria and subsequently rated against each other using a 5-point Likert scale ranging from 1 to 5 (where 5 is excellent and 1 is poor) for each alternative (Dell'Isola, 1982). The function point for each alternative is the weight of the function multiplied by the score of this alternative for this specific function/factor (e.g. fire resistance). The total number of function (performance) points for each alternative is the sum of the points that the factors A through H receive (sum of the row). This is repeated for all alternatives. In order to enhance judgement and minimize bias, the authors of this chapter conducted the analysis separately. Figure 1.27 shows the average values for the criteria scoring as well as the ratings for the alternatives. For example, material savings was rated as important as structural efficiency (AB) but was assigned three times the importance of acoustic insulation (A3). For the alternative scoring, LWC was given a very high rating in materials saving and fire ratings (4 or 5 points). So, it is the combination of criteria importance times the score for the alternative in any given aspects that decides the quality points an alternative can receive. The total (overall) quality points is what finally gives the overall performance (function/quality) for that alterative.

6.2 Estimation of cost

The cost of each alternative was based on the available average prices of the types of concrete considered (Figure 1.27). For the purpose of this chapter, the price ranges (USD per cubic yard, (1 cubic yard = 0.765 cubic metres) were based on quotations received from several prime RMC providers in the north-east of the USA. Both suppliers, who chose to remain anonymous, are well-established mid-tier companies operating in the tristate area. Their services include asphalt concrete and recycled concrete. The price ranges were cross-checked with experts from industry. It should be noted that the prices used represent the initial (present) cost in our case. However, it is vital in large-scale projects to consider the lifecycle cost aspects, such as maintenance, operations, energy, salvage value and other annual costs.

6.3 Determination of value index

Intuitively, the ultimate goal of value engineering is to identify the best alternative. It is the one that can minimize resources, financial expenditure (cost) and maximize the benefits (performance). This can be captured mathematically (Equation (1))using the so-called value index (VI), defined as the ratio of function (performance) to cost (Kelly *et al.*, 2014):

$$Value\ Index\ (VI) = \frac{Performance}{Cost} \times 100 \tag{1}$$

Non-Monetary Criteria

A	Material savings	A							
B	Structural efficiency	AB	B						
C	Carbon emission	AC	B1	C					
D	Energy	A2	D1	D1	D				
E	Thermal conductivity	E1	B2	E2	E1	E			
F	Fire rating	A1	E2	C1	E2	E1	F		
G	Acoustic insulation	A3	E3	C2	E3	G2	G2	G	
H	Durability	A4	B1	H2	D1	H1	F2	GH	H

	A	B	C	D	E	F	G	H	Quality Points	Cost (USD/cyd)			Measured Value			Normalized VI
Weight Points	5	5	4	3	5	5	3	5		Min.	Max.	Avg.	Min.	Max.	Avg.	
Weight %	14	14	11	9	14	14	9	14								
Non-Structural NWC	2	1	4	4	2	2	1	3	231	110	125	**118**	210	185	**197**	80
	28.6	14.3	45.7	34.3	28.6	28.6	8.6	42.9								
SDC	2	3	2	3	3	3	2	3	266	115	175	**145**	231	152	**183**	75
	28.6	42.9	22.9	25.7	42.9	42.9	17.1	42.9								
Semi LWC	3	4	4	3	2	3	2	5	331	155	185	**170**	214	179	**195**	79
	42.9	57.1	45.7	25.7	28.6	42.9	17.1	71.4								
Structural LWC	5	2	3	4	4	4	4	4	374	155	185	**170**	241	202	**220**	90
	71.4	28.6	34.3	34.3	57.1	57.1	34.3	57.1								
High-strength LWC	5	5	3	2	4	4	3	5	406	135	195	**165**	301	208	**246**	100
	71.4	71.4	34.3	17.1	57.1	57.1	25.7	71.4								
High-strength NWC	3	5	2	2	4	2	3	5	337	140	200	**170**	241	169	**198**	81
	42.9	71.4	22.9	17.1	57.1	28.6	25.7	71.4								
Structural NWC	2	3	3	2	2	2	3	5	277	115	175	**145**	241	158	**191**	78
	28.6	42.9	34.3	17.1	28.6	28.6	25.7	71.4								

Figure 1.27 *WEM for the alternative concrete types, based on average scoring given by three experts: (a) setting the criteria and relative importance of the factors; (b) weight points based on setting the criteria, assigning a value score for each alternative considering aspects (A–H), and quality points calculation and estimation of value index VI. (Weight points: 5 excellent, 4 very good, 3 good, 2 fair, 1 poor; white cells are user/evaluator entries in the matrix, and gray cells are calculated automatically).*

The procedure is automatically programmed using the analysis matrix (Figure 1.27). The VIs shown are based on the average cost. Accordingly, the optimum alternative is the one that receives the highest VI (high-strength LWC). It is important to realize that the criteria, weights and alternatives in this procedure should preferably be set by several experts to enhance objectivity.

6.4 Discussion

The authors have obviously considered personal preferences and judgements in performing value engineering analysis – truly, there is no way to entirely eliminate subjectivity. Ranking the alternatives is accordingly based on the scores they received (VI). Despite the differences in the VI values between the three evaluators, the ranking of alternatives came out as quite consistent. In descending order the ranking was as follows: high-strength LWC, structural LWC, high-strength NWC, non-structural NWC, semi LWC, structural NWC and finally SDC. One can still debate whether this agreement was a coincidence or was, rather, driven by the related background and mindset of the evaluators. The truth is that setting the criteria plays a major role in the end result, which is why 'Identify decision maker – first step' is pivotal in this process. Put simply: Who makes the final call? In this example, the set criteria follow many sustainability aspects in construction, which reflects the background and inclination of the group setting it. With a revenue-driven mentality, many of those aspects may not find their way to this list. It is therefore important to ensure diversity in the team performing value engineering.

7. Success stories

The use of LWC has been shown to be feasible in many structures, particularly bridges and high-rise buildings. The versatility of LWC has made possible the successful construction of new and rehabilitation projects. Using LWC seems to be the answer to building high-rise structures in narrow areas and upgrading deteriorated bridge decks in congested junctures. It was a distinct challenge to find recent references collecting the characteristics and features of LWC structures, either in the USA or elsewhere. There are more than 400 simple and long-span LWC bridges throughout the world, mainly in the USA and Canada (Holm *et al.*, 1984). The current number of LWC bridges is expected to be much larger than this figure. The Federal Highway Administration (FHWA) conducted a comprehensive study on the available LWC bridges and concluded that most of these bridges are in a good condition (FHWA, 1985).

In view of the limited resources, the information on LWC bridges and structures was gathered from the websites of local and public entities involved in bridge and high-rise construction. LWC bridge construction and bridge replacement in the USA involves spans ranging from a couple of hundred to thousands of feet (1 foot = 0.048 m) with total lengths up to several miles (1.61 km) (Table 1.4). Several examples of LWC bridge replacement projects from other countries are shown in Table 1.5. Mid-rise and high-rise LWC buildings in the USA have been shown to perform adequately since their construction, some of which were built in the 1970s (Table 1.6). Several low-rise to high-rise buildings made of LWC from other countries are presented in Table 1.7. While many of these LWC buildings are business centers, the applications extend to hospitals, museums and residential properties. As shown in Tables 1.4–1.7, LWC is more popular in cosmopolitan areas, where there is a high need for major transportation projects and high-rise buildings to combat limited land availability and challenging accessibility. In this section, we single out a few examples of the successful use of LWC in bridges and high-rise buildings in the USA and discuss the main features of these projects in light of the merits of LWC.

Table 1.4 *Length and span of selected LWC bridges in the USA.*

Bridge	State	Year	Length (ft)	Max. Span (ft)
[1]Governor William Preston Lane Jr. Memorial Bridge Eastbound	MD	1952	21,279	1,600
[2]Coronado Bridge	CA	1969	11,179	660
[1]Governor William Preston Lane Jr. Memorial Bridge Westbound	MD	1973	21,051	1,500
[3]Antioch Bridge	CA	1978	9,437	460
[4]Parrotts Ferry Bridge	CA	1979	1,293	640
[5]Arthur Ravenel Jr Bridge (Cooper River Bridge)	SC	1991	16,450	800
[6]Brooklyn Bridge	NY	1999	5,989	1,596
[7]Neuse River Bridge	NC	1999	10,560	3,200
[8]James River Bridge Restoration	VA	2002	4,185	415
[9]Benicia-Martinez Bridge	CA	2007	8,976	528
[10]Sam White Bridge	UT	2011	354	177
[11]Skagit River Bridge	WA	2013	160	757
[12]Thaddeus Kosciusko Bridge (Interstate 87)	NY	2013	764	550
[13]I-40 Bridge*	TN	2015	2,411	312
[14]Pulaski Skyway	NJ	2016	18,491	550

*Rehabilitation over the French Broad River.
[1] Maryland Department of Transportation, 2020.
[2] OPAC Consulting Engineers (n.d.(a))
[3] Metropolitan Transportation Commission, 2013.
[4] OPAC Consulting Engineers (n.d.(b))
[5] McCabe, R., 1993.
[6] L. B. Foster Company, 2018.
[7] Sigmon, 2000.
[8] Kozel, 2004.
[9] Metropolitan Transportation Commission, 2009.
[10] NSBA, 2021.
[11] Baker *et al.* (2014).
[12] Martin and Blabac, 2009.
[13] Bell & Associates, 2015.
[14] FHWA, 2017.

Table 1.5 *Selected LWC bridge replacement projects from other countries.*

Bridge	Location	Year	Length (ft)	Max. Span (ft)
[1]Pont de Beaumont-sur-Oise	Île-de-France, France	1997	630	394
[2]Rugsund Bridge	Sogn og Fjordane, Norway	2001	991	623
[3]Sundøy Bridge	Nordland, Norway	2003	1,765	978
[4]Sandsfjord Bridge	Rogaland, Norway	2015	1,903	951

[1]Structurae, 2016a.
[2]Mailhot and Eng.
[3]Structurae, 2016b.
[4]Structurae, 2018.

Table 1.6 *Selected mid- and high-rise LWC buildings in the USA.*

Project	State	Year	Floors	Height (ft)	Total footage (sq. ft.)
[1]One Shell Plaza	TX	1971	52	714	1,300,000
[2]North Pier Apartment Tower	IL	1991	61	581	550,000
[3]Bank of America Corporate Center	NC	1992	60	871	1,400,000
[3]First National Center	NE	2002	45	633	730,000
[4]Goldman Sachs Tower	NJ	2004	42	781	1,600,000
[5]Comcast Tower	PA	2008	60	1,176	1,399,997
[6]Duke Energy Center	NC	2010	48	786	1,500,000
[7]Panasonic Headquarters	NJ	2013	20	unknown	340,000
[8]Prudential Tower	NJ	2014	20	313	744,000
[9]Wilshire Grand Center	CA	2017	73	1,100	1,500,000

[1]Colaco, 2004.
[2]ACI, 2005.
[3]ESCSI, 2003.
[4]Council on Tall Buildings and Urban Habitat, no date(a).
[5]Schwing America Inc., no date.
[6]Council on Tall Buildings and Urban Habitat, no date(b).
[7]Schwartz, 2013
[8]KPF, no date.
[9]Arcosa Lightweight, no date.

Table 1.7 *Selected low- and high-rise LWC buildings from other countries.*

Project	Location	Year	Floors	Height (ft)	Total footage (sq. ft.)
[1]Standard Bank Centre	Johannesburg, South Africa	1968	39	456	322,917
[2]Central Square Building	Sydney, Australia	1972	26	302	Unknown
[3]Guy's Hospital (400 beds)	London, England	1974	34	488	Unknown
[4]Picasso Tower	Madrid, Spain	1988	51	515	1,302,000
[5]Guggenheim Museum Bilbao	Abando, Spain	1997	3	187	350,000
[6]CityLife Residential Buildings	Milan, Italy	2013	13	194	409,029

[1]Vosloo, 2020.
[2]Clarke, 1993.
[3]Gutenberg, no date.
[4]Per-gestora, no date.
[5]Emporis GMBH, no date.
[6]EXCA, 2011.

7.1 Bridge rehabilitation

7.1.1 Pulaski Skyway

The 3.5-mile Pulaski Skyway in New Jersey used 41.37 MPa (6000 psi) precast LWC deck panels for a major rehabilitation after 85 years of service. The bridge accommodates an average daily traffic of 67,000 vehicles that commute across this highly congested area (Cheng, 2013). The elevated bridge includes a series of superstructures: two 168-m (550-ft) through-truss main spans with 107-m (350-ft) flanking-truss spans over the Passaic River and the Hackensack River, three steel through trusses over railroads in Jersey City (east), deck trusses between the two through trusses, and deck trusses and girder bridges in the remaining spans (Figure 1.28). Replacing and upgrading the deteriorated decks and girders using conventional concrete would not have been possible either financially or logistically. Thanks to the low weight of the expanded shale, the use of LWC decks significantly reduced the superstructure loads. Such a decision allowed a smart transition from full replacement of the bridge to optimized rehabilitation (only the deck), which

Figure 1.28 *Pulaski Skyway, New Jersey.*

consequently enabled substantial cost savings, accelerated construction, with minimal traffic impact. The use of LWC segments allowed construction within 2 years and extended the lifecycle of the structure to an additional 75 years at an estimated cost of approximately $1.5 billion.

7.1.2 Thaddeus Kosciusko Bridge (I-87)

The Thaddeus Kosciusko Bridge (also known as the Twin Bridges), built in 1959, is located in upstate New York and spans the Mohawk River on Interstate 87. It consists of a pair of identical through steel arch bridges with concrete decking carrying three northbound and three southbound lanes of traffic. The New York State Department of Transportation (NYSDOT) foresaw the inevitable deterioration in the bridge's lifespan and the need for future repairs. That being said, NYSDOT had to perform the rehabilitation taking into consideration the potentially dramatic impact of repair work on the 110,000 vehicles per day using this busy corridor. In order to minimize road closure and traffic impact, a precast LWC deck replacement was accordingly selected to increase the load ratings of the refurbished structures. The northbound bridge deck was removed and replaced over seven weekends in the fall of 2012 and the southbound bridge deck was removed and replaced over five weekends in the spring of 2013. The schedule included lane construction to perform the work and complete the project in only 12 weekend closures with minimal impact to the commuters. The LWC mix design achieved a compressive strength of 27.58 MPa (4000 psi) within 18 hours with an average 28-day strength greater than 48.26 MPa (7000 psi) (PCI, n.d.). The specified 1842.00 kg/m^3 (115 pcf) density allowed the LWC segments to be 26% lighter, which – in addition to the structural efficiency – encouraged the use of smaller mobile cranes.

7.2 New bridge construction

Although now less common for new construction, a few authorities in USA elect to utilize LWC in bridges from the start. In 2002, Virginia Dare Bridge was built using LWC to connect the Manns Harbor and Roanoke Island in North Carolina (Figure 1.29). At 8.37km (5.2 miles), the bridge is one of the longest concrete bridges ever built on the East Coast (Stalite Lightweight Aggregate, 2020). It consists of 268 spans with precast, prestressed bulb-tee concrete girders and a cast-in-place LWC riding deck surface. The high performance LWC deck has a specified 28-day compressive strength of 31.02 MPa (4500 psi) and a maximum equilibrium density of 1842.00 kg/m^3 (115 pcf) (Castrodale, 2006; MBP, n.d.). The deck occupies more than 172,387.15 m^2 (1,855,560 square feet) consuming approximately 33,640.41 m^3. (44,000 cubic yards) of concrete (KCI Technologies, 2017; Stalite Lightweight Aggregate. 2020;). With a targeted service life of 100 years, the bridge was designed to resist a highly corrosive coastal ecosystem, coastal storm surges, navigable clearances, vessel collision forces and scour impacts. More than 300 concrete mixtures were developed and tested to ensure the durability of all the bridge components needed to warrant a 100-year service life, twice as long as the preceding generation of bridges (Prokopy, 2004).

Figure 1.29 *The Virginia Dare Bridge, Dare County, North Carolina.*

7.3 High-rise building

7.3.1 150 North Riverside Plaza

This gravity-defying skyscraper is located right on the Chicago River in downtown Chicago. The 226.16-m (742-foot) tower (54 floors) was completed in 2017. Approximately 10,733,77 m³ (14,000 cubic yards) of steel fiber reinforced LWC was used in its construction (Morris, 2017). The concrete mix utilized expanded shale to render a low density of 1762 kg/m³ (110 pcf) and a characteristic strength of 27,58 MPa (4000 psi). Despite the height, the mix was pumpable with adequate workability. The designed LWC provided the optimum strength and desired fire rating. In designing the building the architects considered a number of sustainable aspects. The project utilized expanded shale material to open and enrich the planting soil in a uniquely designed green roof and also maximized the open space within the building. The light weight combined with the specially designed core allowed the tall building to be narrower at the bottom and wider at the upper floors – such an unusual 'gymnastic' architectural figure. In fact, the building received a LEED-CS Gold Pre-Certification for the wide spectrum of sustainable features it brings.

7.3.2 The Tower at First National Center

The First National Center is a 40-story, 192.94-m (633-ft) tall office building located in downtown Omaha. Built in 1997, the tower is still the tallest building in the upper Midwest between Minneapolis and Denver. The tower utilizes steel frames and LWC floor slabs on a metal deck to minimize dead loads. The floor slabs comprise 8410.10 m³ (11,000 cubic yards) of structural LWC made of vacuum-saturated expanded shale LWA. The floors enjoy a high compressive strength of 55.18 MPa (8000 psi,), which provides the rigidity needed for a high-rise. The mere 13.34-cm (5.25-inch) slabs allow a 2-hour rating and eliminate the need to spray-fireproof the metal deck, which results in considerable savings given the total floor area.

7.4 Commentary

The diversity of projects utilizing LWC is indicative of the material's success – not to mention that the projects are geographically located in different climates and vary in size and requirements. The adequate performance of LWC in harsh environments particularly cold weathers is evident. As such, the performance of LWC has proved beyond doubt that the technical myths often brought up as potential concerns are a complete fallacy. The examples presented in this section have shown that LWC can provide attractive building solutions from both a practical and technical viewpoint. It is the authors' observation that LWC is rarely chosen as the preferred building material under normal circumstances. In reality, however, LWC is a material with special characteristics that make it an excellent alternative for rehabilitation or to combat challenging site conditions and construction limitations.

8. Prospective

There is a clear bias toward NWC in construction, which is clearly manifest by its use in almost all projects. The indiscriminate use of NWC as the mainstream building material in USA is certainly uneconomical, unsustainable and often unjustifiable. It is the authors' observation, that the values that LWC can bring are either not well appreciated by end-users or are possibly concealed by the major market players. To change this situation would require a game changer. Accordingly, the authors propose tackling this situation from the business and the management ends which, when combined, can enhance the presence of LWC in construction.

8.1 Business intervention

Based on the survey presented earlier in this chapter, there are a number of clear barriers facing LWC in the USA: construction culture, the market dominance of NWC, level of awareness, and technical myths. Despite the declining sales trends for LWC and a customer mindset fixated on NWC, it our assessment that buyers and suppliers probably have equal opportunity to reshape this competition bias and bring balance to the market mechanics. The observations and findings of this study could be further supported if LWC production records made available. Consideration should be given to Kotter's model as a strategic guide for stakeholders to initiate and sustain a paradigm change in stagnant construction cultures (Mousa, 2015b). As shown in Figure 1.30, the model embraces successful organizational change through three transformation phases: namely, unfreeze, change, and lock (freeze) (Kotter, 1995).

8.1.1 Unfreeze

Building the sense of urgency, creating a coalition and developing a vision are necessary for initiating the unfreeze phase (unlock). Calling for a prompt change in construction culture can better be accomplished by identifying the potential threats

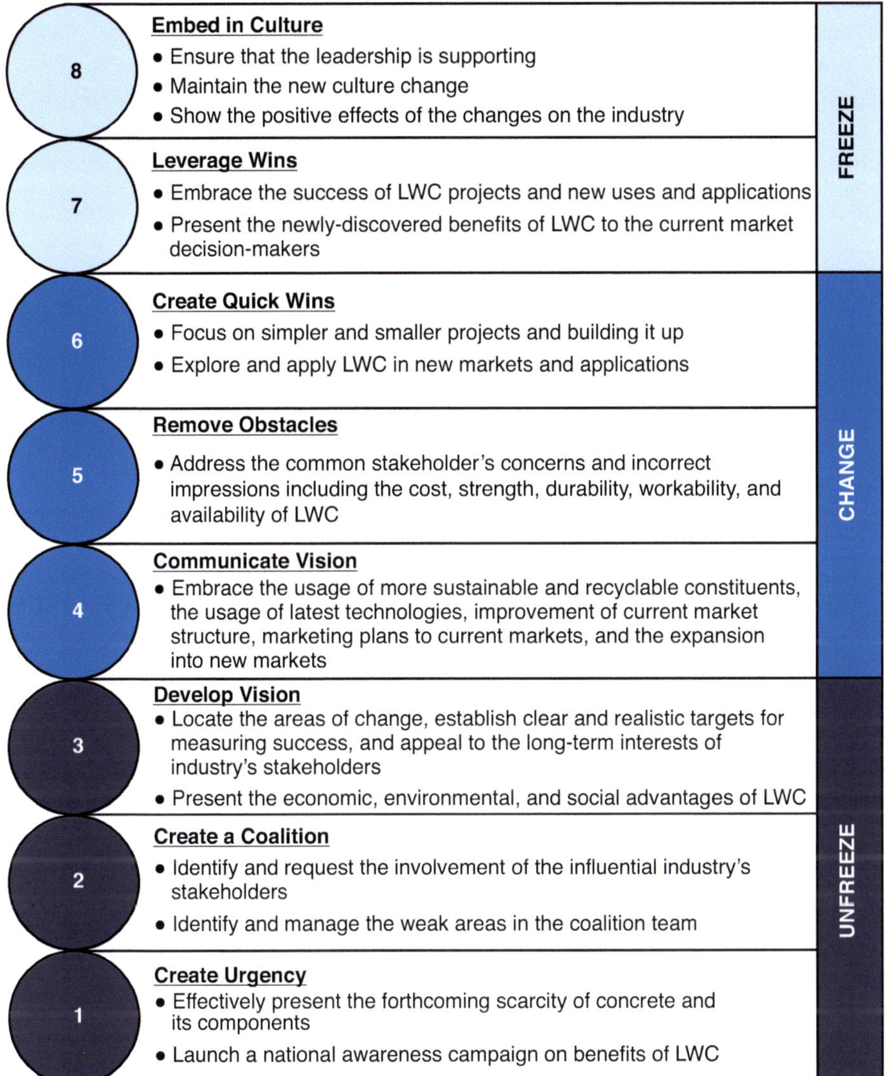

Figure 1.30 *Kotter's 8-step process for potential change in the LWC market (adapted from Mousa et al., 2018).*

to the presence of LWC and simultaneously highlighting its benefits (as discussed earlier). The multidimensionality of the transformation requires the formation of a formidable coalition with different market players: green and sustainable construction certifiers, government officials, end-users/owners, structural engineers, architects, contractors, concrete suppliers, building materials manufacturers, and suppliers of the LWA. This coalition of stakeholders should be at the forefront of the launched change initiative in order to achieve the desired outcomes. Federal and state organizations – e.g. the Portland Cement Association (PCA), the ACI, NRMCA and Departments of

Transportation – are expected to promote the use of LWC in local projects on the basis of its sustainable benefits and also to encourage concrete producers to embrace and develop new LWC technologies. Architects and main consultants, as the most knowledgeable group of LWC value, are in appropriate positions to recommend and specify the use of LWC rather than NWC where applicable. It is paramount to bring about an integrated understanding of the benefits of LWC – including its cost-effectiveness, durability, LEED credits, and green construction incentives – to the public and private sectors. The major obstacles facing LWC must be equally addressed.

8.1.2 Change

Effectively communicating the developed vision and strategy to stakeholders is the first step in bringing about any desired change to the status quo – it lays the seed for bringing about the desired change. Stakeholders should advocate the potential long-term benefits – technical and financial – of using LWC for the construction industry as a result of the required change. For positive change to occur, it is essential to eliminate, or at least minimize, the barriers identified earlier, while following a practical, pre-established plan for prioritizing which issues to address first. Stakeholders likely to backup and counter the change must be identified. The rationale behind their positions should be carefully collected and analyzed. To this end, conflict of interest/ market rivalry must be considered in order to create the good consensus needed for a smooth transition in a construction culture that is centered around NWC by default. Several legislative changes will be required to advance the position of LWC in the construction market. Consideration will need to be given to providing incentives, such as environmental regulations, alternative fuel sources, or rates and tax reliefs, that will appeal to and attract market players. These proposed legislative changes would be expected to draw the attention of a wide audience in the construction industry to consider LWC as a sustainable and affordable building material. Adequate research funding from governmental and industrial entities should be allocated to enhance LWC properties and performance in order to warrant the competitiveness of LWC.

8.1.3 Lock

It is paramount to lock (freeze) the change and consistently evaluate and praise the short-term wins and achieved successes produced by the change (Kotter, 1995). Premature cessation of the efforts needed to maintain the achieved improvements has been responsible for failure of positive market changes that were originally successful. Losing grip on a new change typically happens upon fulfillment of easy or quick gains under the false hope that the market can sustain the change and that the mechanics have integrated the boosting effort (e.g. promoting the use of LWC in our case). For this to be avoided, the master vision must be routinely revisited in view of the observed industrial trends (e.g. concrete, LWA and cement sales). For example, the tax shield and favorable fuel rates provided to suppliers should be re-evaluated in response to the targeted increase in the market share of LWC. In order to incorporate the transformation into the core practices of the industry, there should be a measure for ensuring that the changes are well received and fairly benefit all market players.

Approaches that could favor some stakeholders are prone to failure in the long term. Therefore, monitoring the industrial changes, and ensuring that the leadership is supporting and maintaining the new culture, are necessary.

Promoting the sustainability benefits of LWC presented earlier could inspire a necessary transformation in construction procedures and practices. Obviously, promoting the technical arguments alone in order to support LWC use falls short of making this a reality. To realize the hoped-for transformation, appropriate marketing tools will be needed to enhance the presence of LWC in construction. Of a wide range of business solutions, segmentation/targeting/positioning is probably the most sustainable marketing framework for LWC (Proctor, 2000). The authors believe that LWC is well suited for certain major projects (e.g. bridge construction and repair) and particular clients (targets, e.g. DOTs), which should allow positioning the product in a more competitive place in the construction industry. Should those clients embrace LWC as a viable concrete product, its presence in the market would be expected to grow.

8.2 Value engineering implementation

In addition to the proposed business moves, there is a parallel need to seed awareness of LWC characteristics, applications and – as important – the added value that LWC brings amongst key market players. The answer to the limited familiarity and appreciation of LWC in the construction industry in the USA is to support and enhance business decision making. A quick management approach to achieve this should include lifecycle costing for construction projects considering LWC as a candidate. The evaluation must consider the performance and long-term sustainability aspects. The long-term value of any building material is predicated upon a combination of cost, durability, functionality and aesthetics. ESCSI has emphasized that any increase in the up-front cost of LWC is more than offset by the cost savings in labor, lower dead loads, higher fire resistance resulting in reduced concrete thickness, and less rebar required in building frames, girders, piers and footings. This extends to the long-term savings in heating and cooling costs due to the higher insulating properties and the overall superior thermal performance of building made using LWC.

VE encompasses the foregoing factors. However, for proper implementation, the obstacles need to be understood and overcome. First, regulations should be designed to embrace and enforce VE procedure in construction projects. Aigbavboa *et al.* have argued that stakeholders' 'canned' views and preconceptions could contribute to the challenges of accepting the VE umbrella in construction (Aigbavboa *et al.*, 2016). They further flagged the difficulty in conducting VE analysis and the evaluation of functions and alternatives. As such, 'coaching' construction stakeholders is unavoidable. To start with, they need to consider construction materials that have never been used in their projects (e.g. LWC), believing that new alternatives could outperform – as applicable – the traditional materials they constantly rely on (e.g. NWC). For this to happen, the dialogue between stakeholders on VE analysis should be enhanced (Aghimien *et al.*, 2018). Jaapar *et al.* (2009) indicated that even among construction stakeholders, those who apply VE are likely to disregard the impact of value (performance) and praise the impact of cost for a selected alternative. In a local study in Nigeria, Hayat pointed out that the lack of incentives to support the use of VE in public projects is another major challenge, which requires revisiting the guidelines of policymaking (Hayatu, 2015). Thus, Al-Yami suggested that support from top management is essential to the successful adoption of VE procedures in construction projects (Al-Yami, 2008). To this end, consideration of LEED certification could largely sway the mind of a wide spectrum of owners and clients.

Acknowledgements

The survey included in this chapter was made possible only with the involvement of the Concrete Industry Management (CIM) program and its students. The authors are indebted to the consultants, industrial partners, officials and academics who participated in the questionnaire and to the individual whose interviews are presented here. Their input and views provided us with insights into the culture of the construction market in USA. The interviewees requested to remain anonymous for reasons of confidentiality and to avoid possible conflicts of interest and/or market rivalry issues that their collective technical views might trigger. The help of Mr Ahmed Farouk of Universiti Teknologi Petronas was indispensable to us for the value engineering analysis presented in this chapter.

Appendix A. Participants' responses to the distributed questionnaire

I. BACKGROUND & LEVEL OF ENGAGEMENT

(1) Which general industry description most applies to your current role?

(a) Architect — 27%
(b) Manufacturer, Ready-Mix — 18%
(c) Concrete contractor — 15%
(d) Lightweight aggregate concrete supplier — 13%
(e) Structural Engineer — 10%
(f) Manufacturer, Precast — 7%
(g) Manufacturer, Concrete Products — 5%
(h) Other — 5%

(2) How would you classify your role within your company?

(a) Owner/Partner/Executive Management — 57%
(b) Designer/Specifier — 19%
(c) Business Development/Sales — 12%
(d) Production—foreman or crew — 10%
(e) Contractor/Installer—foreman or crew — 2%

(3) What percentage of your company's business would be considered work/ projects involving lightweight concrete?

(a) Less than 10% — 56%
(b) 10–25% — 22%
(c) None — 12%
(d) 26–50% — 5%
(e) Over 50% — 5%

(4) What concrete construction markets do you currently participate in?*

(a) Commercial — 25%
(b) Residential — 19%
(c) Government — 17%
(d) Institutional — 15%
(e) Green building — 13%
(f) Retrofit — 11%
(g) Other — 0%

(5) If you answered "Other" to question 4, please specify

Open Ended

(6) Where does most of your company's work take place?
- (a) U.S.-Mid West — 25%
- (b) U.S.-South West — 19%
- (c) U.S.-South East — 17%
- (d) U.S.-North East — 16%
- (e) U.S.-North West — 9%
- (f) Total U.S./National — 8%
- (g) North America — 6%

(7) How do you learn about new developments in lightweight concrete construction?*
- (a) Trade magazines/websites — 29%
- (b) Peer/Word-of-mouth — 20%
- (c) CE Courses — 13%
- (d) Trade Shows — 12%
- (e) Trade Associations — 11%
- (f) Direct mail from manufacturers — 8%
- (g) Sales calls — 6%

(8) Where do you see the most growth opportunities for your company in the concrete industry?
- (a) Green building — 27%
- (b) Adding new/different markets — 21%
- (c) High structural concrete — 20%
- (d) Geographic expansion — 16%
- (e) Lightweight concrete — 12%
- (f) Other — 4%

(9) If you answered "Other" to question 8, please specify

Open Ended

(10) Thinking about the entire concrete industry, how much growth potential do you see for projects using lightweight concrete in the next two years?
- (a) 6–10% — 33%
- (b) 5% or less — 31%
- (c) 11–20% — 20%
- (d) No increase — 14%
- (e) Decrease — 2%

(11) What types of lightweight materials are readily available in your area now?*

- (a) No Answer — 26%
- (b) Naturally occurring aggregates (such as slag stone, pumice, scoria, etc.) — 23%
- (c) Industrial byproducts (such as recycled glass, cinders, fly ash, etc.) — 22%
- (d) Manufactured aggregates (such as processed clay, shale, slate, perlite, etc.) — 17%
- (e) Lightweight synthetic particles — 11%
- (f) None — 2%

(12) What types of lightweight materials available in your area shall remain good supply in the next 5 years?*

- (a) Manufactured aggregates (such as processed clay, shale, slate, perlite, etc.) — 23%
- (b) Naturally occurring aggregates (such as slag stone, pumice, scoria, etc.) — 21%
- (c) Industrial by-products (such as recycled glass, cinders, fly ash, etc.) — 21%
- (d) No Answer — 16%
- (e) Lightweight synthetic particles — 16%
- (f) None — 2%

(13) How many of the following lightweight structural concrete technologies are you aware of?*

- (a) Traditional lightweight aggregates — 29%
- (b) Air entrained concrete — 27%
- (c) Cellular concrete — 16%
- (d) Lightweight synthetic particles — 15%
- (e) Aerated auto-claved concrete — 13%
- (f) Other — 1%

(14) If you answered "Other" to question 13, please specify

Open Ended

(15) How many of the following lightweight structural concrete technologies have you used?*

- (a) Traditional lightweight aggregates — 39%
- (b) Air entrained concrete — 31%
- (c) Lightweight synthetic particles — 12%
- (d) Cellular concrete — 12%
- (e) Aerated auto-claved concrete — 7%
- (f) Other — 1%

(16) If you answered "Other" to question 15, please specify
Open Ended

II. HURDLES

(17) What hurdles do you see designing with structural lightweight concrete now or in the future?*
    (a) Price — 21%
    (b) Performance of placed concrete (likelihood cracking, delamination, etc.) — 13%
    (c) Structural strength — 10%
    (d) Life cycle cost — 10%
    (e) Consistency batch to batch — 10%
    (f) Durability — 9%
    (g) Ability to adhere to codes — 9%
    (h) Placement of concrete (ease of pumping, placing and finishing) — 9%
    (i) Supply — 6%
    (j) Other — 2%

(18) If you answered "Other" in question 17, please specify
Open Ended

(19) Which hurdle(s) if overcome you are likely to use/specify/recommend LWC?*
    (a) Price hurdle — 16%
    (b) Consistency batch to batch hurdle — 14%
    (c) Ability to adhere to codes hurdle — 14%
    (d) Life cycle cost hurdle — 14%
    (e) Performance of placed concrete (likelihood cracking, delamination, etc.) hurdle — 14%
    (f) Placement of concrete (ease of pumping, placing and finishing) hurdle — 13%
    (g) Supply hurdle — 12%
    (h) Other hurdles — 4%

(20) What projects have you supplied, designed or constructed using structural lightweight concrete?*
    (a) Elevated decks steel frame building — 33%
    (b) Slabs on grade — 17%
    (c) Precast/Tilt-up panels/Cast-in-place — 16%
    (d) Manufactured concrete products (columns, facades) — 16%
    (e) Prestressed or post-tensioned concrete elements — 12%
    (f) Bridge decks — 6%

(21) Which are likely to be the most important to your company in the future?*
- (a) Slabs on grade — 24%
- (b) Elevated decks steel frame building — 20%
- (c) Precast/Tilt-up panels/Cast-in-place — 17%
- (d) Manufactured concrete products (columns, facades) — 15%
- (e) Prestressed or post-tensioned concrete elements — 14%
- (f) Bridge decks — 11%

III. PERCEPTION

(22) What are the most important benefits of structural lightweight concrete in projects you have constructed?*
- (a) Fire resistance — 17%
- (b) Design flexibility — 14%
- (c) Longer spans — 14%
- (d) Thermal performance — 13%
- (e) Cost savings — 13%
- (f) Smaller columns in steel structures — 10%
- (g) Decreased story height — 9%
- (h) Material source reduction — 9%

(23) If you answered "Other" to question 22 please specify
Open Ended

(24) Compared to normal weight concrete are there any consistency issues with lightweight concrete?
- (a) Consistency batch to batch — 33%
- (b) Meeting structural specifications job to job — 27%
- (c) Meeting applicable codes — 27%
- (d) Meeting structural specifications country-to country — 14%

(25) Are there any are pumping, placing, finishing issues with lightweight concrete?
- (a) No — 52%
- (b) Yes (If so, please specify) — 24%
- (c) N/A — 24%

(26) Please specify any pumping, placing or finishing issues with lightweight concrete.
Open Ended

(27) Do you consider lightweight concrete as having any of the following green or sustainable impact?*

    (a) Transportation savings   37%

    (b) Thermal performance   27%

    (c) Saving resources   21%

    (d) Labor savings   13%

    (e) Other   2%

(28) If you answered "Other" to question 27, please specify

    Open Ended

(29) How many of the following lightweight technologies would you recommend?*

    (a) Traditional lightweight aggregates   35%

    (b) Air entrained concrete   31%

    (c) Lightweight synthetic particles   13%

    (d) Cellular concrete   11%

    (e) Aerated auto-claved concrete   9%

    (f) Other   1%

(30) If you answered "Other" to question 29, please specify

    Open Ended

(31) What is your perception of the price of lightweight structural concrete vs. traditional concrete?

    (a) Lightweight concrete is more expensive   69%

    (b) Don't know   16%

    (c) Cost is the same   7%

    (d) Lightweight concrete is less expensive   7%

(32) What is your perception of the quality of lightweight structural concrete vs. traditional concrete?

    (a) Quality is the same   39%

    (b) Traditional concrete had greater quality   35%

    (c) Don't know   15%

    (d) Lightweight concrete had greater quality   12%

IV. PERCEIVED FUTURE

(33) How much do you see yourself using lightweight concrete in the future?

    (a) About the same   67%

    (b) More   25%

    (c) None   5%

    (d) Less   3%

(34) What would you like to see happen in the industry to increase your consideration of lightweight concrete?
Open Ended

(35) Do you foresee any legislative changes that would impact the lightweight concrete industry?
(a) No   91%
(b) Yes (If yes, please specify)   9%

(36) If you answered "Yes" to question 35, please specify the legislative changes that would impact the lightweight concrete industry.
Open Ended

*question allows multiple answers (normalized by number of answers- not by number of respondents)

References

Abdullahi, M., Al-Mattarneh, H.M.A., Hassan, A.A., Hassan, M.H. and Mohammed, B. (2008) Trial mix design methodology for Palm Oil Clinker (POC) concrete. International Conference on Construction and Building Technology in Kuala Lumpur.

Abidin, N.Z. and Pasquire, C.L. (2005) Delivering sustainability through value management: Concept and performance overview, *Engineering, Construction and Architectural Management*, Vol. 12 No. 2, pp. 168–180.

ACI. (2005) *Specifications for Structural Concrete, ACI 301-05 with Selected ACI References: Field Reference Manual.* Michigan: American Concrete Institute.

ACI. (2007) *Code Requirements for Determining Fire Resistance of Concrete and Masonry construction assemblies.* Report by Joint ACI-TMS Committee 216 (ACI 216.1-07/TMS-0216-07). Michigan: American Concrete Institute.

ACI PRC-213-14 Guide for Structural Lightweight-Aggregate Concrete.

Acoustics.com (2004) STC ratings for masonry walls. STC Ratings. Acoustics.com. Available at www.stcratings.com/masonry.html.

Aghimien, D.O., Oke, A.E. and Aigbavboa, C.O. (2018) Barriers to the adoption of value management in developing countries. *Engineering, Construction and Architectural Management* **25**(7), 818–834.

Aginam, C., Chidolue, C. and Nwakire, C. (2013) Investigating the effects of coarse aggregate types on the compressive strength of concrete. *International Journal of Engineering Research and Applications* **3**, 1140–1144.

Aigbavboa, C.O., Oke, A.E. & Mojele, S. (2016), Contribution of value management to construction projects in South Africa, in Proceedings of the 5th Construction Management Conference, Port Elizabeth, South Africa, pp. 226–234.

Akers, D.J., Gruber, R.D., Ramme, B.W., Boyle, M.J., Grygar, J.G., Rowe, S.K., Bremner, T.W., Kluckowski, E.S., Sheetz, S.R., Burg, R.G. and Kowalsky, M.J. (2014) *Guide for Structural Lightweight-aggregate Concrete.* ACI 213R-14. Committee 213. Michigan: American Concrete Institute.

Alhaddi, H. (2015) Triple bottom line and sustainability: A literature review. *Business and Management Studies* **1**(2), 6–10.

Al-Khaiat, H. and Haque, N. (1999) Strength and durability of lightweight and normal weight concrete. *Journal of Materials in Civil Engineering* **11**(3), 231–235.

Al-Yami, A.M. (2008) An integrated approach to value management and sustainable construction during strategic briefing in Saudi construction projects. Doctoral dissertation, School of Architecture, Building and Civil Engineering, Loughborough University.

Arcosa Lightweight. (n.d.) Hydrolite lightweight aggregate plays key role in record-breaking structure. Frazier, CA: Arcosa Lightweight. Available at https://arcosalightweight.com/case-studies/structural-lightweight-concrete/wilshire-grand.

ASCE. (2009) Report card for America's infrastructure. Reston, VI: American Society of Civil Engineers.

Aslam, M., Shafigh, P., Jumaat, M.Z. and Lachemi, M. (2016) Benefits of using blended waste coarse lightweight aggregates in structural lightweight aggregate concrete. *Journal of Cleaner Production* **119**, 108–117.

ASTM. (2020) ASTM C70-20: Standard test method for surface moisture in fine aggregate. West Conshohocken, PA: ASTM International. Available at www.astm.org.

ASTM. (2017) ASTM C330/C330M-17a: Standard specification for lightweight aggregates for structural concrete. West Conshohocken, PA: ASTM International. Available at www.astm.org.

Baker, T., Khaleghi, B. Harrison, T., Fuller, P. and Zeldenrust, R. (2014) I-5 Skagit River bridge collapse and rapid replacement. Olympia, WA: Washington State Department of Transportation. Available at aspirebridge.com/magazine/2014Winter/ABC_Win14_Web.pdf.

Bamforth, P.B. (1987) The relationship between permeability coefficients for concrete obtained using liquid and gas. *Magazine of Concrete Research* **39**(138), 3–11.

Bell & Associates. (2015) I-40 bridge rehabilitation over the French Broad River. Available at balp.com/projects/i-40-bridge-rehabilitation-over-french-broad-river.

Benazzouk, A., Douzane, O., Mezreb, K., Laidoudi, B. and Quéneudec, M. (2008) Thermal conductivity of cement composites containing rubber waste particles: Experimental study and modelling. *Construction and Building Materials* **22**(4), 573–579.

Bernhardt, D. and Reilly, I.I. (2019) *Mineral Commodity Summaries*. Reston, VA: US Geological Survey.

Bilow, D.N. and Kamara, M.E. (2008) Fire and concrete structures. In D. Anderson (ed.) *Structures Congress 2008: 18th Analysis and Computation Specialty Conference*. Reston, VA: ASCE, 5 vols, vol. 4, pp. 2707–2716.

Bogas, J.A. and Gomes, T., 2015. Mechanical and durability behaviour of structural lightweight concrete produced with volcanic scoria. *Arabian Journal for Science and Engineering* **40**(3), 705–717.

Bouvard, D., Chaix, J.M., Dendievel, R., Fazekas, A., Létang, J.M., Peix, G. and Quenard, D. (2007) Characterization and simulation of microstructure and properties of EPS lightweight concrete. *Cement and Concrete Research* **37**(12), 1666–1673.

Boyd, S. R. (1998) The effect of lightweight fine aggregate on alkali–silica reaction and delayed ettringite formation. MS thesis, Department of Civil Engineering, University of New Brunswick, Fredericton, NB, Canada.

Bremner, T.W., Holm, T.A. and Stepanova, V.F. (1994) Lightweight concrete: A proven material for two millennia. In S.L. Sarkar and M.W. Grutzeck (eds) *Advances in Cement and Concrete: Proceedings of an Engineering Foundation Conference Held in Durham, New Hampshire, July 24–29, 1994*. Reston, VI: American Society of Civil Engineers, pp. 37–41.

Bremner, T. W., Boyd, A. J., Holm, T. A. and Boyd, S. R. (1998) Indirect tensile testing to evaluate the effect of alkali–aggregate reaction in concrete. Paper No. T192-2. Structural Engineering World Wide Conference, San Francisco, CA.

Castrodale, R.W. (2006) Lightweight high-performance concrete for bridge decks. In Presentation in Virginia Concrete Conference, March 10, 2006, Richmond, Virginia.

Chandra, S. and Berntsson, L. (2002) *Lightweight Aggregate Concrete: Science, Technology and Applications* (1st edition) (Building Materials Science Series). New York, NY: William Andrew Inc. (Elsevier), pp. 35–63.

Chen, B. and Liu, N. (2013) A novel lightweight concrete fabrication and its thermal and mechanical properties. *Construction and Building Materials* **44**, 691–698.

Cheng, X.H. (2013) Preliminary seismic considerations for Pulaski Skyway rehabilitation project. *In Proceedings of the 29th US–Japan Bridge Engineering Workshop* (Workshop by the Public Works Research Institute (PWRI), Japan), November 11–13.

Clarke, J.L. (1993) *Structural Lightweight Aggregate Concrete*. Boca Raton, FL: CRC Press.

Colaco, J. (2004) One Shell Plaza, Houston. *Practice Periodical on Structural Design and Construction* **9**(2), 79–82.

Council on Tall Buildings and Urban Habitat (CTBUH) (n.d.(a)) 30 Hudson Street, New Jersey. CTBUH Tallest Building Lists. www.skyscrapercenter.com/building/30-hudson-street/1025.

Council on Tall Buildings and Urban Habitat (CTBUH) (n.d.(b)) Duke Energy Center, Charlotte. CTBUH Tallest Building Lists. www.skyscrapercenter.com/building/duke-energy-center/1077.

Cousins, T.E., Roberts-Wollmann C.L. and Brown M.C. (2013) *National Cooperative Highway Research Program (NCHRP) Report 733: High-performance/High-strength Lightweight Concrete for Bridge Girders and Decks.* Washington, DC: Transportation Research Board.

Cowan, J. (2014) Building acoustics. In Thomas D. Rossing (ed.) *Springer Handbook of Acoustics 2nd Edition*. New York, NY: Springer, pp. 403–442.

CRSI. (2016) Vibration and sound control in reinforced concrete buildings, CRSI Technical Note ETN-B-3-16 (8 pp.). Schaumburg, Illinois: Concrete Reinforcing Steel Institute.

Dawood, E.T. and Ramli, M. (2008) Rational mix design of lightweight concrete for optimum strength. In *2nd International Conference on Built Environment in Developing Countries* (ICBEDC, 2008).

Dell'Isola, A.J. (1982) *Value Engineering in the Construction Industry. 3rd Edition.* New York, NY: Van Nostrand & Reinholt.

Demirboğa, R. and Gul, R. (2003) The effects of expanded perlite aggregate, silica fume and fly ash on the thermal conductivity of lightweight concrete *Cement and Concrete Research* **33**, 723–727.

Dulsang, N., Kasemsiri, P., Posi, P., Hiziroglu, S. and Chindaprasirt, P. (2016) Characterization of an environment friendly lightweight concrete containing ethyl vinyl acetate waste. *Materials and Design* **96**, 350–356.

Ellis, R.C., Wood, G.D. and Keel, D.A. (2005) Value management practices of leading UK cost consultants. *Construction Management and Economics* **23**(5), 483–493.

Emporis GMBH (n.d.). Guggenheim Museum Bilbao. www.emporis.com/buildings/112096/guggenheim-museum-bilbao-bilbao-spain.

ESCSI. (2003) Publication #4640: Finishing lightweight concrete floors. Salt Lake City, UT: Expanded Shale, Clay and Slate Institute (ESCSI) https://www.escsi.org/wp-content/uploads/2017/10/4640.0-Finishing-Lightweight-Floors-1.pdf?utm_source=chatgpt.com.

EXCA. (2011) CityLife residential buildings. Brussels, Belgium: European Expanded Clay Association. https://www.exca.eu/light-projects/citylife-residential-buildings/.

FHWA. (1985) *Criteria for Designing Lightweight Concrete Bridges*. Washington, DC: United States Department of Transportation, Federal Highway Administration.

FHWA. (2017) UHPC on the Pulaski Skyway: Owner's perspective. Washington, DC: United States Department of Transportation, Federal Highway Administration. https://www.fhwa.dot.gov/innovation/everydaycounts/edc_4/uhpc_august15_transcript.pdf.

Goeb, E. (1985) Pumping structural lightweight concrete. *Concrete Construction* **30**(6), 505–510.

Guggemos, A.A. and Horvath, A. (2005) Comparison of environmental effects of steel-and concrete-framed buildings. *Journal of Infrastructure Systems* **11**(2), 93–101.

Gündüz, L. and Uğur, İ. (2005) The effects of different fine and coarse pumice aggregate/cement ratios on the structural concrete properties without using any admixtures. *Cement and Concrete Research* **35**(9), 1859–1864.

Gutenberg (n.d.). Guy's Hospital. Available at: self.gutenberg.org/articles/eng/Guy%27s_Hospital. Project Gutenberg self-publishing.

Harmon, K.S. (2007) Engineering properties of structural lightweight concrete. Carolina Stalite Company.

Harris, C.M. and Knudsen, V.O. (1950) *Acoustical Designing in Architecture*. New York & Chichester: John Wiley & Sons.

Hayatu, U.A. (2015) An assessment of the Nigerian construction industry's readiness to adopt value management process in effective project delivery. Unpublished MSc thesis, Department of Quantity Surveying, Faculty of Environmental Design, Ahmadu Bello University, Zaria.

Hockstad, L. and Hanel, L. (2018) Inventory of US greenhouse gas emissions and sinks (No. cdiac: EPA-EMISSIONS). Environmental System Science Data Infrastructure for a Virtual Ecosystem (ESS-DIVE). doi:10.15485/1464240.

Hoff, G.C. (1992) High strength lightweight-aggregate concrete for arctic applications. In T.A. Holm and A.M. Vaysburd (eds) *SP-136 Structural Lightweight Aggregate Concrete Performance.* Farmington Hills, MI: American Concrete Institute.

Holm, T.A. (1980) Performance of structural lightweight concrete in a marine environment. In V.M. Malhotra (ed.) *Performance of Concrete in a Marine Environment.* ACI SP-65. Detroit, MI: American Concrete Institute.

Holm, T.A., and Bremner, T.W. (1994) High Strength Lightweight Aggregate Concrete. In S.P. Shah and S.H. Ahmad (eds), *High Performance Concrete: Properties and Applications.* New York, NY: McGraw-Hill, 341–374.

Holm, T.A. and Bremner, T.W. (2000) *State-of-the-Art Report on High-strength, High-durability Structural Low-density Concrete for Applications in Severe Marine Environments.* US Army Corps of Engineers, Engineer Research and Development Center.

Holm, T. and Ries, J. (2007) *Reference Manual for the Properties and Applications of Expanded Shale, Clay and Slate Lightweight Aggregate.* Chicago, IL: Expanded Shale, Clay and Slate Institute (ESCSI).

Holm, T.A., Bremner, T.W. and Newman J.B. (1984) Concrete bridge decks: Lightweight aggregate concrete subject to severe weathering. *Concrete International* **6**(6), 49–54.

Hossain, K.M.A. and Lachemi, M. (2005) Thermal conductivity and acoustic performance of volcanic pumice based composites. *Materials Science Forum* **480**, 611–616.

Hunag, L.J., Wang, H.Y. and Wang, S.Y. (2015) A study of the durability of recycled green building materials in lightweight aggregate concrete. *Construction and Building Materials* **96**, 353–359.

Jaapar, A., Endut, I.R., Bari, N.A.A. and Takim, R. (2009) The impact of value management implementation in Malaysia. *Journal of Sustainable Development* **2**(2), 210–219.

Jin, R., Chen, Q. and Soboyejo, A. (2015) Survey of the current status of sustainable concrete production in the U.S. *Resources Conservation and Recycling* **105**, 148–159.

Kahn, L.F., Kurtis, K.E., Lai, J.S., Meyer, K.F., Lopez, M. and Buchberg, B. (2004). *Lightweight Concrete for High Strength/High Performance Precast Prestressed Bridge Girders.* Research Report No. 04-1, Georgia Institute of Technology (January 2004).

Kan, A. and Demirboğa, R. (2009) A novel material for lightweight concrete production. *Cement and Concrete Composites* **31**(7), 489–495.

KCI Technologies (2017) Innovative safety inspection of the Virginia Dare Memorial Bridge. Safety inspection assesses condition of longest bridge in North Carolina, March 22, 2017.

Keeton, J.R. (1970) Permeability studies of reinforced thin-shell concrete. Technical Report R692 YF51.42.001, 01.001. Port Hueneme, CA: Naval Civil Engineering, Laboratory.

Kelly, J., Male, S. and Graham, D. (2014) *Value Management of Construction Projects.* New York & Chichester: Wiley.

Kim, H.K., Jeon, J.H. and Lee, H.K (2012) Workability, and mechanical, acoustic and thermal properties of lightweight aggregate concrete with a high volume of entrained air. *Construction and Building Materials* **29**, 193–200.

Kogel, J.E., Trivedi, C.N., Barker, J.M., Krukowski, S.T. (2006). *Industrial Minerals and Rocks: Commodities, Markets, and Uses. 7th Edition.* Littleton, CO: Society for Mining, Metallurgy, and Exploration.

Kotter, J.P. (1995) Leading change: Why transformation efforts fail. *Harvard Business Review.* 59–67. Reprint No. 95204.

Kozel, S. (2004) James River Bridge (US-17). Highway and Transportation History Website. www.roadstothefuture.com/US17_JRB.html.

KPF (n.d.). Prudential Newark Receives LEED Gold Certification, www.kpf.com/projects/prudential-newark Wilshire Grand Center.

Kumar, N., Sharma, A., & Agarwal, A. (2017) Title of the paper. IOP *Conference Series: Materials Science and Engineering*, Volume 377, Article 012345. https://doi.org/10.1088/1757-899X/377/1/012345.

Kurpinska, M., Grzyl, B. and Kristowski, A. (2019) Cost analysis of prefabricated elements of the ordinary and lightweight concrete walls in residential construction. *Materials* **12**(21), 3629.

L.B. Foster Company (2018) Case History: Brooklyn Bridge Redecking. www.lbfoster-fabricatedproducts.com/pdf/Brooklyn%20Bridge%20new.pdf.

Lo, T. and Cui, H. (2004) Properties of green lightweight aggregate concrete. In Kejin Wang (ed.) *Proceedings of the International Workshop on Sustainable Development and Concrete Technology, Beijing, China, May 20–21, 2004*. Ames, IO: Center for Transportation Research and Education, Iowa State University, pp. 113–117.

Lo, T.Y., Tang, W.C. and Nadeem, A. (2008) Comparison of carbonation of lightweight concrete with normal weight concrete at similar strength levels. *Construction and Building Materials* **22**(8), 1648–1655.

Mailhot, G. and Eng, M. Développement de mélanges et utilisation du béton léger haute performance dans le cadre de travaux visant le remplacement de tabliers de ponts. http://www.bv.transports.gouv.qc.ca/mono/0981440/08_Developpement_melanges_utilisation_beton_haute_performance.pdf.

Marceau, M., Nisbet, M.A. and Van Geem, M.G. (2006) *Life Cycle Inventory of Portland Cement Manufacture* (No. PCA R&D Serial No. 2095b). Skokie, IL: Portland Cement Association.

Martin, Barney T. and Blabac B.A. (2009) Replacing the suspender ropes of a tied arch bridge using suspension bridge methods. In Masahiro Shirato (ed.) *Proceedings of the 25th US–Japan Bridge Engineering Workshop*, October 19, 20, 21, 2009. Japan: Center for Advanced Engineering Structural Assessment and Research Public Works Research Institute.

Maryland Department of Transportation (2020) William Preston Lane Jr. Memorial Bridge (Chesapeake Bay Bridge) 1952 and 1973. https://web.archive.org/web/20200625041520/https://roads.maryland.gov/mdotsha/pages/index.aspx?PageId=271.

MBP (n.d) North Carolina Department of Justice, Virginia Dare Memorial Bridge. MBP, www.mbpce.com/project/north-carolina-department-of-justice-virginia-dare-memorial-bridge/.

Meyer, C., Egosi, N. and Andela, C. (2001) "Concrete with Waste Glass as Aggregate" in "Recycling and Re-use of Glass Cullet", Dhir, Dyer and Limbachiya, editors, Proceedings of the International Symposium Concrete Technology Unit of ASCE and University of Dundee.

McCabe, R. (1993) Closing the loop: Cooper River Bridge in Charleston, South Carolina. *Civil Engineering* **63**(7), 64–66.

Medgar, L. (2007) *Life Cycle Inventory of Portland Cement Concrete*. PCA R&D Serial No. 3007. Skokie, IL: Portland Cement Association.

Metropolitan Transportation Commission (2009) Benicia-Martinez Bridge. mtc.ca.gov/about-mtc/what-mtc/bay-area-toll-authority/benicia-martinez-bridge.

Metropolitan Transportation Commission (2013) Antioch Bridge. mtc.ca.gov/about-mtc/what-mtc/bay-area-toll-authority/antioch-bridge.

Morris, T. (2017) 54 Stories of pumped lightweight concrete. ESCSI, 20 Oct. 2017. Chicago, IL: Expanded Shale, Clay and Slate Institute. www.escsi.org/e-newsletter/54-stories-of-pumped-lightweight-concrete/.

Mortensen, A. (ed.) (2006) *Concise Encyclopedia of Composite Materials*. New York, NY: Elsevier Science.

Mousa, A. (2015a) A business approach for transformation to sustainable construction: An implementation on a developing country. *Resources, Conservation and Recycling* **101**, 9–19.

Mousa, A. (2015b) Six sigma DMAIC for shaking stagnant construction cultures: A conceptual perspective. *Journal of Civil Engineering and Environmental Sciences* **1**(1), 13–20.

Mousa, A., Mahgoub, M. and Hussein, M. (2018) Lightweight concrete in America: Presence and challenges. *Sustainable Production and Consumption* **15**, 131–144.

NASA. (2019) Global climate. Overview: Weather, global warming and climate change. NASA, August 28, 2019. www.climate.nasa.gov/resources/global-warming-vs-climate-change/.

Nisbet, M.A., Marceau, M.L. and VanGeem, M.G. (2002) *Life Cycle Inventory of Portland Cement Concrete.* PCA R&D Serial No. 2095b. Skokie, IL: Portland Cement Association (PCA).

Norlite Lightweight Aggregate. (2019) Mix designs. Retrieved September 6, 2025, from https://www.norliteagg.com/mix-designs.

Norton, B.R. and McElligott, W.C. (1995) *Value Management in Construction: A Practical Guide.* London: Macmillan.

NRMCA. (2003) CIP 36-Structural lightweight concrete. Silver Spring, MD: National Ready Mixed Concrete Association (NRMCA).

NRMCA (2008) Concrete CO2 fact sheet. Silver Spring, MD: National Ready Mixed Concrete Association (NRMCA).

NSBA (2021) Sam White Bridge. National Steel Bridge Alliance. www.aisc.org/nsba/prize-bridge-awards/prize-bridge-winners/sam-white-bridge.

OPAC Consulting Engineers (2010a) San Diego – Coronado Bay Bridge. www.opacengineers.com/projects/Coronado.

OPAC Consulting Engineers (2010b) Parrotts Ferry Bridge, Vallecito, CA. OPAC. www.opacengineers.com/projects/ParrottsFerry.

Ozyildirim, C. (2008) Durability of structural lightweight concrete. Proceedings of 2008 Concrete Bridge Conference, St Louis, MO, paper 142.

Pade, C., Guimaraes, M., Kjellsen, K. and Nilsson, A. (2007) The CO_2 uptake of concrete in the perspective of life cycle inventory. *International Symposium on Sustainability in the Cement and Concrete Industry.* Presented at the International Symposium on Sustainability in the Cement and Concrete Industry, Aalborg, Denmark.

PCI (Precast/Prestressed Concrete Institute) (n.d.). I-87 Twin Bridges deck replacement (Thaddeus Kosciuszko Bridge) over the Mohawk River. www.pci-projects.interactivetwist.com/project/i-87-twin-bridges-deck-replacement-thaddeus-kosciuszko-bridge-over-the-mohawk-river/?search=%2Fproject-profiles-pci-pc%2F&pctemplate=one_column_pc.

Per Gestora Inmobiliaria. (n.d.). Torre Picasso. Retrieved September 6, 2025, from https://www.torre-picasso.com/en/.

Porter, M.E. (1979) How competitive forces shape strategy. *Harvard Business Review*, March–April 1979, 1–10.

Proctor, T. (2000) *Strategic Marketing: An Introduction.* Hove: Psychology Press.

Prokopy, J.G. (2004) Next performance. Stronger concrete use on the rise as industry tests future generation. *Roads & Bridges* **42**(1), 29–31. www.roadsbridges.com/next-performance.

Raithby, K.D. and Lydon, F.D. (1981) Lightweight concrete in highway bridges. *International Journal of Cement Composites and Lightweight Concrete* **3**(2), 133–146.

Ries, J.P. and Holm, T.A. (2004) A holistic approach to sustainability for the concrete community: Lightweight concrete – Two millennia of proven performance. ESCSI Information Sheet 7700. Chicago, IL: Expanded Shale, Clay and Slate Institute.

Ries, J., Speck, J. and Harmon, K. (2010) Lightweight aggregate optimizes the sustainability of concrete. Concrete Sustainability Conference, April 13–15, 2010, Tempe, Arizona. National Ready Mixed Concrete Association.

Roberz, F., Loonen, R.C., Hoes, P. and Hensen, J.L. (2017) Ultra-lightweight concrete: Energy and comfort performance evaluation in relation to buildings with low and high thermal mass. *Energy and Buildings* **138**, 432–442.

Russell, H.G. (2008) Lightweight concrete – Material properties for structural design. Proceedings of the Concrete Bridge Conference, 4–7 May 2008, St Louis, MO.

Sabnis, G.M. (2015) *Green Building with Concrete: Sustainable Design and Construction. Second Edition.* Boca Raton, FL: CRC Press.

Structural Building Components Association. (2002). Metal Plate Connected Wood Truss Handbook. Madison, WI: Wood Truss Council of America. Retrieved March 11, 2020, from https://www.sbcindustry.com/sites/default/files/uploads/attachments/node/424/section18.pdf.

Schwartz, H. (2013) Panasonic celebrates grand opening of new headquarters in Newark, NJ. Red Bank, NJ: Business Facilities. www.businessfacilities.com/2013/09/panasonic-celebrates-grand-opening-of-new-headquarters-in-newark-nj/.

Schwing America Inc. (n.d.) New Comcast Tower will be Philadelphia's tallest building. www.schwing.com/new-comcast-tower-will-be-philadelphias-tallest-building/.

Sengul, O., Azizi, S., Karaosmanoglu, F. and Tasdemir, M.A. (2011) Effect of expanded perlite on the mechanical properties and thermal conductivity of lightweight concrete. *Energy and Buildings* **43**(2–3), 671–676.

Shafigh, P., Nomeli, M.A., Alengaram, U.J., Mahmud, H.B. and Jumaat, M.Z. (2016) Engineering properties of lightweight aggregate concrete containing limestone powder and high volume fly ash. *Journal of Cleaner Production* **135**, 148–157.

Sigmon, R.G. (2000) Steel spans the Neuse River. *Modern Steel Construction* (October 2000) www.aisc.org/globalassets/modern-steel/archives/2000/10/2000v10_neuse_river.pdf.

Simons, K.J. (2010) Affordable lightweight high performance concrete (ALWHPC) – Expanding the envelope of concrete mix design. Master's thesis, University of Nebraska.

Skoyles, E.R., Bardhan-Ro, B.K., Craig, A., Coulthard, R.D., Ralph, D.W. and Williams, J. (1979) A comparative study of the economics of lightweight structural concrete floor slabs in building: The concrete society-lightweight concrete committee. *International Journal of Cement Composites and Lightweight Concrete* **1**(1), 9–27.

Stalite Lightweight Aggregate. (2019) Sample mix designs. www.stalite.com/lwa-mix-designs.

Stalite Lightweight Aggregate. (2020) Virginia Dare Bridge. https://static1.squarespace.com/static/59c91fb8f7e0ab097112fbc4/t/5b552bae758d4613de0b19a1/1532308399719/5.+VA+Dare+Bridge.pdf.

Structurae. (2016a) Pont De Beaumont-sur-Oise (Beaumont-sur-Oise/Bernes-sur-Oise, 1997). February 5, 2016. structurae.net/en/structures/pont-de-beaumont-sur-oise.

Structurae. (2016b) Sundøy Bridge (Leirfjord, 2003). Structurae, April 5, 2016. structurae.net/en/structures/sundoy-bridge.

Structurae. (2018) Sandsfjord Bridge (Suldal, 2015). Structurae, October 5, 2018. structurae.net/en/structures/sandsfjord-bridge.

Teo, D.C.L., Mannan, M.A. and Kurian, V.J. (2010) Durability of lightweight OPS concrete under different curing conditions. *Materials and Structures* **43**(1–2), p. 1.

Ünal, O., Uygunoğlu, T. and Yildiz, A. (2007) Investigation of properties of low-strength lightweight concrete for thermal insulation. *Building and Environment* **42**(2), 584–590.

US-EIA. (2019) World Energy Projection System Plus (2019), run r_190808_161601, and EIA, *International Annual Energy Outlook 2019* (January 2019), Washington, DS: U.S. Energy Information Administration (EIA). www.eia.gov/aeo.

US EPA. (1995) *AP 42 - Compilation of Air Pollutant Emission Factors, Volume I: Stationary Point and Area Sources.* Washington, DC: Environmental Protection Agency.

US EPA. (2011) *Municipal Solid Waste in the United States: 2011 Facts and Figures.* Washington, DC: United States Environmental Protection Agency. https://archive.epa.gov/epawaste/nonhaz/municipal/web/pdf/mswcharacterization_fnl_060713_2_rpt.pdf.

US EPA. (2016) Why build green? Washington, DC: US Environmental Protection Agency. February 20, 2016. https://archive.epa.gov/greenbuilding/web/html/whybuild.html.

US EPA. (2017) Understanding global warming potentials. Washington, DC: US Environmental Protection Agency. https://www.epa.gov/ghgemissions/understanding-global-warming-potentials.

USGBC. (2017) LEED Credits, prerequisites and points: How are they different? Atlanta, GA: U.S. Green Building Council. www.usgbc.org/articles/whats-difference-between-leed-credit-leed-prerequisite-and-leed-point.

USGBC. (2020) LEED credit library. Atlanta, GA: U.S. Green Building Council. www.usgbc.org/credits.

USGS. (2014) *Mineral Commodity Summaries 2014*. Reston, VA: U.S. Geological Survey.

USGS. (2016) *Mineral Commodity Summaries 2017*. Reston, VA: U.S. Geological Survey.

USGS. (2018) *Minerals Yearbook, Volume II, Area Reports – Domestic*. Reston, VA: U.S. Geological Survey. https://doi.org/10.3133/mybvII.

USGS. (2020) *Mineral Commodity Summaries 2020*. Reston, VA: U.S. Geological Survey. https://doi.org/10.3133/mcs2020.

Van Vliet, K., Pellenq, R., Buehler, M.J., Grossman, J.C., Jennings, H., Ulm, F.-J. and Yip, S. (2012) Set in stone? A perspective on the concrete sustainability challenge. *MRS Bulletin* **37**(4), 395–402.

Vosloo, C. (2020) Sustainable skyscrapers: The Standard Bank Centre, Johannesburg. *Athens Journal of Architecture* **6**(1), 53–78.

Wagner, A.M. (2015) Durability of lightweight concrete for bridge decks. Master's thesis, University of Tennessee – Knoxville.

World Commission on Environment and Development. (1989) *Our Common Future.* London: Oxford University Press.

Yazdani, N. and Goucher, E. (2015) Increasing durability of lightweight concrete through FRP wrap. *Composites Part B: Engineering* **82**, 166–172.

Yu, Q.L., Spiesz, P. and Brouwers, H.J.H. (2015) Ultra-lightweight concrete: Conceptual design and performance evaluation. *Cement and Concrete Composites* **61**, 18–28.

Zhang, B. and Poon, C.S. (2015) Use of furnace bottom ash for producing lightweight aggregate concrete with thermal insulation properties. *Journal of Cleaner Production* **99**, 94–100.

Zhang, M.H. and Gjorv, O.E. (1991) Characteristics of lightweight aggregates for high-strength concrete. *Materials Journal* **88**(2), 150–158.

Zidan, A., Mousa, A. and Mahgoub, M. (2013) A survey-based vision for restructuring the concrete business in new residential communities in Egypt. *Industrial and Systems Engineering Review* **1**(2), 162–172.

Chapter 2
The mix design of structural lightweight aggregate concrete

José Alexandre Bogas

1. Introduction

Although reinforced structural lightweight aggregate concrete (SLWAC) has been used in buildings and special structures since the first half of the last century, there is still no consensus on a universal concrete mix design method when considering the type of light weight aggregate (LWA). Most mix design methods have been based on prior experimental experience for a given type of LWA and concrete composition. As such, the validity of these methods is limited to the specific concretes used in those trials. Moreover, these methods are time sensitive and not easily adjusted to new technological advances in SLWAC production. In fact, with the development of more powerful high-performance admixtures, the development of high-fluid concrete and the pressing need for more eco-efficient additions and alternative binders over the last few years, empirical or semi-empirical methods became useless.

According to ACI 213R (ACI, 2014), as per normal weight concrete (NWC) the SLWAC mix design must be defined based on the economical combination of the constituents used, taking into account the required properties of the concrete fresh and hardened. The same principles of the mix design adopted for NWC applied to SLWAC. However, the behavior of SLWAC may be affected differently. Some of its properties, such as compressive strength, durability, workability, density, and thermal conductivity, are affected by the mix design in opposite ways. Furthermore, compared to NWC, the concrete density, the increased risk of segregation, and the water uptake (absorption) of LWA, are additional parameters that must be considered in SLWAC mix design (Dreux, 1986; Bogas and Gomes, 2013b; ACI 213R, 2014). This makes SLWAC mix design more complex than that considered for NWC.

This chapter presents a brief overview of mix design methods for SLWAC and its key differences from NWC. In this chapter, only the most common SLWAC with natural sand is covered, although in general the same main principles may be applied to all-lightweight concrete (ASLWAC) with coarse and fine LWA. Most of the significant advances and typical mix design methodologies are covered. At the end of this chapter, an expeditious method for SLWAC mix design is proposed, one that aims to be easily implemented in practice, regardless of the type of LWA and concrete composition. Only structural lightweight aggregate concrete is considered. The mix design of non-structural lightweight concrete follows a different philosophy depending on the type of concrete (no-fines concrete, cellular, non-structural lightweight aggregate concrete).

2. Constituents of structural lightweight concrete

SLWAC utilizes the same constituents as NWC except that the natural aggregate is partially or totally replaced with LWA. The direct replacement, however, is not simple because LWA can have a very distinct porosity, which controls the respective concrete mix proportion and subsequently affects the concrete properties.

European standards EN 13055 (2016) and EN 206 (2016) define LWA as aggregates of mineral origin, having a particle density not exceeding 2000 kg/m^3 or a loose bulk density not exceeding 1200 kg/m^3. Various types of aggregates meet these requirements and are suitable for the production of lightweight concrete. However, despite the wide range of available aggregates, only some inorganic LWA can be used to produce structural lightweight concrete – namely, natural volcanic aggregates (pumice, slag), sintered fly ash, expanded slag, clay, shale, and slate aggregates (Dolby, 1995; Bogas, 2011; ACI 213R, 2014). LWA produced from by-products and recycled materials have also been explored to ensure SLWAC sustainability but their commercial use is still scarce (Mousa *et al.*, 2018; Hussein *et al.*, 2021). Other aggregates are only suitable for low-strength or insulating lightweight concrete.

Natural LWA usually exhibits greater variability than artificial (manufactured/byproducts) LWA and, therefore, can only be used to produce low- to moderate-strength SLWAC. Manufactured aggregates can, in turn, be divided into two subgroups: (1) - those made from natural materials; (2) - those made from industrial by-products, according to the source of their raw materials: those made by thermal processes using either natural materials, such as clay, shale, slate, or industrial by-products, such as glass, fly ash or expanded slag (EuroLightConR2, 1998; Kockal and Ozturan, 2011; Bogas *et al.*, 2017; Ting *et al.*, 2019). Expanded clay, slate, and shale are the most common LWA incorporated into SLWAC. These LWAs have a honeycombed porous structure surrounded by a denser outer shell, resulting from their manufacturing process, which improves their mechanical properties (Faust 2000; Ke *et al.*, 2009; Kockal and Ozturan, 2011; Bogas *et al.*, 2012a). On the other hand, fly ash LWA is sintered without expansion and without forming a dense outer shell, which reduces the mechanical strength compared to expanded LWA of the same porosity. Moreover, LWA may be conformed in distinct ways (pelletized; extruded; pre-molded inside the kiln, crushed after burning) and with different expansion properties and sizes (Clark, 1993; Bogas, 2011; Bogas *et al.*, 2012a). Therefore, LWA may be produced with a very distinct pore structure and porosity (Bogas *et al.*, 2012a). This structure affects the absorption, bulk density and hardened properties, such as workability, strength, durability, shrinkage and creep (e.g. Zhang and Gjørv, 1989; Lo *et al.*, 1999; Moravia *et al.*, 2006). This unique structure must be taken into account for SLWAC mix design.

SLWAC can be produced with any type of water and standard cement (Thienel *et al.*, 2020). Like in NWC, there are no restrictions on the use of water, including recycling water (DIN EN 1008, 2002). However, the type of cement has a different impact on SLWAC compared to NWC. As discussed in Section 4, once concrete strength exceeds a certain threshold, the influence of LWA becomes significant, and the effect of cement strength on this property is reduced (FIP, 1983). In addition, SLWAC is usually produced with a higher binder content and a lower water/binder (w/b) ratio compared to NWC of equal strength. The lower w/b ratio partially compensates for the reduction in the aggregate strength of the SLWAC (see Section 4). The high proportion of binder increases fresh concrete cohesion – reducing the risk of segregation during handling, pouring and vibration (Pankhurst, 1993; Holm and Bremner, 2000). Moreover, it allows the production of more fluid concrete, thereby reducing the energy of vibration. Excessive vibration is less effective and can impair the homogeneity of lightweight concrete. Richer mixes ensure a better involvement and protection of LWA, reducing their effect on the transport properties and improving SLWAC durability. On the other hand, since the porosity of LWA usually increases with aggregate size, especially in expanded aggregate, the maximum aggregate size is limited to 12–19 mm (for strength and workability reasons) (FIP, 1983; Dreux, 1986; Holm and Bremner, 2000; Chandra and Berntson, 2003). Therefore, more volume of paste is necessary to compensate for the higher surface area of LWA. For the same reasons, the volume of coarse LWA is typically lower than 400 l/m^3 (FIP, 1983; Holm and Bremner, 2000; Chandra and Berntson, 2003; Bogas and Gomes, 2013a).

Based on a large number of existing SLWAC bridges and buildings from the 1950s, the amount of cement binder varied between 350 and 500 kg/m^3 and the w/b ratio between 0.3 and 0.6 (fib8, 2000; Bogas 2011; ACI 213R, 2014). With the modern generation of admixtures and the increased durability requirements, a w/b ratio of less than 0.45 in SLWAC is commonly used for compressive strengths greater than 35-40 MPa. This obviously depends on the type of LWA. In Norway, high-performance SLWAC with more than 400 kg/m^3 of binder and a w/b ratio of 0.3–0.4 was widely applied in special structures (Helgesen, 1995). In SLWAC exposed to marine environments, Holm and Bremner (2000) reported binder contents 25–80 kg/m^3 higher than those in NWC, resulting in total binder contents exceeding 360 kg/m^3. More than 450 kg/m^3 of fine material (lower than 250 μm) is recommended in FIP (1983) to avoid problems of concrete segregation. Fine aggregate content of 420–525 kg/m^3 is recommended for SLWAC with coarse aggregate having a maximum size of 8–25 mm (FIP, 1983). For typical binder contents, approximately 10% of the sand content should be up to 250 μm in size. Paste volumes of the order of 0.28–0.35 m^3 are common (Chandra and Berntson, 2003; Bogas and Gomes, 2013b).

Greater amounts of cement in SLWAC increase the shrinkage, creep, density, heat of hydration and unit cost, and comes with an environmental impact (FIP, 1983). Accordingly, the use of large amounts of mineral additives is recommended in SLWAC as necessary to counter the urge to use high cement content. Mineral additions – such as fly ash, calcined clay and blast furnace slag – are also recommended to ensure a modest hydration heat release. In fact, the higher cement content combined with the insulating properties of SLWAC may lead to an excessive rise of temperature in thick elements. This may lead to a delayed ettringite formation or generation of thermal gradients and subsequent cracking (Bogas, 2011; Holm and Bremner, 2020; Thienel *et al.*, 2020). Naturally, the need for richer mixes depends on the type of aggregate. The more porous and weaker the aggregate, the higher the cement content. There is, therefore, a unique relation between the cement content or w/b ratio and the concrete compressive strength for a given type of LWA (see Section 4).

As with NWC, the optimal composition of SLWAC depends on a variety of factors, including the type of application, targeted compressive strength, density class, economical restrictions, required workability and the characteristics of the concrete's constituents. Consequently, countless examples of SLWAC compositions are reported in the literature, ranging from low-to high-strength or high-performance concrete. Selected buildings and bridges are presented in Table 2.1.

Table 2.1 *SLWAC composition in selected structures.*

Materials	Picasso Tower Madrid 1988 (fib8 2000)	Sandhornøya Bridge, Norway 1989 (Holm 2000)	Nations Bank Charlote 1991 (ACI 213)	Heidrun TLP Norway 1995 (fib8 2000)	Portugal pavilion Expo-Lisboa 1998 (fib8 2000)	National Hospital, Oslo 2000 (fib8 2000)	Nordhordland bridge, Norway 1994 (Heimdal 1995)
Cement (kg/m^3)	320	400	385	420	420	375	430
Fly ash (kg/m^3)	120	–	83	–	100	–	–
Silica fume (kg/m^3)	–	25	–	20	15	11	35
water, (l/m^3)	190	144	180	163	196	174	144
LWA, (kg/m^3)	345(Arlita F7)	650(Leca)	534(Solite)	561(Liapor)	130(Leca2–4)	254(Leca4–10)	570(Leca/Liapor)
Sand, (kg/m^3)	900	575	763	720	838	900	630
w/b$_{effect}$	≅0.43	≅0.33	≅0.40	≅0.37 ± 0,02	≅0.37	≅0.45	≅0.31
Slump (mm)	230	200	130	250	–	150	150
Density (kg/m^3)	1855	1900	1890	1943	1850	1716	1893
$f_{cm,28d}$ (MPa)	30.1	59.8	47	79	LC25/28	30	69.9

Chemical admixtures are equally effective in SLWAC as they are in NWC. However, dry or partially wet LWA may absorb part of these admixtures during mixing, which reduces their efficiency. This may slightly increase the plasticizers' saturation point and may promote a more rapid loss of workability over time (Bogas *et al.*, 2012c). The use of pre-saturated LWA, or the delayed addition of admixtures, can combat this concern (ACI 304R, 2002)

3. Principles of mix design

Despite the particularities, the same mix design principles of NWC may be adopted for SLWAC (Dreux, 1986; EuroLightConR14, 2000; ACI 211.2, 2004). According to Maage *et al.* (2000), the water absorption by LWA and the risk of LWA segregation are the main factors distinguishing SLWAC mix design. Regarding absorption, water in the mix design has to be compensated by the water uptake by LWA. This is discussed in more detail in Section 6.

Concerning the risk of segregation, sufficient binder and extra fines are necessary to counteract the rise of LWA in the mix. This is attributed to the high mismatch between the densities of the LWA and the mortar (Sandvik and Hammer, 1995). According to Beris *et al.* (1985) the sedimentation of a heavy particle in a Bingham fluid occurs when Y_{GY} is lower than 0.143 (Equation (1)):

$$Y_G = \frac{3\tau_0}{D.(\rho_p - \rho_{fl}).g}$$

(1)

where g is the acceration due to gravity, D is the particle diameter, τ_0 is the fluid yield stress and ρ_p and ρ_{fl} are the densities of the particle and fluid, respectively. Therefore, assuming that this behavior is approximately valid in fresh concrete, the risk of segregation is governed by the mortar yield stress and the density difference between phases (mortar and aggregate). In SLWAC, although the movement of particles is in the opposite direction, the principle is the same. Since LWA density is usually 700–1400 kg/m^3 and the density of the mortar is greater than 2000 kg/m^3, the mismatch is higher than when normal-weight aggregates are used (typical density 2600 kg/m^3). From Equation (1), τ_0 for SLWAC would have to be increased by more than twice (richer pastes of low w/b ratio) to minimize the risk of segregation. Moreover, the greater the size of the LWA and the lower their density the higher the risk of segregation. The segregation phenomenon is more relevant in concrete with normal weight sand (NWS) than with lightweight sand (LWS), since the density of the mortar is reduced (ACI 213R, 2014). For the same reason, the incorporation of air entrainment also reduces this effect (Johnsen *et al.*, 1995). On the other hand, using pre-wetted or pre-saturated aggregates reduces the risk of segregation because it increases the LWA density.

Besides strength and workability, density is also a determining factor in SLWAC mix design (Dreux 1986; ACI 213R, 2014). In fact, according to ACI 213R (2014) and EN 206 (2016), SLWAC must be specified according to its density class in addition to the strength, durability and workability requirements. The fresh and hardened properties of SLWAC are affected by the type and amount of LWA, and so is the mix design (EuroLightConR2, 1998; Videla and López, 2002; Chandra and Berntsson, 2003;

Lijiu *et al.*, 2005). In fact, the mechanical strength of SLWAC depends on the w/b ratio, the quality of the aggregate–paste interface transition zone (ITZ) and the type and volume of LWA. Therefore, mix design methods that rely only on the w/b ratio, as per NWC, are not adequate for SLWAC. In addition, as discussed in Section 4, the influence of aggregate depends on the strength grade of the SLWAC, which further increases the complexity of the concrete mix design. When lighter LWA is used in high-strength SLWAC, or when denser LWA is used in low-density SLWAC, less efficient and less attractive solutions are obtained.

The accurate mix design of SLWAC should be performed on a volume basis, defined as the optimal arrangement of the volume of concrete constituents, including the air content. This should be determined according to the volumetric method in ASTM C173 (2016), since pressurized methods are affected by the air entrapped in the LWA (EuroLightConR14, 2000; Holm and Bremner, 2000; ACI 213R, 2014). Typical air content in the range of 20–40 l/m^3 is reported for SLWAC (Heimdal, 1995; Videla and López, 2002). The mix design must also take into account the casting conditions (ACI 213R, 2014).

SLWAC may be produced using natural sand, lightweight sand or a combination of both (FIP, 1983; EuroLightConR14, 2000; ACI 213R, 2014). Differences are mainly related to the more difficult control of water absorption and w/b ratio, as well as the more complex prediction of SLWAC strength. The use of natural sand allows better control of mixing water and workability, reduces the shrinkage and concrete cost, and increases the strength and modulus of elasticity of the concrete. Therefore, only when a reduction in the thermal conductivity or density is the main objective, and the compressive strength is less relevant (such as in some rehabilitation work), is all-lightweight concrete (ASLWAC) justified. Besides the high porosity, LWA fines are frequently composed of crushed particles, which greatly increases the water absorption, the irregularity of the particle shape and the surface roughness, increasing the water demand (Bogas, 2011; Thienel *et al.*, 2020). The water absorption of lightweight sand (LWS) may be as high as 40% (EuroLightConR2, 1998; Bogas, 2011). However, as mentioned, when natural sand is replaced by fine LWA the mortar density becomes more compatible with the coarse aggregate density and the risk of segregation is reduced. Replacing natural sand by lightweight sand may reduce the dry density by about 200 kg/m^3 but there is a commensurate reduction in mechanical strength (Thienel *et al.*, 2020).

4. Compressive strength

As is well known, when aggregates have less stiffness than the surrounding mortar, as may occur in SLWAC, forces are mainly transferred through the matrix, leading to transverse stresses in the aggregates and the mortar (Figure 2.1). Since the ITZ is stronger than the LWA, the fracture path tends to travel through them and the compressive strength of the concrete becomes lower than that of the mortar (Bogas and Gomes, 2013a). In this case, the failure path occurs in the areas with the greatest number of LWA (Virlogeux, 1986). This behavior may be also found in high-strength concrete (Aïtcin, 1998).

As with NWA, if the stiffness of the LWA is higher than that of the paste, the failure path travels around the aggregate – usually across the weaker ITZ – and the

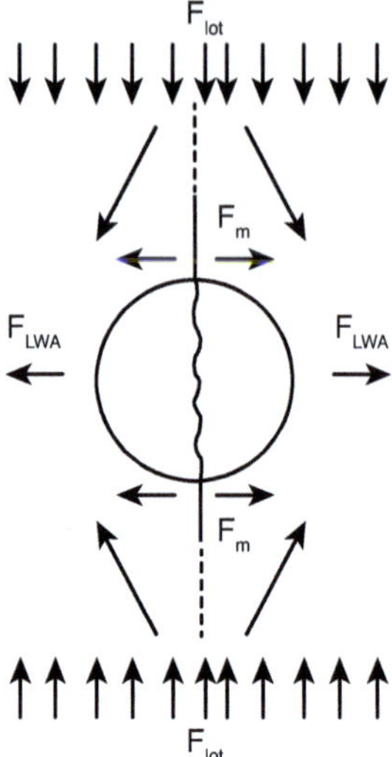

Figure 2.1 *The force transfer mechanism in SLWAC utilizing LWA having less stiffness than the surrounding mortar (adapted from Gerritse, 1981; FIP, 1983; Bogas and Gomes, 2013a).*

strength of the concrete is close to the strength of the mortar (Bogas and Gomes, 2013a). The compressive strength behaviour of SLWAC is depicted in Figure 2.2. When the concrete reaches a limit strength, f_L, beyond which the modulus of elasticity of the LWA is lower than that of the mortar, then the compressive strength of the concrete is affected by both the LWA and the mortar (Chen *et al.*, 1999). This transition

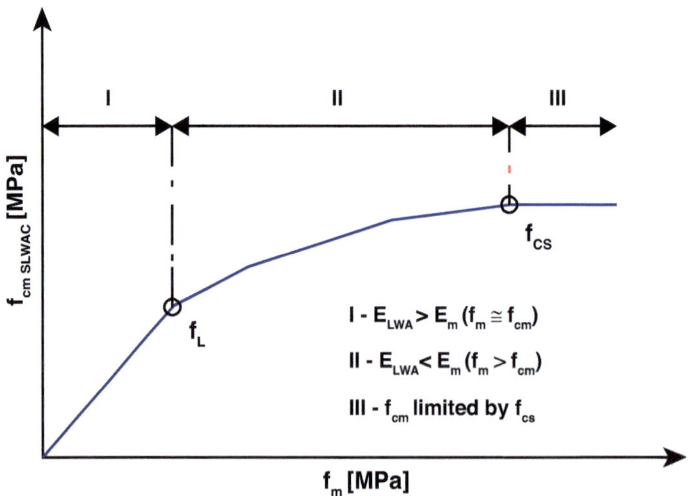

Figure 2.2 *The compressive strength f_{cm} of SLWAC versus the compressive strength f_m of the mortar. In region I, SLWAC presents the same behavior as NWC (adapted from Bogas and Gomes, 2013a).*

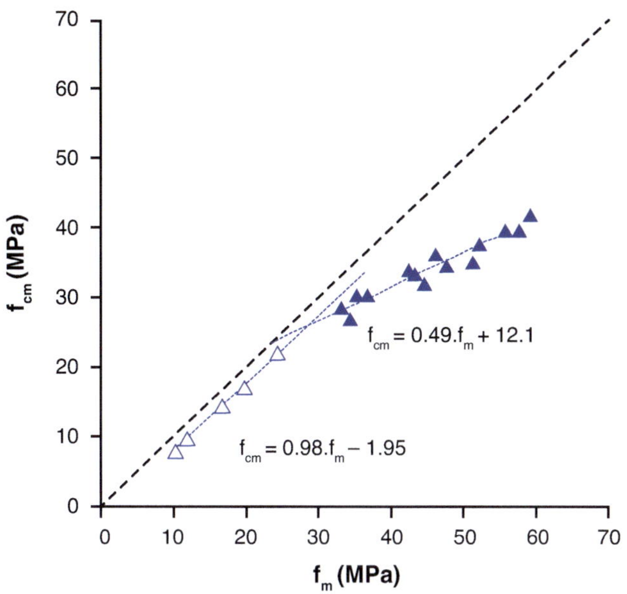

Figure 2.3 *The compressive strength of mortar, f_m, versus the compressive strength of SLWAC, f_{cm}, using expanded LWA (ρ_p = 1070 kg/m³). Regression is based on data from Bogas et al., 2017.*

has been documented by various authors (Chen *et al.*, 1995; Holm and Bremner, 2000; Ke *et al.*, 2009). The value of f_L for a given type of aggregate may be estimated by comparing the strength of concrete and that of mortar of equal composition (Chen *et al.*, 1995; Adámek, 2000). Figure 2.3 illustrates one example where f_L for a certain type and volume of LWA was determined from two experimental mixtures tested over time and under similar conditions (curing, specimen type, age), one with concrete and the other with just the mortar used to produce that concrete. In this case f_L was estimated as 25–27 MPa.

In general, below f_L the compressive strength of the concrete is governed by the mortar, and the SLWAC shows the same performance as NWC of equal composition. This typically occurs in SLWAC with dense LWA, a low w/b ratio or tested at an early age (Bogas and Gomes, 2013a). Above f_L, the relationship between the compressive strength of the mortar and that of the concrete is parabolic (Figure 2.2), tending to a maximum ceiling strength, f_{cs}, beyond which further improvement of the mortar quality has little influence on the compressive strength of the concrete (Chen *et al.*, 1999; Faust, 2000; Bogas *et al.*, 2017). In Figure 2.3, this corresponds to a value close to 42 MPa. This parameter differs for each type of LWA and provides an idea of its maximum domain of application. For the estimation of f_{cs}, production of concrete with a high-quality paste (a water–cement ratio of about 0.3), especially for dense LWA, is recommended (EuroLightCon, 2000; Faust, 2000; Bogas and Gomes, 2013a). This parameter also varies with the volume of LWA, although with minimal impact within the typical range of 350–400 l/m³. The latter value may be conservatively adopted. In general, f_{cs} can be increased by reducing the maximum size of the coarse aggregate (ACI 213R, 2014).

Contrary to NWC, in which the compressive strength is mainly determined by the cement matrix composition (type of binder and w/b ratio), the mechanical behavior of SLWAC is also controlled by the type and volume of aggregate and the strength level of the concrete (Gerritse, 1981; FIP, 1983; Alduaij *et al.*, 1999). This means that the influence of the w/b ratio or binder content on the SLWAC strength will depend on the type of aggregate, being less relevant than in NWC (FIP, 1983; Virlogeux, 1986; Zhang and Gjørv, 1991b; Faust, 2000). In fact, the lower the w/b ratio, the lower the influence of the cement matrix properties on the SLWAC compressive strength (Zhang and Gjørv, 1991b; Faust, 2000; Bogas and Gomes, 2013a). For example, Zhang and Gjørv (1991b) reported that the strength of high-strength SLWAC was almost unaffected by the cement content and the w/b ratio. There is a ceiling strength and above this the compressive strength is not significantly changed by a further improvement in the cementitious matrix quality (Chen *et al.*, 1999; Faust, 2000; Bogas *et al.*, 2017). Thus, the compressive strength of the SLWAC cannot be predicted from the w/b ratio alone. Also, the w/b ratio only takes into account the effective water (i.e. the water not absorbed by the LWA) that immediately affects the cementitious matrix microstructure (see Section 6).

Various studies have been published concerning the compressive strength of SLWAC. However, most of them involve SLWAC with a given type of aggregate, and so their conclusions are only valid for the specific case studied. One exception is the comprehensive research of Bogas and Gomes (2013a) and Bogas *et al.* (2017). Covering SLWAC in a wide range of strength classes (LC12/13 to LC60/66), the authors analyzed the failure modes and the influence of the main constituents of concrete on compressive strength, namely the cement content, amount of water and the type and volume of aggregates. It was found that the cement content and the sand/cement ratio had little influence on the SLWAC strength. Moreover, varying the LWA volume, even above f_L, led to a smaller decrease in strength than altering the w/b ratio, for the same changes in concrete density.

Denser LWAs ($\rho_p > 1400$ kg/m^3) have proven suitable for the production of high-strength concrete, aggregates of intermediate density (1000 kg/m$^3 < \rho_p < 1400$ kg/m^3) for moderate-strength concrete, and more porous aggregates ($\rho_p < 1000$ kg/m^3) for low-strength concrete, or for applications in which the reduction of density is a deciding factor. Up to 60 MPa, SLWAC with denser LWA exhibited the same behavior as NWC. It was concluded that the influence of LWA on the SLWAC strength is more important the lower the density, the greater the mortar strength and the older the concrete. This influence is highly dependent on the strength level and concrete composition, corroborating the schematic relation shown in Figure 2.2. Moreover, the mechanical properties of LWA are also affected by their production process and not only by their total porosity, which also complicates the generalization of the compressive behavior of SLWAC. For example, the lower efficiency of SLWAC with fly ash LWA is related to the omission of the dense outer shell in these aggregates (Hammer and Smeplass, 1995; Bogas *et al.*, 2017). Thus, SLWAC should not be specified according to the relative strength against NWC, and the methodology for analyzing the compressive strength behavior must be other than that considered in NWC.

Several studies have attempted to relate compressive strength to density (Virlogeux, 1986; Zhang and Gjørv, 1991b; Lazarus, 1993; Hammer and Smeplass, 1995) because

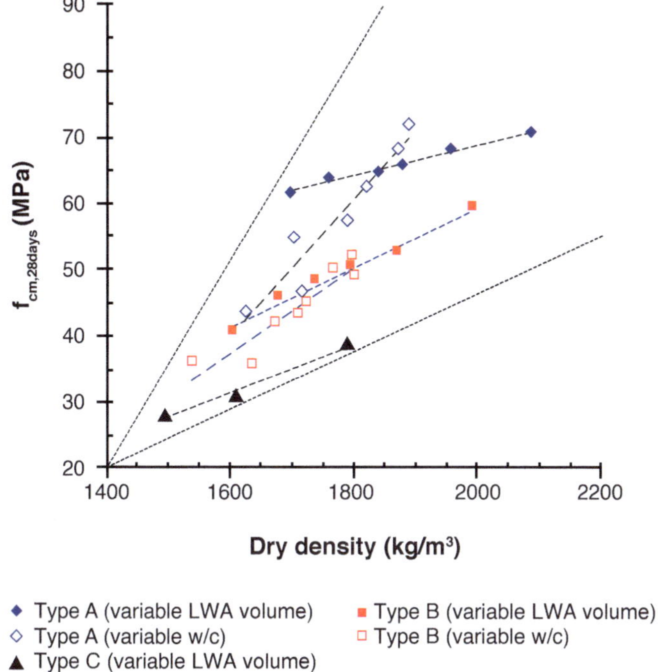

Figure 2.4 *Dry density versus the 28-day average compressive strength ($f_{cm,28days}$) for different types (densities) of LWA and variable w/c ratio (based on Bogas and Gomes, 2013a).*

both properties are affected by concrete porosity. However, the relation between these properties may widely vary with the concrete composition and the strength level (Figure 2.4). Therefore, density cannot be a criterion for predicting the compressive strength either.

To sum up, the mechanical properties of SLWAC vary with the type of aggregate and the strength level of the concrete. Therefore, the prediction of SLWAC compressive strength and the definition of a general mix design methodology can be more challenging and must take into account the biphasic mechanical behavior of concrete illustrated in Figures 2.2 and 2.3.

5. Common structural lightweight concrete

Until the late 1980s, SLWAC was adopted in many countries. Compressive strengths were generally below 45 MPa, limited by the aggregate and concrete characteristics at that time. However, with the introduction of more efficient admixtures and the technological development of high-performance SLWAC, higher compressive strengths have been progressively attained. In the 1980s Wilson and Malhotra (1988) expounded the possibility of producing SLWAC with a compressive strength of 70 MPa, and Hoff (1985) settled the challenge of producing SLWAC with a density lower than

1800 kg/m³ and a compressive strength of 60–80 MPa, using superplasticizer and silica fume. Some years later, Zhang and Gjørv (1991c) produced laboratory high-strength SLWAC with a dry density of less than 1735 kg/m³ and 100 mm cube strengths of 102 MPa at 28 days. These uneconomical concretes were produced with a w/c ratio of 0.28 and as much as 600 kg/m³ of cement and 60 kg/m³ of silica fume. Since the late 1980s, high-strength SLWAC for offshore platforms and wide-span bridges was produced in Norway, with concrete classes LC55-LC60 and a density of 1900 kg/m³. Nowadays, fluid SLWAC up to about 70 MPa can be easily produced with cement content below 450 kg/m³ (Harmon, 2003; Haque *et al.*, 2004; Bogas *et al.*, 2013; ACI 213R, 2014). Moreover, by choosing the correct type of LWA, SLWAC may be produced with a compressive strength in the range of common NWC, which increases its competitiveness in the construction industry. Nevertheless, most structures do not require higher strengths than about 40–50 MPa, avoiding the excessive enrichment of mortars.

Today, mix design methods must be adequate for common SLWAC, as specified by the relevant standards. According to EN 206 (2016), SLWAC is defined for oven-dry densities between 800 and 2000 kg/m³ and is divided into density subclasses differing by 200 kg/m³. According to this standard, the strength class of common SLWAC may vary between LC 8/9 and LC 55/60, with that of high-strength SLWAC up to LC 80/88. According to the American standard ACI 213R (2014), SLWAC is specified by a minimum 28-day compressive strength of 17.2 MPa (in cylinders) and a maximum equilibrium density of 1920 kg/m³. This standard was essentially designed and prepared for SLWAC in the narrow range of 20–35 MPa, high-strength SLWAC being designated a compressive strength of more than 40 MPa. Table 2.2 gives the density and strength classes of current SLWAC and ASLWAC.

6. Water absorption

One of the main challenges in SLWAC production is the accurate estimation of the high water absorption by LWA, which makes control of the mixing water difficult, affecting the mixing, workability, transportation, placement and compaction of SLWAC and, as a consequence, its mechanical properties and durability (Smeplass, 2000; Bogas *et al.*, 2012b; ACI 213R, 2014). However, water absorption by the LWA may also be beneficial, since it promotes internal curing and increases the paste–aggregate ITZ quality, which in turn improves the mechanical properties and durability of SLWAC.

Table 2.2 *Common ranges for compressive strength and density in SLWAC (Bogas and Gomes, 2013b; Thienel et al., 2020).*

Compressive strength classes	Density classes	
	SLWAC	*ASLWAC*
LC16/18 – LC30/33	>D1.6	>D1.4
LC35/38 – LC55/60	>D1.8	>D1.6
LC60/66 – LC70/77	>D2.0	>D1.8
LC80/88	>D2.0	>D2.0

The water absorption of LWA is mainly affected by its microstructure, surface characteristics, the number of crushed particles, the initial water content and the characteristics of the fresh mortar (Hoff, 1992; Wasserman and Bentur, 1997; Bogas *et al.*, 2012b). Lightweight aggregates adequate for structural concrete are characterized by up to about 25% of water absorption, with most of that occurring in the first minutes, followed by a marked drop in the rate of absorption (Swamy and Lambert, 1981; Lo *et al.*, 1999; Harmon, 2003; Holm and Bremner, 2000). Zhang and Gjørv (1991a) reported that 60–80% of the 24-hour absorption by expanded LWA occurred after one hour. In turn, Bogas *et al.* (2012a) found that in the first three minutes the water absorption was 53–79% of that at one hour. For fly ash aggregates without outer shell, 70% water absorption is reported after 30 seconds (Swamy and Lambert, 1981). In fact, the denser outer shell of expanded clay aggregates has an important role in delaying the water uptake (Lo *et al.*, 1999; Chandra and Bernstsson, 2003).

The water absorption is inversely proportional to the initial water content (EuroLightConR20, 2000; Bogas *et al.*, 2012a, 2012b; ACI 213R, 2014). Moreover, the water absorption is not only affected by the water content but also by the moisture distribution inside the particles (EuroLightConR20, 2000; Smeplass, 2000; Holm *et al.*, 2003). In Figure 2.5, cases 1 and 4 represent unusual extreme situations of completely dry (case 1) or fully saturated (case 4) particles immersed for long periods. Case 5, with the core of the particle saturated and the outside dry, is the result of drying previously saturated particles at room temperature. Case 2 represents the opposite situation, in which previously dried aggregate is immersed for a short period of time, keeping the core unsaturated. Cases 2 and 3 are the most common. Case 3 corresponds to an aggregate that starts with a similar distribution to that of case 2, which is then dried at room temperature. All the cases may imply distinct water absorption behavior (Bogas *et al.*, 2012b).

The fresh mortar involved in the aggregate also affects the rate of absorption. Muller-Rochholz (1979) confirmed that absorption depends on both the characteristics

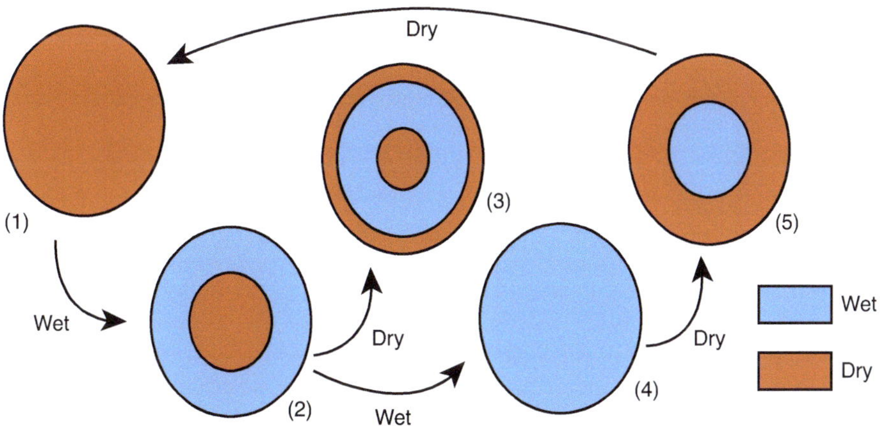

Figure 2.5 *The distribution of moisture in the aggregate before concrete mixing (based on Bogas et al., 2012b).*

of the aggregate and the w/b ratio of the surrounding mortar. Punkki and Gjørv (1995) reported that two types of LWA absorbed 6.0–7.0% after 60 minutes in pure water but only 3.5–4% and 4.5–5.5%, respectively, in fresh concrete. In addition, crushed particles have a faster water uptake because of the partial or total loss of their outer shell and the increase in surface area (Bogas, 2011; Bogas *et al.*, 2012b). Some authors have reported differences of 30–50% in water absorption between broken and intact particles (Zhang and Gjørv, 1991a; Lo *et al.*, 1999). Bogas *et al.* (2012b) found experimentally that, depending on the type of aggregate and percentage of broken particles, absorption in mortars with a w/b ratio of 0.3 could be 20–50% of the absorption in water. The absorption was higher for more porous aggregates with a higher percentage of broken particles.

For all the foregoing reasons, the water absorption by LWA is still only roughly estimated, usually based on a rule of thumb or on previous experience. According to EN 206 (2016), the water absorption during concrete mixing may be roughly considered as equivalent to the 1-hour absorption of as-used wet LWA. However, according to Hammer (Hammer and Smeplass, 1995), this rule is only valid for w/b ratios of about 0.4, decreasing for lower ratios. Absorption values that are 10% higher are mentioned for concretes with a w/b ratio of 0.55. In Germany, the recommended estimated absorption by aggregate in mortar is twice the 30-minute absorption in water (EuroLightConR2, 1998). Chandra and Berntsson (2003) suggested that the aggregate absorption in fresh concrete is 75–100% of the first 30 minutes to 1 hour of absorption in water. Besides the inaccurate nature of these approaches, using aggregates with different initial water content and determining the absorption properties of the LWA at every hour is not practical.

To overcome this issue, SLWAC is usually produced with pre-wetted or pre-saturated aggregate. In pre-wetting, LWA is previously mixed with part of the mixing water. For better control of the effective water or for when concrete has to be pumped, the LWA may be pre-saturated by immersing it for at least 24 hours in water or through vacuum saturation (Pankhurst, 1993). Nevertheless, to control the final mixing water, the water absorption of the LWA after saturation must be previously determined. The LWA may also be used in a dry state, easily predicting the absorption from standard tests in water. Nevertheless, bringing the aggregates to any of these conditions is expensive.

Based on a comprehensive experimental program and by introducing new parameters that take into account the microstructure and water content of the aggregates, a new methodology for predicting the LWA absorption in concrete was suggested by Bogas *et al.* (2012b). The purposed procedure can provide a more accurate estimation of the amount of mixing water to be specified in the SLWAC mix design, regardless of the type of aggregate.

To better understand the influence of water content on LWA absorption two new parameters were defined: Rel_w, corresponding to the quotient between the absorption in a given aggregate with a certain water content and the absorption of the same aggregate, initially dry; and w_{R48h}, corresponding to the relative water content at 48 h and which may be determined by the quotient of the initial water content and the saturation water content of the aggregate at 48 hours. These parameters take into account the initial water content and the microstructural characteristics of the

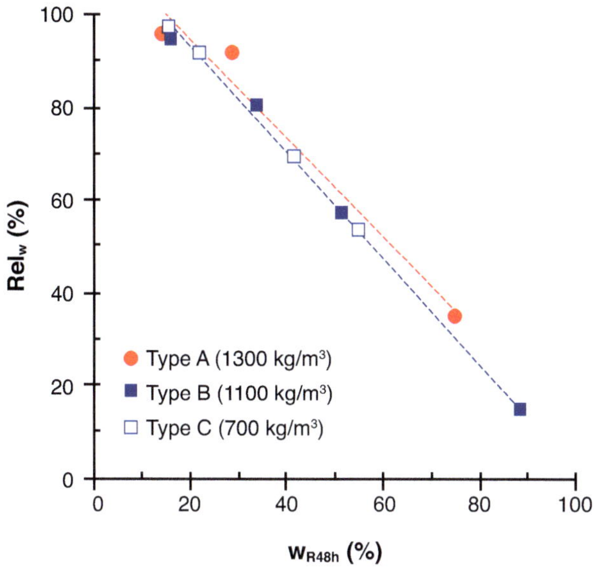

Figure 2.6 *The relationship between Rel$_w$, and the relative water content at 48 hours, w$_{R48h}$, for LWA with different porosities/densities (Bogas et al., 2012b).*

aggregate, respectively. As shown in Figure 2.6, a very good correlation is obtained between these two parameters, regardless of the type of LWA, even considering aggregates of very different porosity. Using these parameters, a general expression was suggested to estimate the water absorption, which is almost independent of the type of aggregate.

The suggested expression was further corrected with a dimensionless empirical factor, K_{LWA}, that depends on the type of LWA and assumes that aggregates are surrounded by fluid mortars with w/b ratios of up to 0.4. Depending on the percentage of broken particles and the porosity of the LWA outer shell, K_{LWA} varies between 1.10 and 1.25. To reduce this imprecision and to avoid the significant mortar influence, the authors suggested pre-wetting the aggregate for 2–5 minutes because the LWA absorption is greatly reduced after the first few minutes and further absorption after mixing was found negligible (Bogas *et al.*, 2012b). Equation (2) was evaluated experimentally by producing several concrete mixtures, leading to reasonable estimates of the total absorption in the concrete (Bogas *et al.*, 2012b).

$$W_{abs,LWA} = K_{LWA} \times \frac{116 - 1.14 \times W_{R48h}}{100} \times W_{abs,ag,w_0=0\,(\%)} \tag{2}$$

Nevertheless, it must be clear that, as with recycled aggregates, the high water absorption of LWA does not impair the mechanical strength as long as the effective water is not altered. Indeed, as already mentioned, the water absorbed by aggregates may even contribute to the mechanical strength through internal curing (Bentz *et al.*, 2006; ACI 213R, 2014).

7. Mix design methodologies

Various SLWAC mix design methodologies have been suggested in the literature, essentially based on preliminary experimental work within a narrow range of SLWAC compositions and for a certain type of LWA (FIP, 1983; Dreux, 1986; EuroLightConR9, 2000; Videla and López, 2000; ACI 211.2, 2004; Lijiu *et al.*, 2005; Yang *et al.*, 2014). The methodology is only valid for the specific materials considered in the experimental program and hence significant differences may be expected when these methodologies are adopted in other types of SLWAC. As noted, the characteristics of LWA may greatly vary, depending on the raw material, the process of manufacture and the level of expansion or incorporated voids. Moreover, depending on the type of LWA, SLWAC properties are also affected by the type and proportion of other constituents (cement, additions, admixtures). Therefore, the implementation of a universal methodology for SLWAC mix design has been more complicated (Videla and López, 2000).

ACI 211.2 (2004) suggests two simple mix design methods supported by significant preliminary experimental work: the weight method and the volumetric method. The weight method, usually adopted in NWC mix design, is based on the fact that the volume of fresh concrete corresponds to the sum of the absolute volumes of the concrete constituents (binder, aggregates, water, admixtures) and air voids. The method also takes into account that the volume of mortar must be sufficient to fill the gaps between the aggregate particles and must ensure adequate workability.

The amount of water is defined based on listed results from previous experimental work, depending on the maximum size of the aggregate, the required slump and the eventual incorporation of chemical admixtures, such as air-entraining. The same tables defined for NWC are adopted for SLWAC, although the slump is affected by the fresh density of the concrete (EuroLightCon R2, 1998; Bogas *et al.*, 2012c). This method also requires the determination of the water absorption and density of the LWA, taking into account their initial water content before mixing. The cement content is determined from abacuses relating the compressive strength to the w/c ratio. However, as mentioned, since the SLWAC strength depends on the type and amount of LWA for a given w/b ratio, a new relationship must be defined when a new type of LWA is adopted. Otherwise, ACI 211.2 (2004) provides conservative values for concrete with cement type I and a maximum aggregate size between 19 mm and 25 mm. Finally, the dosage of coarse aggregate is obtained from empirical relationships, depending on the sand fineness modulus and the maximum aggregate size, in order to achieve concrete with adequate workability in usual applications. For SLWAC requiring higher workability, a reduction of 10% in the volume of LWA is recommended. This method only provides an idea of a first SLWAC composition, which then has to be adjusted further. This method is limited to concretes that have been subjected to trial experimental tests and it cannot be extended to other concretes. Thus, the great limitation of this methodology may be inferred immediately.

The volumetric method, explained in greater detail in ACI 211.2 (2004), starts with the production of an experimental mixture, designed from estimated volumes of binder and aggregate with sufficient water added to achieve the required concrete workability. Afterward, the concrete mixture is assessed in terms of workability, fresh density and finishing quality, and the necessary corrections are implemented. The SLWAC composition is defined taking into account that the sum of the apparent volume of wet aggregates varies between 1.04 and 1.26 m^3/m^3 and the apparent volume of fine aggregate is around 40–60% of this amount. In this method, the w/b ratio cannot be estimated with sufficient accuracy due to the inherent difficulty in determining the real amount of water absorbed by the LWA. Therefore, the compressive strength is inaccurately estimated from the amount of cement and air content, instead of the w/b ratio. This implies the production of various mixtures with different cement content in order to cover a wide range of strength classes. This method is almost exclusively based on experimental work and is also limited to tested SLWAC. Therefore, both weight and volumetric methods are laborious, time-consuming, essentially based on empirical relationships and have a limited range of applications. Based on an extensive experimental campaign, Jafari and Mahini (2017) found that the results obtained for compressive strength and density were far from those assumed in ACI 211.2 (2004). They considered that the ACI method is only adequate for a narrow range of concrete compositions.

The FIP (198)) suggests a mix design method also based on extensive experimental work. The type of aggregate is selected taking into account the intended density and compressive strength of the SLWAC. The amount of cement and water are estimated from pre-established curves defined for each type of aggregate, relating the compressive strength to the w/b ratio and the consistency to the water content. The volume of coarse aggregate may be obtained from different procedures: the absolute volume method, the volumetric method, the specific gravity factor method; the weight method; and the effective w/c ratio method. These methods are discussed in detail in FIP (1983) and their detailed analysis is outside the scope of this chapter. Finally, the aggregate grading respects the reference curves established by the German standard DIN 1045 (1988).

Alternatively, the FIP (1983) also suggests a mix design method based on the relative performance of SLWAC and NWC of the same composition. This type of methodology requires various trial mixes in order to build detailed diagrams comparing the compressive strength of NWC and SLWAC with distinct types of LWA. Besides being laborious and time-consuming, the method is highly sensitive to small variations in the characteristics of the concrete constituents (EuroLightCon R2, 1998).

Dreux (1986) adapted his proposed design method for NWC to low-strength SLWAC, in which the concrete density and the cement content were defined according to the properties of the LWA, essentially based on empirical relations. The method is applicable to SLWAC with natural sand up to 2 mm in size, since larger fine aggregates may have little influence on the workability but a significant impact on the reduction of the concrete density. The w/b ratio is estimated based on the intended compressive strength, the type of cement, and a correction coefficient depending on the characteristics of the LWA. However, the method does not take into account that the compressive strength of SLWAC is affected by the proportion of LWA and the strength range of concrete (Section 4).

A more rational SLWAC mix design method was suggested in EurolightconR5 (2000) but it is only valid for concrete with a w/c ratio of 0.25–0.35. For a given slump, the volume of paste is determined from s-curve diagrams. In general, it was found that 30% paste content provides good workability.

Videla and López (2000) proposed one method in which the mixing water is determined from empirical expressions as a function of the slump, aggregate fineness modulus and intended fresh density of concrete. It is assumed that the workability is little affected by the type of aggregate. Concrete is considered a two-phase material in which compressive strength depends on the strength and proportion of each phase. The strength of the aggregate is estimated from the ten percent fines value method of BS 812 (British Standards Institute, 1990), which consists of determining the load required to produce 10% of fines.

Due to the difficulty in determining the effective water, some authors consider that the SLWAC strength should be related to the cement content (Dreux, 1986; Amiri *et al.*, 1994; Lijiu *et al.*, 2005). Lijiu *et al.* (2005) considered the influence of mortar based on the cement content and the influence of the LWA based on the dry density of the concrete. The compressive strength of concrete was estimated as a function of the cement content to the dry density ratio. However, as shown by Bogas and Gomes (2013a), the cement content has little influence in compressive strength for the same w/b ratio.

Chandra and Berenson (2003) proposed a semi-empirical method that accounts for the influence of LWA type and volume on the compressive strength of SLWAC using Equation (3),

$$\log(f_{cm}) = v_{LWA} \cdot \log(f_{LWA}) + v_m \cdot \log(f_m) \tag{3}$$

where f_{cm} is the mean compressive strength of the concrete and f_m is the mean compressive strength of the mortar with the same composition as that used for concrete; v_m and v_{LWA} are the relative volumes of the mortar and LWA in the mix; f_{LWA} is the strength of the aggregate in concrete, estimated from empirical relations. This equation takes into account that the density of SLWAC is determined by the density of concrete constituents and that the density of each phase is related to the logarithm of its compressive strength. Basically, it is assumed that concrete behaves as a two-phase material, composed only of LWA and mortar, a reasonable assumption given the high quality of the ITZ in SLWAC (Videla and López, 2002). The strength of the aggregate in concrete, f_{LWA}, is estimated as a function of the LWA density and two empirical coefficients.

However, this mix design method does not take into account that the SLWAC compressive strength depends greatly on the strength level and composition of the concrete (Bogas and Gomes, 2013a). As discussed in Section 4, the compressive strength behaviour of SLWAC depends on the limit and ceiling strength, which must be considered in the mix design (Chen *et al.*, 1999; Bogas *et al.*, 2017). Moreover, the mix design method does not consider the concrete density and requires the predetermination of empirical parameters.

Some authors have considered the limit strength, f_L, in their estimation of compressive strength (Chen *et al.*, 1995, 1999). In this case, f_L is previously determined from concrete and mortar samples of equal composition, as discussed in Section 4. It was found that the f_L varies slightly with the volume and maximum size of the LWA (Chen *et al.*, 1999). However, the difference in f_L was lower than ±5% for LWA volumes varying from 350 l/m^3 to 300 and 450 l/m^3, respectively (Chen *et al.*, 1999). Therefore, it is reasonable to assume a constant value for this parameter within the narrow range of common LWA volumes used in SLWAC (Bogas and Gomes, 2013b). Knowing f_L, the strength of the mortar is related to the strength of the concrete according to two regression lines determined below and above f_L. However, as shown in Figure 2.2, above f_L the relation between the concrete and mortar strength is not linear. This assumption may not be a reasonable one, especially for concrete strengths near the celling strength. Finally, the cement content is determined from conventional relations between the mortar strength and the w/b ratio, as done for conventional NWC.

Mix design optimization depends on various factors, such as environmental and application conditions, as well as the fresh and hardened properties required (Chandra and Berntson, 2003; Bogas and Gomes, 2013b). Thus, concrete mix design is always a trial-and-error process. However, the lower the number of iterations the more robust the mix design method. Based on previous contributions, Bogas and Gomes (2013b) suggested the more rational method described in Section 9, which is valid for different types of LWA and takes into account the key factors that affect SLWAC performance, namely the LWA water absorption, the concrete dry density and the compressive strength behaviour.

More recently, Yang *et al.* (2014) proposed an empirical SLWAC trial mix design methodology based on regression analysis from various results obtained in expanded clay and fly ash SLWAC. From the data provided in ACI 211.2 (2004), Abdullahi *et al.* (2011) suggested a computer-model approach for SLWAC mix design. This method aimed to implement more expeditiously the ACI methods but no relevant contribution in the design fundamentals was provided. Nadesan and Dinakar (2017) suggested a mix design procedure for sintered fly ash SLWAC, valid in the wide w/c ratio range 0.25–0.75. The authors considered the reference grading curves of DIN 1045 (1988) to enhance the packing density of aggregates, with the aim of achieving an improvement in the workability and mechanical properties of the concrete. The compressive strength is predicted from empirical curves, only valid for the specific concretes studied by the authors. In general, no relevant advances were introduced with this mix design method. Jafari and Mahini (2017) used gene expression programming to predict the SLWAC compressive strength as a function of some design parameters. The method was strictly based on curve-fitting regression analysis from various experimental results. Therefore, this study is also limited to the specific cases of expanded clay and pumice LWA used by the authors.

8. Aggregate grading optimization

As with conventional NWC, the aggregate packing optimization of SLWAC allows the more rational determination of mixing water and volume of paste for a given workability. In NWC, this optimization may contribute to a reduction in porosity, shrinkage, creep and unit cost, as well as an increase in strength and durability. However, contrary to NWC, the strength and density of SLWAC is directly affected by the volume of LWA, which usually has a higher influence on these properties than the grading optimization, assuming that workability is not adversely affected (Bogas and Gomes, 2013a, 2013b). Moreover, the risk of segregation increases with the volume of LWA.

As noted, SLWAC needs a greater volume of mortar than NWC to ensure the sufficient involvement of the LWA. In addition, the strength and stiffness of LWA tend to decrease with the aggregate size, especially in expanded LWA, which also limits the optimization of aggregate grading in SLWAC. Therefore, SLWAC mix design cannot be based only on packing optimization criteria. However, taking these specificities into account – namely limiting the maximum volume and size of the LWA – packing models adopted for NWC may also be considered for SLWAC mix design.

Continuous and discrete packing methods may be used for aggregate grading optimization. The first group includes theoretical reference grading curves, such as those of Fuller, Bolomey, Joisel and Faury. One issue with these original curves is that they underestimate the amount of fines less than 250 mm in size, resulting in lean mixtures. To overcome this problem, original curves have been modified. For example, the Fuller curve, given in Equation (4)

$$P = \left(\frac{d}{D} \right)^{0.5} \tag{4}$$

was modified to become Equation (5)

$$P = \left(\frac{d^r - D^r_{min}}{D^r - D^r_{min}} \right) \tag{5}$$

where D is the maximum size of aggregates, d is the minimum size of the aggregate fraction and r is a variable exponent. This modification is based on the initial Andreasen (1930) model, where r varies between 0.33 and 0.5, and on the later contribution of Funk and Dinger (1994), which introduces a minimum particle size (D_{min}).

According to Cánovas and Gutiérrez (1992) the Faury method is backed by a long record of adequate application in practice and showed to be extended to high-performance concrete, provided that some adaptations are made.

More complex discrete models are an alternative to reference curves, such as the compressive packing model of Larrard (1999). This model assumes that interaction between aggregate particles is possible and that part of the particles occupy the space left by others. In addition, two relevant restrictions to packing – namely, the wall effect and loosening effect – are considered, as well as a compaction index that takes into account the compaction energy used.

EuroLightConR9 (2000) considered the use of Fuller curves and a computer program called 'Europack' to optimize the aggregate packing density in SLWAC, depending on the grading properties and loose bulk density of the LWA. The composition and characteristics of the surrounding paste was determined separately, taking into account its required strength, stability and rheology. Chandra and Berntsson (2003) proposed consideration of the Fuller curve (Equation (4)) for SLWAC, where r is equal to 0.5 for d up to 2 mm and variable from 0.55 to 0.7 for d greater than 2 mm.

Most design methods that have been proposed for SLWAC do not consider the aggregate packing optimization, although they may be adjusted to also take this aspect into account. The packing models may be more adequate when the LWA are supplied in different fractions, which is not the usual case. In the more usual circumstances, when the aggregate is composed only of natural sand and one type of coarse LWA, the aggregate composition may be determined directly by setting the maximum content of coarse LWA or experimentally maximizing the packing volume of both types of aggregates. Grading curves according to the B16 (DIN 1045, 1988) should be followed/considered to provide adequate workability and strength (FIP, 1983). As noted, to avoid segregation a minimum amount of fine particles under 250 µm is recommended. As with NWC, continuous grading curves usually reduce the risk of segregation, contrary to gap grading curves (FIP, 1983). One difficulty in the aggregate packing optimization and mix design of SLWAC is that the grading curves of the LWA need to be measured by volume because the density of the LWA usually varies with the particle size (FIP, 1983; Hoff, 1992; ACI, 213R, 2014).

To sum up, packing models may be used to improve the overall granular compactness of SLWAC but they should not be considered alone. Within a narrow range, the proportion of each aggregate fraction may need to be tuned by packing optimization. The packing methodologies aim at maximizing the aggregate compactness without compromising the applicability of fresh concrete (ICAR-105-1, 2003). Indeed, the maximum aggregate packing may not correspond to the optimal concrete mix design, as it may lead to rough and discontinuous mixtures with subsequent problems of segregation, bleeding and applicability (Neville, 1995; ICAR-105-1, 2003).

9. SLWAC mix design methodology

A novel method for the mix design of SLWAC produced with coarse LWA and natural sand was proposed in Bogas and Gomes (2013b). The method takes into account the contributions of Chandra and Berntsson (2003), Chen *et al.*, (1995, 1999) and Bogas and Gomes (2013a), regarding the modelling of SLWAC compressive strength behaviour. Contrary to other methods, the proposed mix design methodology is not based extensively in experimental work or empirical relations, limiting its range of application. The method considers simultaneously the requirements of concrete density, the more accurate prediction of LWA absorption, the influence of the proportion and strength of the concrete constituents and the dependence of the compressive behaviour of SLWAC on the level of compressive strength. The objective is to provide a simple and easily implementable method, requiring reduced experimental effort and being valid for any type of LWA and mortar composition, within the common strength and density classes of SLWAC. Basically, the idea is to provide a starting composition for the production of SLWAC, which should be only subjected to further minor adjustments to tune it for the intended workability and final hardened properties. The general procedure for the suggested mix design is shown schematically in Figure 2.7.

9.1 Initial mix design data

Previous knowledge of the intended workability, density, compressive strength and durability, as well as information about the main properties of the concrete ingredients

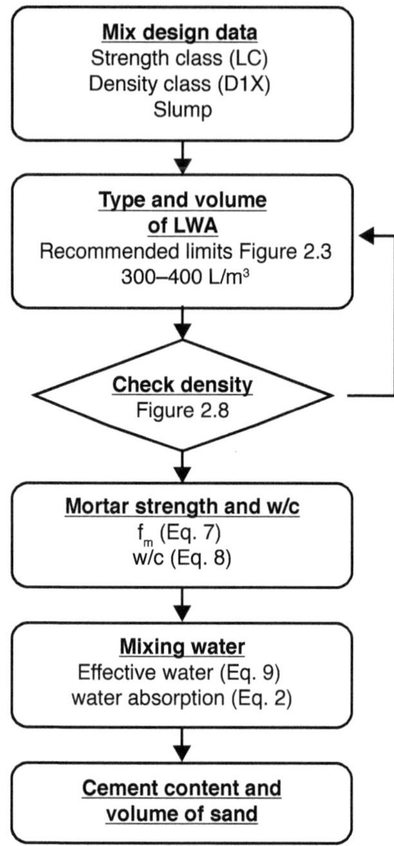

Figure 2.7 *SLWAC mix design procedure suggested by Bogas and Gomes (2013b).*

is necessary. Therefore, the following preliminary information is required (Bogas and Gomes 2013b):

- The strength, density, durability and workability of the SLWAC, in accordance with the current specification of concrete, as required in EN 206:2013+A1 (2016). For example, the following concrete may be specified by the designer: EN 206: LC30/33·XC3·Cl 0.2·D$_{max}$12·D1.8·S3.
- The properties of the concrete constituents according to the standards: density, absorption and grading of LWA; density and grading of natural sand; type and characteristics of the binder and admixtures.
- The strength parameters and initial water content of LWA. This is unusual information to need but it can be provided easily by the supplier.

In order to predict the compressive strength of SLWAC, it is necessary to first determine the strength of the LWA in the concrete, f_{LWA}, the limit strength, f_L and the ceiling strength of the SLWAC, f_{cs}. Values for f_L and f_{cs} may be easily determined as discussed in Section 4. At most, only two experimental mixtures are needed to determine these parameters: a concrete mix with the LWA that has to be characterized, and a mix with only the mortar fraction used to produce that concrete. An additional mixture may be needed if the ceiling strength of the SLWAC made with a given type of LWA is too high.

A knowledge of f_{LWA} is necessary in order to estimate the compressive strength of the SLWAC above f_L, according to Equation (3). The strength of the LWA must be measured in concrete, since it is increased by the confinement effect provided by the surrounding mortar (Zhang and Gjørv, 1991b; Bogas and Gomes, 2013a). Thus, one experimental mixture is necessary to estimate f_{LWA} according to Equation (6),

$$f_{LWA} = 10^{\left(\dfrac{\log(f_{cm}) - v_m \cdot \log(f_m)}{1 - v_m} \right)} \tag{6}$$

knowing the relative volume, v_m, and the compressive strength, f_m, of the mortar. If f_m is not known, one more mixture with only mortar has to be tested. However, the same mixtures used to determine f_L may be also adopted to estimate f_{LWA}.

In addition, f_{LWA} may be conveniently estimated from empirical expressions suggested in Bogas *et al.* (2017), which are valid for different types of coarse LWA (Figure 2.8). This allows a first reasonable approximation of f_{LWA} to be made, avoiding additional experimental tests (Bogas *et al.*, 2017).

Only two experimental mixtures are therefore needed to estimate all the necessary parameters (f_{LWA}, f_L, f_{cs}). Moreover, these trial mixtures are only needed if a new type of LWA, whose properties are not known, is used in the SLWAC production. Nevertheless, LWA suppliers can easily provide the necessary values of f_{LWA}, f_L and f_{cs} in

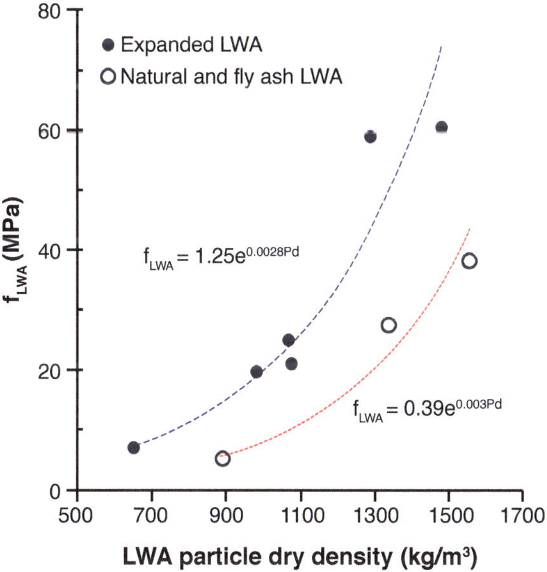

Figure 2.8 *Estimation of f_{LWA} from the LWA dry density for different types of expanded clay or slate LWA (Bogas and Gomes, 2013a; Bogas et al., 2017), fly ash LWA (Bogas et al., 2017), pumice (Videla and López, 2002) and slag LWA (Bogas and Gomes, 2015a) without outer shell. Adapted from Bogas et al., 2017.*

their technical sheet. Therefore, the concrete designer can define the starting SLWAC composition with reasonable precision and without the need of previous experimental mixtures. Some values of f_L and f_{LWA}, are indicated in Table 2.3 for different types of aggregate. The suggested method was intended for SLWAC produced with natural sand and a maximum aggregate size of up to about 12 mm, although it may be easily extended to other grades of LWA with minor adjustments.

9.2 Selection of lightweight aggregate

The type of aggregate may be selected based on the application, which is governed by its ceiling strength, f_{cs} (Section 4) and the expected range of compressive strength and density of the SLWAC. Recommended ranges for common types of expanded LWA and fly ash LWA are based on the works of Bogas and Gomes (2013a) and Bogas *et al.* (2017) (Table 2.3). The recommended limits were defined as a function of the type and density of LWA, taking into account the production and testing of various concrete compositions with 250–400 l/m³ of LWA of distinct porosities and w/c ratios of 0.3–0.55. Outside the indicated ranges, SLWAC becomes uneconomical and less efficient, with a low strength-to-density ratio. Table 2.3 clearly shows that the most appropriate LWA should be chosen depending on the desired/targeted characteristics of SLWAC.

As noted, the volume of coarse LWA must be limited due to the risk of concrete segregation and an excessive reduction of its strength and durability. On the other hand, larger volumes of LWA are necessary for effectively reducing the SLWAC density. Therefore, the best compromise may be attained for LWA volumes within the range 300-400 l/m³. The exact value may be adjusted based on the required workability and density. The specified dry density of the SLWAC can be checked approximately with the aid of abacuses (see Figure 2.9). In this case, for a given target dry density range, the required volume of LWA with a specific particle density (ρ_p) can be estimated as a function of the effective mixing water. The area delimited by the dashed lines in Figure 2.9 represents most common SLWAC.

Table 2.3 *Recommended strength and density for different types of LWA.*

Type of LWA	LWA density, ρ_p	f_L	f_{cs}	f_{LWA}	Recommended limits	
	(kg/m³)	(MPa)	(MPa)	(MPa)	Strength class	Density class
	ρ_p>1300 (Stalite, Arlita)	59–62	70–75	60	LC40/44-LC55/60 LC25/28-LC40/44	D2.0 D1.8
expanded LWA	1000<ρ_p<1300 (Leca)	27	45–50	25	LC25/28-LC35/38 LC20/22-LC25/28	D1.8 D1.6
	ρ_p<1000 (Argex)	<10	30–35	7	LC20/22-LC25/28 LC12/13-LC20/22	D1.8 D1.6
Fly ash LWA	ρ_p>1300 (Lytag)	29	50–55	27	LC25/28-LC40/44 LC20/22-LC25/28	D1.8 D1.6

f_L limit strength
f_{cs} ceiling strength
f_{LWA} strength of the aggregate in concrete

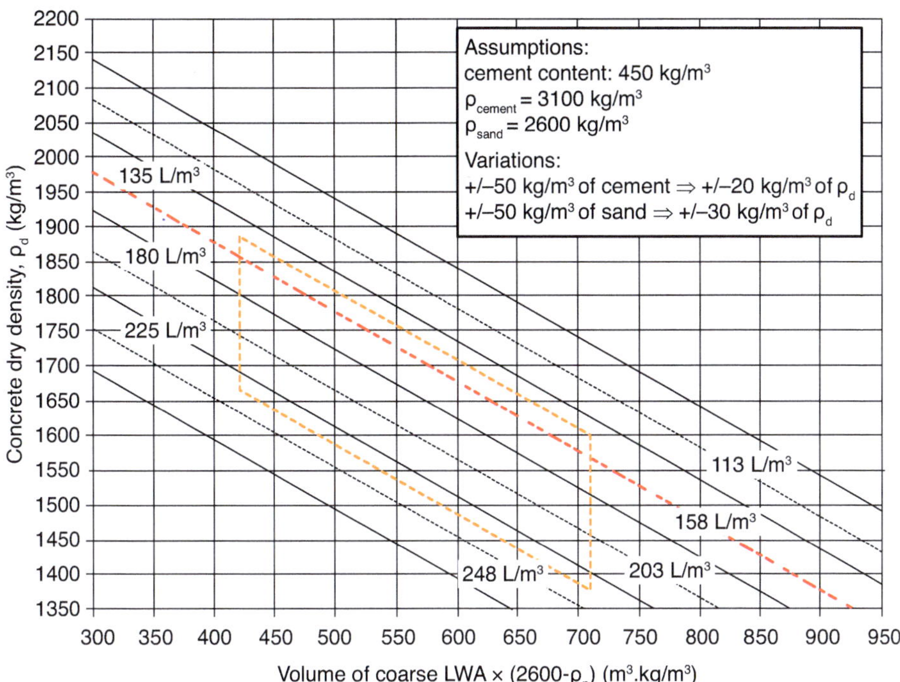

Figure 2.9 *Dry density, ρ_d, as a function of the LWA volume and particle density, ρ_p, for different amounts of effective water (adapted from Bogas and Gomes, 2013b). The narrower dashed region indicates the most common SLWAC.*

As noted, the volume of LWA and the relative proportion of sand may be tuned from reference grading curves or other packing methods, aiming for maximum aggregate compactness. However, the optimal aggregate grading may be incompatible with the optimal strength and applicability of SLWAC. Moreover, these methods tend to underestimate the volume of fines in fluid concrete, leading to less cohesive mixtures and increasing the risk of segregation. Therefore, a good compromise is to consider the packing density optimization within the limit range of the recommended LWA volume.

9.3 Mortar strength and water–cement ratio

The mortar compressive strength, f_m, for a given required concrete compressive strength, f_{cm}, can be estimated from Equation (7), when $f_{cm} > f_L$. For $f_{cm} < f_L$, the mortar strength for each type of LWA can be estimated using a regression line for f_L (as shown in Figure 2.3).

$$f_m = 10^{\left(\dfrac{\log(f_{cm}) - \upsilon_{LWA}\,\log(f_{LWA})}{1 - \upsilon_{LWA}}\right)} \tag{7}$$

Based on two comprehensive experimental works, covering different compositions and types of LWA, Bogas and Gomes (2013a) and Bogas *et al.* (2017) found

95

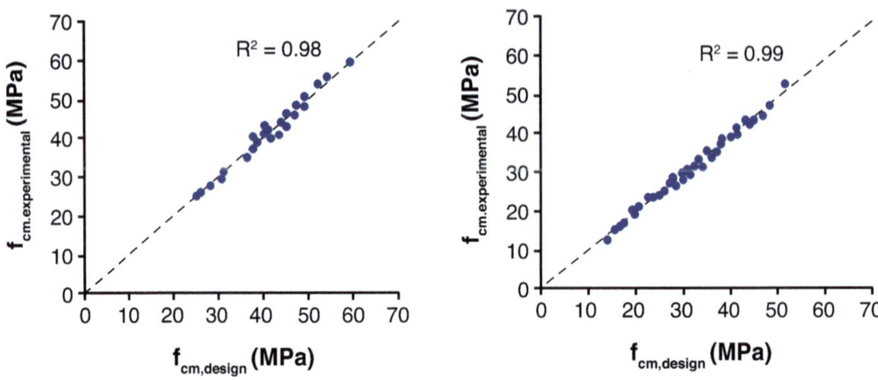

Figure 2.10 *Experimental versus design compressive strength for different types of expanded and fly ash LWA (a) results from Bogas and Gomes (2013a); (b) results from Bogas et al. (2017).*

that Equation (3) can adequately predict the compressive strength of SLWAC. Within the validity domain of the model ($f_{cm} > f_L$), high correlations were obtained between design and experimental values (Figure 2.10), confirming the suitability of the biphasic model for estimating the compressive strength of SLWAC. These studies support the use of this model in the new expeditious SLWAC mix design proposed by Bogas and Gomes (2013b).

Knowing the mortar strength, the w/b ratio can be estimated from common expressions relating these two parameters, such as the Abrams' law (Equation (8)),

$$\frac{w}{b} = \frac{ln\left(\dfrac{K_1}{f_{cm}}\right)}{ln(K_2)} \tag{8}$$

where K_1 and K_2 are constants depending on the type of binder (Videla and López, 2002; Bogas and Gomes, 2013b). New relations must be built for each type of binder.

9.4 Mixing water and binder content

The volume of paste must be sufficiently high to provide the required fluidity and applicability, without segregation. Bearing this in mind, for LWA up to a maximum size of 12 mm, the volume of paste, V_p, should be around 0.29–0.34 m³/m³ (Chandra and Berntsson, 2003; Bogas, 2011). Higher volumes of paste are needed for higher slumps. An initial V_p of 0.3 m³/m³ may be considered. Then, knowing the required w/c ratio and arbitrating a volume of paste in this range, the amount of effective water, w_{eff}, can be easily estimated from Equation (9),

$$w_{eff} = \frac{V_p \cdot \rho_b \cdot \left(\dfrac{w}{b}\right)}{1000 + \rho_b \cdot \left(\dfrac{w}{b}\right)} \cdot 1000 \, (kg/m^3) \tag{9}$$

where ρ_b is the density of binder. Finally, the amount of binder can be directly calculated from w_{eff} and w/b.

The mixing water is given by the sum of w_{eff} and the water due to LWA absorption, $w_{abs,LWA}$ (section 6). This $w_{abs,LWA}$ may be estimated from Equation (2), as suggested by Bogas *et al.* (2012b). In a simpler approach, the $w_{abs,LWA}$ can be roughly estimated as equivalent to 1 hour of absorption in pure water (EN 206, 2016). Alternatively, the LWA may be pre-saturated, enabling better control over the mixing water. This is in line with the North American approach (ACI 211.2, 2004). However, considering cost and durability, the use of non-saturated LWA is suggested (Nadesan and Dinakar, 2017). Moreover, the use of pre-wetted LWA during mixing was shown to be sufficient in providing efficient internal curing effect (Bogas *et al.*, 2014). The suitability of any of these approaches depends on the available means on the construction site and on the required level of accuracy.

9.5 Volume of sand

The volume of sand, V_s, is directly determined as the difference between the unit volume of concrete and the sum of the volumes of LWA, binder, water, admixtures and air. For the maximum size of LWA of 9.5–12.5 mm, it is reasonable to assume a volume of air of 30 l/m³ (Bogas and Gomes, 2013a, 2013b). The final composition may then be adjusted slightly after producing the first trial mixture, i.e. with regard fine-tuning the workability, density and compressive strength. It was found that design values led to differences in strength and density of less than 4% and 1%, respectively (Bogas and Gomes, 2013b).

9.6 Mix design example

As an example, the suggested method was applied to the mix design of one SLWAC with the following specification: EN 206: LC30/33·XC3·Cl 0.2·D$_{max}$12·D1.8·S3. It was assumed that SLWAC was produced with cement type I 42.5 (EN 197-1, 2011), and the LWA was pre-saturated before mixing.

From Table 2.3 it can be seen that only the more porous aggregate (ρ_p<1000 kg/m³) is inadequate for SLWAC production, since the required compressive strength class is outside the recommended domain for this LWA. The other types of aggregates may be used, knowing that denser LWA leads to leaner and cheaper mortars and that more porous LWA may better meet the low-density requirement. In this case, however, the option taken was to select the aggregate with intermediate density (1000 kg/m³<ρ_p< 1300 kg/m³, Leca).

The volume of paste and aggregate may then be assumed as 300 l/m³ and 350 l/m³, respectively. Assuming that the Leca has 1070 k/m³ of density, from Figure 2.9 the dry density ranges between 1600 kg/m³ and 1800 kg/m³ (within the required D1.8 class), for expected amounts of mixing water whiting 140–220 l/m³. Assuming a minimum average compressive strength of 37 MPa for LC30/33 (in 150 mm cubic specimens), the target strength exceeds the f_L of LWA (27 MPa in Table 2.3). Therefore, from Equation (7), the estimated mortar compressive strength, f_m, is 46 MPa, where f_{cm}, v_{LWA} and f_{LWA} are 37 MPa, 0.35 and 25 MPa, respectively.

For cement type I 42.5, K_1 and K_2 in Equation (8) may be considered as 182.7 and 14.4, respectively (Bogas and Gomes, 2013b). Therefore, the calculated w/c ratio is 0.52. Now, based on Equation (9) and taking 3100 kg/m³ for the cement density, the effective water is 185 l/m³, which is within the above-mentioned expected range. This corresponds to 356 kg/m³ of cement. Finally, knowing that the volume of cement, LWA and water is 115 l/m³, 350 l/m³ and 185 l/m³, respectively, and assuming 30 l/m³ of air, the volume of sand becomes 320 l/m³.

10. Conclusions

In this chapter, a brief overview of the composition and mix design of common SLWAC has been presented. The more complex nature of the SLWAC mix design, compared to that of NWC, has been emphasized, with more parameters, such as the density and strength level of the concrete, the absorption, volume and strength of the LWA, needing to be closely considered. Particular attention has been given to the distinct compressive strength and absorption characteristics of SLWAC, and current approaches for taking this into account in concrete design have been discussed. The lower volume and lower particle size of aggregates are other aspects that distinguish the mix design of SLWAC and NWC.

A simplified mix design method has been proposed with the aim of defining the composition of a first trial mixture more expeditiously. This new mix design is valid for SLWAC with natural sand and can be easily implemented in practice regardless of the type of LWA. In addition to the common mix design aspects, such as the density of materials, absorption of aggregates, and the cement type, three other parameters were considered: namely, the strength of LWA in the concrete, f_{LWA}, the limit strength, f_L, and the potential strength of the SLWAC. Only two experimental mixtures are needed to determine these parameters and only for new types of LWA. One example of an application was presented to better show the ease of implementation of the suggested method.

Future advances should involve the extension of this method to all lightweight concrete and the introduction of more powerful packing optimization techniques, based on computational tools. New suggested mix design methodologies should be validated for any type of LWA, such as the one proposed in this chapter. Such actions will contribute to a more efficient use and better acceptance of SLWAC.

Acknowledgements

The authors wish to thank the Portuguese Foundation for Science and Technology (FCT) for funding this research through project PTDC/ECM-COM1734/2012 and CERIS for supporting this research through project UIDB/04625/2020.

References

ACI (2014) *ACI 213R-14: Guide for Structural Lightweight-Aggregate Concrete*. Farmington Hills, MI: American Concrete Institute.

ACI (2004) *ACI 211.2: Standard Practice for Selecting Proportions for Structural Lightweight Concrete*. Farmington Hills, MI: American Concrete Institute.

ACI (2002) *ACI 304R: Guide for Measuring, Mixing, Transporting, and Placing Concrete.* Farmington Hills, MI: American Concrete Institute.

ASTM (2016) ASTM C173: *Standard Test Method for Air Content of Freshly Mixed Concrete by the Volumetric Method.* West Conshohocken, PA: American Society for Testing & Materials.

Abdullahi, M., Hashem, M., Al-mattarneh, A., Mohammed, B.S. and Sadiku, S. (2011) M-file for mix design of structural lightweight concrete using developed models. *Journal of Engineering Science and Technology* **6**(4): 520–531.

Adámek, J. (2000) Strength of lightweight concrete influenced by strength of lightweight aggregate. In S. Helland, I. Holand and S. Smeplass (eds) *Second International Symposium on Structural Lightweight Aggregate Concrete, 18–22 June, 2000, Kristiansand, Norway.* Oslo: Norwegian Concrete Association, pp. 425–430.

Ahmad, H., Hilton, M. and Noor, N.M. (2007) Physical properties of local palm oil clinker and fly ash. 1st Engineering Conference on Energy & Environment (ENCON2007), December 27-28, 2007, Kuching, Sarawak, Malaysia.

Aïtcin, P.C. (1998) *High Performance Concrete.* Modern Concrete Technology series, vol. 5. London: E & FN Spon.

Alduaij, J., Alshaleh, K., Haque, M.N. and Ellaithy, K. (1999) Lightweight concrete in hot coastal areas. *Cement and Concrete Composites* **21**: 453–458.

Amiri, B., Krause, G.L. and Tadros, M.K. (1994) Lightweight high-performance concrete masonry-block mix design. *ACI Materials Journal* **91(5)**, 495–501 (Technical Paper 91-M50).

Andreasen, A. (1930) Uber die Beziehung zwischen Kornabstufung und Zwischenraum in Produkten aus losen Kornern (mit einigen Experimenten). *Kolloid-Zeitschrift* **50**, 217–228.

Bentz, D., Halleck, P., Grader, A. and Roberts, J. (2006) Water movement during internal curing: Direct observation using X-ray microtomography. *Concrete International* **28**(10), 39–45.

Beris, A.N., Tsamopoulos, J.A., Armstrong, R.C. and Brown, R.A. (1985) Creeping motion of a sphere through a Bingham plastic. *Journal of Fluid Mechanics* **158**, 219–244.

Bogas, J.A. (2011) Characterization of structural lightweight expanded clay aggregate concrete. PhD thesis, Instituto Superior Técnico, Universidade de Lisboa, Lisbon, Portugal.

Bogas, J.A. and Gomes, A. (2013a) Compressive behavior and failure modes of structural lightweight aggregate concrete – Characterization and strength prediction. *Materials & Design* **46**(4), 832–841.

Bogas, J.A. and Gomes, A. (2013b) A simple mix design method for structural lightweight aggregate. *Materials and Structures* **46**(11),1919–1932.

Bogas. J and Gomes, T. (2015) Mechanical and durability behaviour of structural lightweight concrete produced with volcanic scoria. *Arabian Journal for Science and Engineering* **40**(3), 705–717.

Bogas, J.A., Maurício,. A. and Pereira, M.F.C. (2012a) Microstructural analysis of Iberian expanded clay aggregates. *Microscopy & Microanalysis* **18**(5), 1190–1208.

Bogas, J.A., Gomes, A. and Gloria, M.G. (2012b) Estimation of water absorbed by expanding clay aggregates during structural lightweight concrete production. *Materials and Structures* **45**(10), 1565–1576.

Bogas, J.A., Gomes, A., Pereira, M.F.C. (2012c) Self-compacting lightweight concrete produced with expanded clay aggregate. *Construction and Building Materials* **35**, 1013–1022.

Bogas, J.A., Nogueira, R. and Almeida, G. (2014) Influence of mineral additions and different compositional parameters on the shrinkage of structural expanded clay lightweight concrete. *Materials & Design* **56**, 1039–1048.

Bogas, J.A., Ferrer, B., Pontes, J. and Real, S. (2017) Biphasic compressive behavior of structural lightweight concrete. *ACI Materials Journal* **113**(1–6), 151–163.

British Standards Institute (1990) *BS 812 P111:1990 Testing Aggregates. Part III: Methods for Determination of ten per cent fines value (TFV)*. London: British Standards Institute.

Cánovas, M.F. and Gutiérrez, P.A. (1992) Composición y dosificación de los hormigones de alta resistencia. El hormigón de altas resistencias y sus aplicaciones, *Cemento-Hormigon* **709**, 971–990.

Chandra, S. and Berntsson, L. (2003) *Lightweight Aggregate Concrete: Science, Technology and Applications*. Norwich, NY: Noyes publications/Wiliam Andrew Publishing.

Chen, H.J., Yen, T. and Lia, T.P. (1995) A new proportion method of light-weight aggregate concrete based on dividing strength. In I. Holand, T. Arne and F. Flune (eds) *International Symposium on Structural Lightweight Aggregate Concrete, 20–24 June, 1995, Sandefjord, Norway*. Oslo: Norwegian Concrete Association, pp. 463–471.

Chen, H.J., Yen, T., Lia, T.P. and Huang. Y.L. (1999) Determination of the dividing strength and its relation to the concrete strength in lightweight aggregate concrete. *Cement and Concrete Composites* **21**(1), 29–37.

Clarke, J.L. (1993) Design requirements. In J.L.Clarke (ed.) *Structural Lightweight Aggregate Concrete*, London: Chapman & Hall, pp. 45–74.

DIN 1045. (1988) *Beton und Stahlbeton: Bemessung und Ausführung* [*(Structural Use of Concrete: Design and Construction*). Berlin: Beton Verlag for the Deutsches Institut für Normung (DIN), the German Institute for Standardisation.

DIN EN 1008. (2002) Zugabewasser für Beton-Festlegung für die Probenahme, Prüfung und Beurteilung der Eignung von Wasser, Einschließlich bei der Betonherstellung Anfallendem Wasser, als Zugabewasser für Beton [Mixing water for concrete-specification for sampling, testing and assessing the suitability of water, including water recovered from processes in the concrete industry, as mixing water for concrete]. Berlin: Beton Verlag for the Deutsches Institut für Normung (DIN), the German Institute for Standardisation.

Dolby, P.G. (1995) Production and properties of Lytag aggregate. In I. Holand, T. Arne and F. Flune (eds) *International Symposium on Structural Lightweight Aggregate Concrete, 20–24 June, 1995, Sandefjord, Norway*. Oslo: Norwegian Concrete Association, pp. 326–336.

Dreux, G. (1986) Composition des bétons légers. In M. Arnould and M.Virlogeux (eds) *Granulats et Betons Legers-Bilan de Dix Ans de Recherches*. Paris: Presses de l'école nationale des ponts et chaussées, pp. 425–437.

EN 197-1. (2011) *Cement Part 1: Composition, Specifications and Conformity Criteria for Common Cements*. Brussels: European Committee for Standardization (Comité Européen de Normalisation, CEN).

EN 206:2013+A1. (2016) *Concrete. Specification, Performance, Production and Conformity*. Brussels: European Committee for Standardization (Comité Européen de Normalisation, CEN).

EN 13055. (2016) *Lightweight Aggregates*. Brussels: European Committee for Standardization (Comité Européen de Normalisation, CEN). English version.

EuroLightConR2. (1998) *LWAC Material Properties, State-of-the-Art*. European Union – Brite EuRam III, Document BE96-3942/R2, December.

EuroLightConR5. (2000) *A Rational Mix Design Method for LWAC Using Typical UK Materials*. European Union – Brite EuRam III, Document BE96-3942/R5.

EuroLightConR9. (2000) *Technical and Economic Mixture Optimisation of High Strength LWAC*. European Union – Brite EuRam III, Document BE96-3942/R9.

EuroLightConR14. (2000) *Structural LWAC. Specification and Guideline for Materials and Production*. European Union – Brite EuRam III, Document BE96-3942/R14.

EuroLightConR20. (2000) *The Effect of the Moisture History on the Water Absorption of Lightweight Aggregates*. Europan Union – Brite EuRam III, Document BE96-3942/R20.

Faust, T. (2000) Properties of diferent matrixes and LWAs and their influences on the behaviour of structural LWAC. In S. Helland, I. Holand and S. Smeplass (eds) *Second International Symposium on Structural Lightweight Aggregate Concrete, 18-22 June, 2000, Kristiansand, Norway.* Oslo: Norwegian Concrete Association, pp. 502–511.

fib8. (2000) *Lightweight Aggregate Concrete* [Part 1- Recommended extensions to Model Code 90; Part 2 – Identification of research needs; Part 3 – Application of lightweight aggregate concrete.] Lausanne, Switzerland: International Federation for Structural Concrete (Fédération Internationale du Béton, FIB), CEB/FIP Working Group on Lightweight Aggregate Concrete, Task Group 8.1, Bulletin 08.

FIP (Fédération internationale de la précontrainte). (1983) *FIP Manual of Lightweight Aggregate Concrete. Second Edition.* Guilford, Surrey: Surrey University Press.

Funk, J. and Dinger, D. (1994) *Predictive Process Control of Crowded Particulate Suspension, Applied to Ceramic Manufacturing.* Dordrecht, The Netherlands: Kluwer Academic Press.

Gerritse, A. (1981) Design considerations for reinforced lightweight concrete. *International Journal of Cement Composites and Lightweight Concrete* 3(1), 57–69.

Hammer, T.A. amd Smeplass, S. (1995) The influence of lightweight aggregates properties on material properties of the concrete. In I. Holand, T. Arne and F. Flune (eds) *International Symposium on Structural Lightweight Aggregate Concrete, 20–24 June, 1995, Sandefjord, Norway.* Oslo: Norwegian Concrete Association, pp. 517–532.

Haque, M.N., Al-Khaiat, H. and Kayali, O. (2004) Strength and durability of lightweight concrete. *Cement and Concrete Composites* 26(4), 307–314.

Harmon, K.S. (2003) Recent research on the mechanical properties of high performance lightweight concrete. [Theodore Bremner Symposium on High-performance Lightweight Concrete, June, Thessaloniki, Greece.] In J. Ries and T. Holm (eds) *Sixth CANMET/ACI International Conference on Durability of Concrete.* Farmington, MI: American Concrete Institute, pp. 131–150.

Heimdal, E. and Rønneberg, H. (1995) Production of high strength lightweight concrete. The views of a ready mix concrete producer. In I. Holand, T. Arne and F. Flune (eds) *International Symposium on Structural Lightweight Aggregate Concrete, 20–24 June, 1995, Sandefjord, Norway.* Oslo: Norwegian Concrete Association, pp. 380–389.

Helgesen, H.K. (1995) Lightweight aggregate concrete in Norway. In I. Holand, T. Arne and F. Flune (eds) *International Symposium on Structural Lightweight Aggregate Concrete, 20–24 June, 1995, Sandefjord, Norway.* Oslo: Norwegian Concrete Association, pp. 70–82.

Hoff, G.C. (1985) The challenge of offshore concrete structures. *Concrete International* 7(8), 13–22.

Hoff, G.C. (1992) High strength lightweight aggregate concrete for artic applications. In D. Holm and A.M. Vaysburd (eds) *Structural Lightweight Aggregate Concrete Performance (SP-136).* Farmington, MI: American Concrete Institute, Part I–III, pp. 1–245.

Holm, T. and Bremner, T. (2000) *State-of-the-Art Report on High-strength, High-durability Structural Low-density Concrete for Applications in Severe Marine Environments.* Washington, DC and Vicksburg, MS: US Army Corps of Engineers Structures Laboratory and the Engineer and Research Development Center, ERDC/SL TR-00-3.

Holm, T.A., Ooi, O.S. and Bremner, T.W. (2003) Moisture dynamics in lightweight aggregate and concrete. [Theodore Bremner Symposium on High-performance Lightweight Concrete, June, Thessaloniki, Greece.] In J. Ries and T. Holm (eds) *Sixth CANMET/ACI International Conference on Durability of Concrete.* Farmington, MI: American Concrete Institute, pp. 167–184.

Hussein, M.A., Mahgoub, M.A. and Mousa, A. (2021) Is lightweight concrete a viable product? *Concrete International* 43(6), 41–44.

ICAR-105-1 (2003) *Summary of Concrete Workability Test Methods.* Austin, TX: International Center for Aggregates Research, University of Texas at Austin, Research report ICAR 105-1.

Jafari, S. and Mahini, S.S. (2017) Lightweight concrete design using gene expression programing. *Construction and Building Materials* **139**, 93–100.

Johnsen, H., Helland, S. and Heimdal, E. (1995) Construction of the Støvset free cantilever bridge and the Nordhordland cable stayed bridge. In I. Holand, T. Arne and F. Flune (eds) *International Symposium on Structural Lightweight Aggregate Concrete, 20–24 June, 1995, Sandefjord, Norway*. Oslo: Norwegian Concrete Association, pp. 517–532.

Ke, Y., Beaucour, A.L., Ortola, S., Dumontet, H. and Cabrillac, R. (2009) Influence of volume fraction and characteristics of lightweight aggregates on the mechanical properties of concrete. *Construction & Building Materials* **23**(8), 2821–2828.

Kockal, N.U. and Ozturan, T. (2011) Strength and elastic properties of structural lightweight concretes. *Materials & Design* **32**(4), 2396–2403.

Larrard de, F. (1999) *Strcutures Granulaires et Formulations des Betons*. Paris: Laboratoire Central des Ponts et Chaussees.

Lazarus, D. (1993) Lightweight concrete in buildings. In J.L.Clarke (ed.) *Structural Lightweight Aggregate Concrete*. London: Chapman & Hall, pp. 106–149l.

Lijiu, W., Shuzhong, Z. and Guofan, Z. (2005) Investigation of the mix ratio design of lightweight aggregate concrete. *Cement and Concrete Research*, **35**(5), 931–935.

Lo, Y., Gao, X.F. and Jeary, A.P. (1999) Microstructure of pre-wetted aggregate on lightweight concrete. *Building and Environment* **34**(6), 759–764.

Maage, M., Smeplass, S. and Thienel, K. (2000) Structural LWAC specification and guideline for materials and production. In S. Helland, I. Holand and S. Smeplass (eds) *Second International Symposium on Structural Lightweight Aggregate Concrete, 18–22 June, 2000, Kristiansand, Norway*. Oslo: Norwegian Concrete Association, pp. 802–810.

Moravia, W.G., Oliveira, C.A.S., Gumieri, A.G. and Vasconcelos, W.L. (2006) Caracterização microestrutural da argila expandida para aplicação como agregado em concreto estrutural leve [Microstructural characterization of expanded clay for application as an aggregate in lightweight structural concrete]. *Cerâmica* **52** (322), 193–199.

Mousa, A., Mahgoub, M. and Hussein, M. (2018) Lightweight concrete in America: Presence and challenges. *Sustainable Production and Consumption* **15**, 131–144.

Muller-Rochholz, J. (1979) Investigation of the absorption of water by lightweight aggregate from cement paste. *International Journal of Cement Composites and Lightweight Concrete* **1**(1), 39–41.

Nadesan, M.S. and Dinakar, P. (2017) Mix design and properties of fly ash waste lightweight aggregates in structural lightweight concrete. *Case Studies in Construction Materials* **7**, 336–347.

Neville, A.M. (1995) *Properties of Concrete. Fourth Edition*. Harlow, Essex: Longman.

Pankhurst, R.N.W. (1993) Construction. In J.L.Clarke (ed.) *Structural Lightweight Aggregate Concrete*. London: Chapman & Hall, pp. 75–105.

Punkki, J. and Gjørv, O.E. (1995) Effect of aggregate absorption on properties of high-strength lightweight concrete. In I. Holand, T. Arne and F. Flune (eds) *International Symposium on Structural Lightweight Aggregate Concrete, 20–24 June, 1995, Sandefjord, Norway*. Oslo: Norwegian Concrete Association, pp. 604–616.

Sandvik, M. and Hammer, T. (1995) The development and use of high performance light weight aggregate concrete in Norway. In I. Holand, T. Arne and F. Flune (eds) *International Symposium on Structural Lightweight Aggregate Concrete, 20–24 June, 1995, Sandefjord, Norway*. Oslo: Norwegian Concrete Association, pp. 617–627.

Smeplass, S. (2000) Drying of LWAC. In S. Helland, I. Holand and S. Smeplass (eds) *Second International Symposium on Structural Lightweight Aggregate Concrete, 18–22 June, 2000, Kristiansand, Norway*. Oslo: Norwegian Concrete Association, pp. 833–843.

Swamy, R.N. and Lambert, G.H. (1981) The microstructure of Lytag aggregate. *International Journal of Cement Composites and Lightweight Concrete* **3**(4), 273–282.

Thienel, K.-C., Haller, T. and Beuntner, N. (2020) Lightweight concrete – From basics to innovations. *Materials* **13**(5), 1120. doi:10.3390/ma13051120.

Ting, T.Z.H., Rahman, M.E., Lau, H.H. and Ting, M.Z.Y. (2019) Recent development and perspective of lightweight aggregates based self-compacting concrete. *Construction and Building Materials* **201**(2), 763–777.

Videla, C. and López, M. (2000) Mixture proportioning – Methodology for structural sand lightweight concrete. *ACI Materials Journal* **97**(3), 281–289.

Videla, C. and López, M. (2002) Effect of lightweight aggregate intrinsic strength on lightweight concrete compressive strength and modulus of elasticity. *Materiales De Construcción* **52**(265), 23–37.

Virlogeux, M. (1986) Généralités sur les caractères des bétons légers. In M. Arnould and M. Virlogeux (eds) *Granulats et Betons Legers-Bilan de Dix Ans de Recherches.* Paris: Presses de l'école nationale des ponts et chaussées, pp. 111–246.

Wasserman, R. and Bentur, A. (1997) Effect of lightweight fly ash aggregate microstructure on the strength of concretes. *Cement and Concrete Research* **27**(4), 525–537.

Wilson, H.S. and Malhotra, V.M. (1988) Development of high strength lightweight concrete for structural applications. *International Journal of Cement Composites and Lightweight Concrete* **10**(2), 79–90.

Yang, K.H., Kim, G.H. and Choi, Y.H. (2014) An initial trial mixture proportioning procedure for structural lightweight aggregate concrete. *Construction and Building Materials* **55**, 431–439.

Zhang, M.H. and Gjørv, O.E. (1989) *Characteristics of lightweight aggregates for high-strength LWA concrete.* Materialutvikling Hoyfast Betong. Report N.2.2. STF70 A92022.

Zhang, M.H. and Gjørv, O.E. (1991a) Characteristics of lightweight aggregates for high-strength concrete. *ACI Materials Journal* **88**(2), 150–158 (Technical Paper 88-M19).

Zhang, M.H. and Gjørv, O.E. (1991b) Mechanical properties of high-strength lightweight concrete. *ACI Materials Journal* **88**(3), 240–247 (Technical Paper 88-M29).

Chapter 3
Flexural fatigue of high-strength lightweight aggregate concrete

Kazi M. A. Sohel

1. Introduction

Many concrete structures such as bridge decks, airport runways, concrete pavements, and offshore structures experience a million cycles of repetitive loadings in their services live. This exposure to repetitive loading reduces the stiffness of the concrete structures, which leads to fracture generation due to changes caused by the progressive growth of micro-cracks. When the repeated loads are applied at a high frequency rate over a long period, this can eventually lead to fatigue failure (Ballinger, 1971; Wang and Song, 2011). Thus, increasing attention is being paid to the consideration of the fatigue characteristics of the constituent material in the design of these concrete structures.

The use of lightweight cement composites for civil, offshore and marine structures has long been recognized as a durable method of construction. For almost 2,000 years, lightweight concrete has been utilized in construction, with extensive use especially in the last 100 years. A number of the structures that have survived from the Roman Empire have elements that were made of some form of lightweight concrete (Ries and Holm, 2004). These structures utilized locally available natural volcanic materials as the lightweight constituent (ACI 213R-14, 2014). High strength lightweight concrete (HSLWC) has been used frequently in Norway and other parts of Europe. HSLWC, with its high durability and lightweight characteristics, is a much sought after material in the construction of concrete floating platforms. The first major offshore structure to be built mostly with lightweight concrete was the Glomar Beaufort Sea I (commonly called Concrete Island Drilling System (CIDS)). The CIDS was contracted in 1983–1984 using lightweight aggregate concrete having a density of 1840 kg/m^3 and a 56-days design cylinder strength of 45 MPa. This structure was developed for use in the severe climate of the Arctic. High-strength lightweight concrete was also used to construct the Heidrun Tension Leg platform (TLP) in Norway in 1995. The floating structure was made entirely of HSLWC, including the deck support frame. In 1994 the Troll platform in Norway used HSLWC in portions of its extremely long shafts (Hoff, 1994). The Forth Platform base, built in Scotland in 1994, was constructed completely of lightweight concrete (Hoff, 1994). The Hibernia Platform, constructed in Canada in 1997, used HSLWC (80 MPa) in some of the upper portion of the structure. An extensive study (Hoff, 1993) of high-strength lightweight concrete (HSLWC) for Arctic applications found that this type of concrete was suitable for severe offshore environments.

From the 20th century, lightweight concrete is produced using manufactured lightweight aggregate (LWA) such as expanded shale, expanded clay or foam slag (ESCI, 1971). In recent years, lightweight aggregates such pumice, perlite, cenospheres, polyurethane foam, diatomite earth, expanded glass, aerogel and high-impact polystyrene have been used to produce low thermal conductive structural lightweight concrete. In the late 20th century, high-strength lightweight aggregate concrete was successfully developed with a strength ranging from 57 MPa to 102 MPa and a density ranging from 1595 kg/m^3 to 1880 kg/m^3 (Zhang and Gjørv, 1991a). According to a state-of-the-art review on the development of high-strength lightweight concrete (Wee, 2005), the compressive strength of lightweight aggregate concrete (LWC) typically decreases with the decrease of density, and it is a considerable challenge to produce LWC with density below 1500 kg/m^3 and a compressive strength above 50 MPa. However, LWC with low density ranging from 1440 kg/m^3 to 1840 kg/m^3 can achieve high-strength levels (35–70 MPa) by incorporating various lightweight pozzolans (silica fume, fly ash, metakaolin, volcanic ashes, calcined clays and shales) combined with mid- to high-range water-reducing admixtures or both (Holm and Bremner, 2000).

Recently developed ultra lightweight cement composite (ULCC) is a type of novel composite, which could be categorized as a low-density ($\leq$1450 kg/m^3) with high compressive strength ($\geq$60 MPa) cementitious composite. Originally, this cementitious material was developed to be used in sandwich composite structures for offshore and marine applications (Marshall *et al.*, 2012). To achieve low density, cenospheres were used as a filler material in the ULCC mix. The cenospheres are lightweight, inert, hollow microspheres recycled from the ash pond in coal-burning thermal power plants. The hollow interior of the cenosphere is covered by a thin shell made of aluminosilicate that is thermally stable. Cenospheres are being used in many applications due to their superior physical and chemical properties (Kolay and Singh, 2001).

The mechanical properties of ULCC were documented by Chia *et al.* (2011) and Wu *et al.* (2015). Structures with ULCC (e.g. reinforced slabs, steel–concrete–steel sandwich panels and double-skinned steel tubes) were tested for both static and impact load (Sohel *et al.*, 2012; Wang *et al.*, 2015; Wu *et al.*, 2015; Yan *et al.*, 2016). The use of ULCC in composite slabs with profiled deck was studied by Sohel *et al.* (2021). The shear bond characteristics of composite slabs with ULCC were found to be superior to those of composite slabs with conventional concrete. However, the long-term effect such as the fatigue performance of this newly developed cementitious composite was known until a recent study conducted by Sohel *et al.* (2018). The resistance of a material to repeated loading is obviously an important factor in the design of concrete structures like bridges, building floors, offshore structures and composite decks, which are subjected to such a type of load. Knowledge of the relationship between the number of cycles to the failure and the applied stress is essential. Most of the previous research studies on flexural fatigue were carried out on normal-weight concrete. Such type of research on lightweight concrete is very limited in the literature.

This chapter discusses the flexural fatigue performance of ULCC at different stress levels. The relationship between the number of cycles to failure and the applied stress level (i.e. *S–N* curve) was established based on an extensive experimental investigation. The material coefficients of the fatigue equation were determined to allow accurate prediction of the endurance limit of ULCC under repetitive flexural loads.

The fatigue performance of ULCC was compared to that of high-strength LWC and NWC having comparable compressive and flexural strengths. The chapter also presents the probabilistic analysis concepts necessary to describe the fatigue characteristics of ULCC and LWC. The 2-parameter Weibull distribution was examined to describe the fatigue behavior of ULCC and LWC. Application of the failure probability in the S–N relationship is also presented at the end of the chapter.

2. Fatigue of concrete and cement composites

Concrete or cementitious material is a heterogeneous material that is inherently full of flaws (such as pores, air voids, lenses of bleed water under aggregates and shrinkage cracks). When subjected to repeated load, extensive cracks may develop from the flaws, which eventually lead to failure after a sufficient number of load repetitions. The fatigue-fracture process in concrete can generally be divided into three phases, as shown in Figure 3.1. In the first phase, flaws (cracks) initiate in weak regions within the concrete mass, which is termed the microscale phase. The second phase – the crack propagation phase – during which the initial flaws or cracks grow slowly but progressively to a critical size, is referred to as the micro-cracking or mesoscale phase. The growth of a single micro-crack in lightweight concrete is shown in Figure 3.2 (a, b). In the final phase, the macroscale phase, a sufficient number of continuous and unstable cracks develop to cause failure (Elfgren, 1989; Lee and Barr, 2004).

Different cyclic loading has different mechanisms of failure. In low-cycle fatigue, the failure mechanism is due to the formation of progressive networks of cracks in the mortar. In high-cycle fatigue, cracks initiate in the bond between the aggregate and the cement paste. The progress of this process is slow and gradual (Hsu, 1984). The stresses associated with each of these modes of failure are shown in Figure 3.2(c).

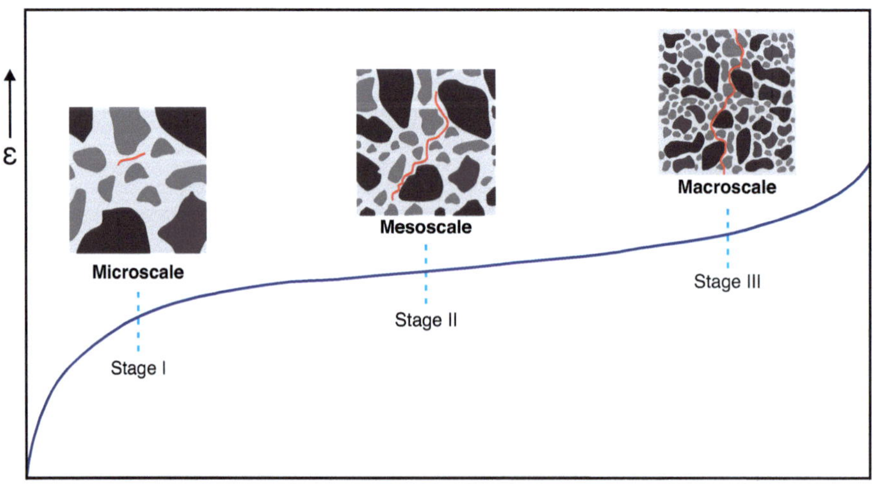

Figure 3.1 *Development of cracks in concrete due to cyclic loading, as a function of the number of cycles and the strain ε.*

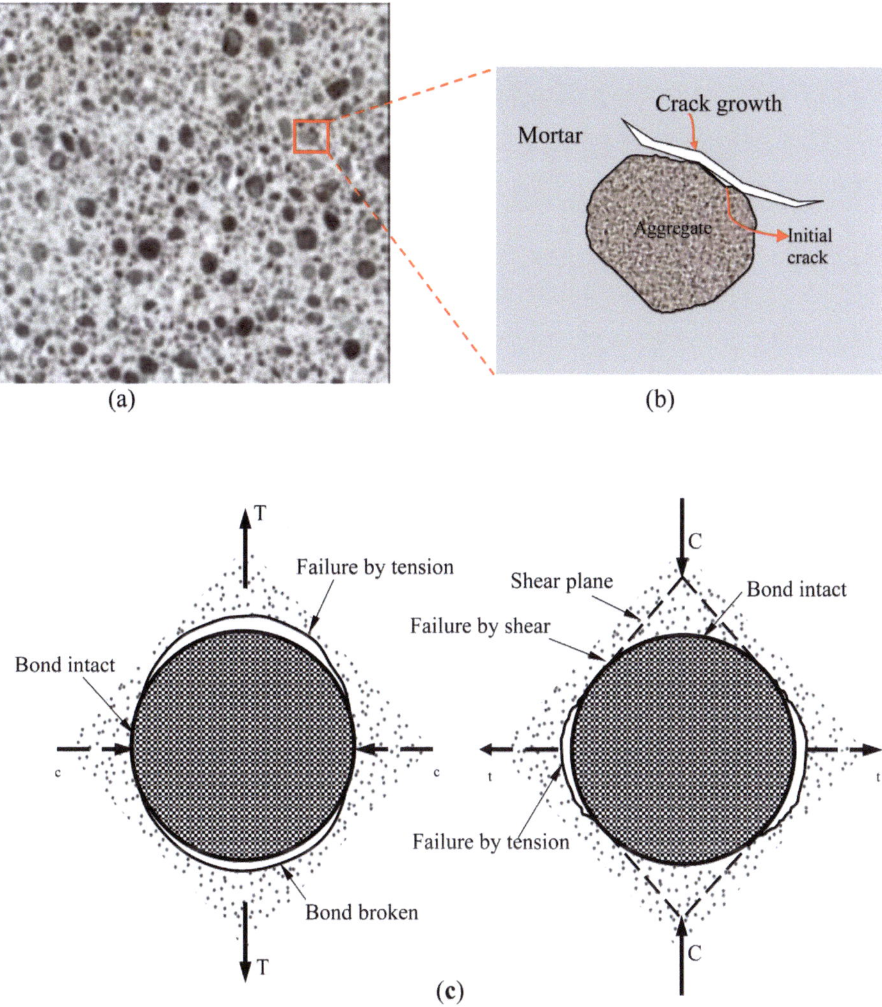

Figure 3.2 *Schematic of single crack growth in concrete: (a) schematic of LWC material; (b) growth of single micro-crack; (c) local stresses around aggregate particle under tensile (T) and compressive (C) loading (Petkovic, 1991).*

The fatigue strength of cementitious material can be defined as the fraction of static strength that can be supported by the material repeatedly for a given number of load cycles. The common factors that influence the fatigue strength of cementitious materials are loading range (or stress ratio), loading rate, matrix composition, mechanical properties, boundary conditions and environmental conditions (Zhang *et al.*, 1996).

As stated previously, the development of fatigue failure in concrete can be divided into three stages. It is feasible to retard and inhibit the growth of the flaws in the second stage by introducing closely spaced and randomly dispersed fibers as reinforcements. In fiber-reinforced cementitious (FRC) material, the action of fiber bridging and fiber

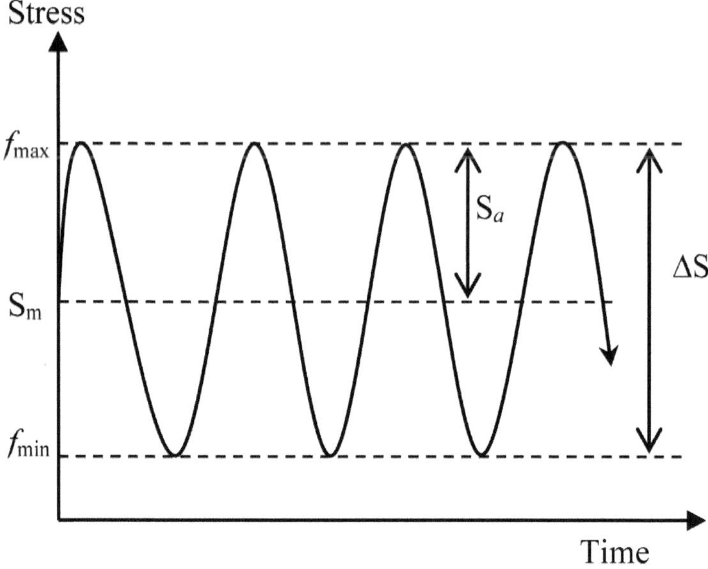

Figure 3.3 *Definition of fatigue loading parameters.*

pullout dissipates energy in the wake of the crack tip. After matrix crack initiation, the stresses are absorbed by bridging fibers. The concrete element does not fail spontaneously when the matrix is cracked. The deformation energy is absorbed and the material becomes pseudo-ductile. This mechanism plays a dominant role in inhibiting crack growth and therefore increases the flexural strength, fatigue strength and fatigue life of FRC specimens.

Cyclic loadings are schematically represented in Figure 3.3. Symbols of reversal stresses and their denotations are summarized in Equations (1)–(6).

$$\text{Stress range, } \Delta S = S_{max} - S_{min} \tag{1}$$

$$\text{Stress amplitude, } S_a = \frac{\Delta S}{2} = \frac{(S_{max} - S_{min})}{2} \tag{2}$$

$$\text{Mean stress, } S_m = \frac{(S_{max} + S_{min})}{2} \tag{3}$$

$$\text{Stress ratio, } R = \frac{S_{min}}{S_{max}} \tag{4}$$

$$\text{Amplitude ratio, } A = \frac{S_a}{S_m} = \frac{1-R}{1+R} \tag{5}$$

$$\text{Stress level, } S = \frac{f_{max}}{f_r} \tag{6}$$

where f_r is the static flexural strength (modulus of rupture) of the corresponding cementitious material determined under static loading conditions.

Table 3.1 *Typical specimen sizes and loading frequency for the fatigue test.*

Investigators	Specimen size (depth×width×length) (mm)	Loading frequency of fatigue test (Hz)
Ganesan *et al.* (2013)	100 × 100 × 500	2
Oh (1991)	100 × 100 × 500	4
Naaman and Hammoud (1998)	100 × 100 × 400	5
Li *et al.* (2007)	100 × 100 × 400	10
Arora and Sing (2016)	100 × 100 × 500	10
Singh and Kaushik (2000)	100 × 100 × 500	12
Ramakrishnan *et al.* (1989)	152 × 152 × 533	20
Ramakrishnan and Panchalan (2003)	76 × 102 × 406	20
Nieto *et al.* (2006)	76 × 76 × 406	20
Singh *et al.* (2005)	100 × 100 × 500	20
Raithby (1979)	102 × 102 × 510	20
Sohel *et al.* (2018)	100 × 76 × 406	5

There is no standard as yet to describe the loading frequency of the fatigue test. Several studies suggested that a loading frequency of between 1 Hz and 15 Hz had a minor effect on the fatigue life when the maximum stress level was less than 75% of the static strength (ACI 215R-74, 1997). In general, four parameters should be taken into account to determine the frequency of the fatigue load. These include the response of the test specimen, the inertia and strain rate effect, the duration of the test, and the strength gain over time. Table 3.1 summarizes the typical loading frequencies of the fatigue tests used in past studies. Based on past studies and laboratory conditions, in this study a sinusoidal load with constant amplitude was applied at 5 Hz.

A number of experimental investigations have been carried out on the fatigue behavior of normal-weight concrete, lightweight concrete and other cement-based composites since the early 20th century. Various prediction models were developed in these studies, to evaluate the fatigue performance of concrete. The relationship between stress level S, the ratio of maximum fatigue stress (f_{max}) to the flexural strength (f_r), and the number of loading cycles (N) that causes failure is commonly used for this purpose (Kesler, 1953; Ballinger, 1971; Hsu, 1981; Sohel *et al.*, 2022). The established relationship is known as the Wöhler fatigue equation (Hsu, 1981; Oh, 1986), and is given by Equation (7):

$$S = \frac{f_{max}}{f_r} = a + b \log_{10}(N) \tag{7}$$

where a and b are the coefficients that can be obtained by a linear regression analysis of the plotted test fatigue lives. Equation (7) shows the relationship between the applied stress level S and the number of cycles N until failure, on a logarithmic scale ($\log(N)$). Therefore, this equation is also known as the single-logarithm fatigue equation.

Oh (1986) obtained values for the coefficients a and b in Equation (7) for plain concrete using the fatigue test data.

A modified version of the Wöhler fatigue equation was used in some studies to estimate the fatigue life of cementitious materials (Tepfers and Kutti, 1979; Hsu, 1981; Oh, 1986). This modified version of the fatigue equation incorporates the fatigue stress ratio R. The fatigue stress ratio R is the ratio of minimum fatigue stress f_{min} to the maximum fatigue stress f_{max} and is included to simulate the loading conditions in actual structures where the minimum value of the repeated stress may not be zero. The modified Wöhler equation takes the form shown in Equation (8):

$$S = \frac{f_{max}}{f_r} = 1 - \beta(1-R)\log_{10}(N) \tag{8}$$

where β is an experimental material coefficient and that can be obtained from the S–N curve. Equation (8) is only suitable when $R = f_{min}/f_{max} \geq 0$ (i.e. no stress reversal). Aas-Jakobsen (1970) obtained the value of β as 0.0640 in Equation (8) for the compression fatigue of concrete. Tepfers and Kutti (1979), however, recommended 0.0685 as a value for β, for both NWC and LWC under compression. Oh (1986) tested Equation (8) for the flexural fatigue of plain concrete and obtained a value of 0.0690 for β. In the case of steel fiber reinforced concrete, the average values of β for 0.5%, 1.0% and 1.5% steel fiber content are 0.0536, 0.0425 and 0.0615, respectively (Singh and Kaushik, 2001).

Another fatigue equation, called the power relationship of fatigue life, was developed for concrete pavements, to estimate the flexural fatigue life of concrete (Treybig et al., 1977; Wirsching and Yao, 1982; Singh and Kaushik, 2001; Koltsida et al., 2018; Sohel et al., 2022). It relates the dimensionless stress level ($S = S_{max}/f_r$) to the number of loading cycles (N) given and is given by Equation (9):

$$N(S)^m = C \tag{9}$$

where C and m are empirical constants, N is the fatigue life, and S is the applied stress level. This equation has wide applicability since the stress level $S = S_{max}/f_r$ is expressed in a dimensionless form. Here, S_{max} is the maximum applied stress and f_r is the concrete modulus of rupture (flexural strength). Taking the logarithm of both sides of Equation (9), the following expression can be obtained:

$$\log(N) = \log(C) - m\log(S) \tag{10}$$

The values of C and m in Equation (10) can be determined from the plot of $\log(N)$ versus $\log(S)$. For instance, using the fatigue test data of normal-weight concrete, the values of m and C are 16.382 and 42.20, respectively, for the stress ratio $R = 0$ (Sohel et al., 2022).

3. Fatigue test on ULCC and high-strength LWC

Conventional high-strength lightweight concrete (density < 1900 kg/m^3) and ultra-lightweight cement composite (density < 1450 kg/m^3) are considered in order to evaluate the fatigue strength of high-strength lightweight concrete by test.

Fatigue test data of these two types concrete are used to evaluate the fatigue performance. In the following sections the detailed properties of these two types of concrete are given.

3.1 Ultra lightweight cement composite (ULCC)

ULCC is a type of fiber-reinforced ultra-lightweight cement composite. The wet mix of ULCC was composed of water, cement, undensified silica fume, cenospheres, chemical admixtures and polyvinyl alcohol (PVA) fibers. Cenospheres, as shown in Figure 3.4(a), were used as microscopic fillers in the production of ULCC. Cenospheres have a hollow spherical shape, obtained as a by-product of thermal power plants, and result from the ash ponds of coal combustion. Approximately 90% of the cenospheres' constituent materials are SiO_2 and Al_2O_3 (Chia *et al.*, 2011; Wu *et al.*, 2015; Yan *et al.*, 2016). Due to their hollow structural form, cenospheres have a very low particle density that generally varies from 600 kg/m³ to 900 kg/m³; consequently, the self-weight of the ULCC was reduced substantially compared to NWC.

The particle size distribution of the cenospheres used to produce ULCC in this study is shown in Figure 3.4(b). The majority of the particle had sizes between 0.1mm and 0.3 mm, with a nominal maximum size of ~0.6 mm. Along with ordinary Portland cement (OPC), silica fume (SF) was used as a supplementary cementitious binder. The resulting mixture became fiber-reinforced cementitious material by adding 6-mm long PVA fibers having a diameter of 0.27 mm. The tensile strength, modulus of elasticity and elongation at fracture of the PVA fiber were 1600 MPa, 39 GPa and 7%, respectively. The specific gravity of the fiber was 1.3. Commercially available shrinkage reducing admixture was used to minimize shrinkage strain and reduce air content in the mix. In order to achieve a workable mix with a low water/cement ratio,

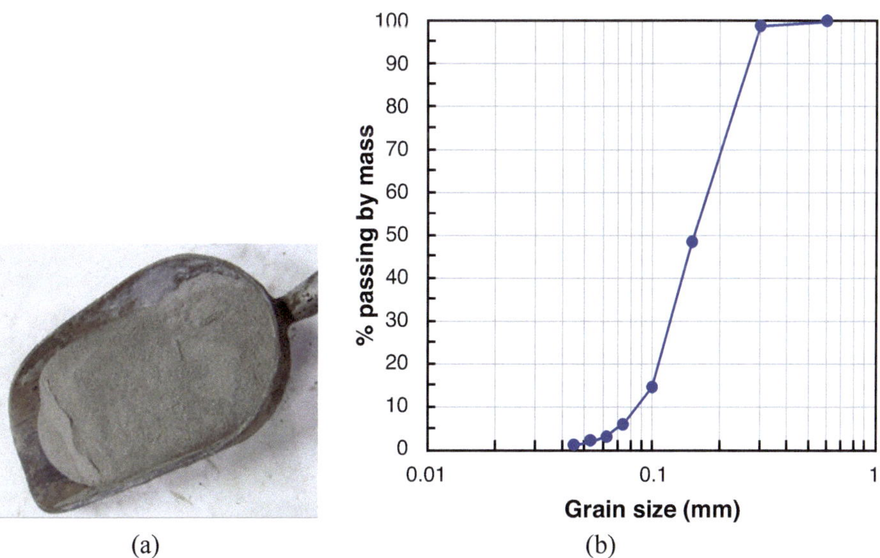

(a)   (b)

Figure 3.4 *(a) Cenospheres (b) sieve analysis of the cenospheres.*

a commercially available superplasticizer was also added. Since no coarse aggregate was added in the mix, the ULCC became highly workable material.

3.2 Materials and mix proportions of ULCC and LWC

The flexural fatigue behavior of the ULCC with 0.9% PVA fiber was evaluated in comparison to that of a high-strength lightweight aggregate concrete (LWC) with comparable compressive and flexural strength. Ordinary Portland cement (OPC) of ASTM Type I, cenosphere, PVA fiber, undensified silica fume (SF), shrinkage reducing admixture (SRA), polymer-based superplasticizer and potable water were used for the ULCC. The characteristics of the cement are given in Table 3.2. The chemical and mineral composition of cenosphere are given in Tables 3.3 and 3.4 (Wang *et al.*, 2012). The major chemical composition of the cenosphere is silica, alumina and iron oxide. These three compounds comprise more than 93% of the chemical constituents of the cenosphere. Therefore, cenosphere in powder form may have pozzolanic properties when combined with Portland cement and water. The X-ray diffraction (XRD) pattern is shown in Figure 3.5 (Wang *et al.*, 2012). From XRD analysis, the major mineralogical compounds of the cenosphere are found to be quartz, mullite and glass. The mix proportion of the ULCC is given in Table 3.5. The flowability of the ULCC mix was measured using the flow table test according to ASTM C1437, as shown in Figure 3.6(a). It was found that the ULCC was a highly flowable cementitious material.

Table 3.2 *Chemical composition of the cement (Wang et al., 2012).*

	Calcium oxide, CaO	64.3
	Silica, SiO_2	20.0
	Aluminium oxide, Al_2O_3	4.7
	Iron oxide, Fe_2O_3	2.9
	Magnesia, MgO	2.1
Chemical composition (%)	Sodium oxide, Na_2O	0.17
	Potassium oxide, K_2O	0.47
	Total alkalinity as $Na_2O+0.658K_2O$	0.48
	Sulphuric anhydride as SO_3	1.9
	Insoluble residue	0.3
	Loss on ignition (LOI)	2.2
Mineral composition according to X-ray diffraction (%)	Tricalcium silicate, C_3S	54.1
	Dicalcium silicate, C_2S	24.8
	Tricalcium aluminate, C_3A	7.5
	Tetracalcium alumninoferrite, C_4AF	7.5

Table 3.3 *Chemical composition (%) of cenosphere across various size ranges (Wang et al., 2012).*

Compound	0–100 μm	100–150 μm	150–300 μm
SiO_2	58.11	60.87	60.11
$Al2O_3$	32.00	28.21	28.42
Fe_2O_3	3.67	4.47	4.77
CaO	0.87	0.73	0.76
MgO	1.38	1.51	1.54
Na_2O	0.81	0.86	0.88
K_2O	3.12	3.34	3.49
SO_3	0.04	0.01	0.02
Loss on ignition (LOI)	1.04		

Table 3.4 *Mineral composition of the cenosphere by weight (%) (Wang et al., 2012).*

	Quartz	Mullite	Amorphous = glass
Particle size: 0–100 μm			
1	4.7	12.8	82.5
2	4.9	14.1	81.0
3	3.9	10.8	85.3
Particle size: 106–150 μm			
1	4.6	9.2	86.2
2	6.3	12.1	81.6
3	5.9	10.4	83.7
Particle size: 150–300 μm			
1	5.8	8.6	85.6
2	6.5	9.3	84.2
3	5.6	7.9	86.5

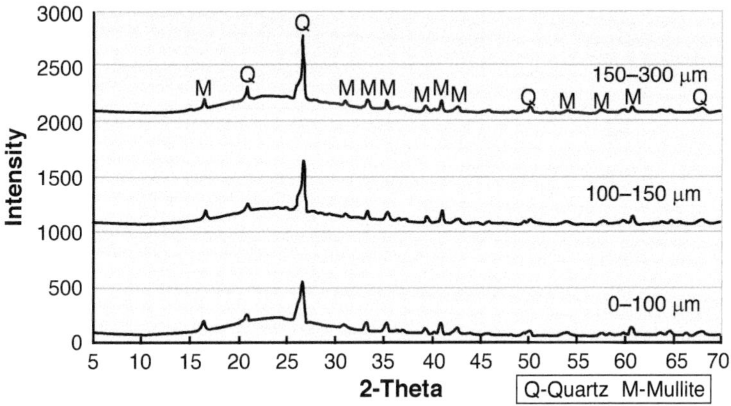

Figure 3.5 *X-ray diffraction patterns of the cenosphere (Wang et al., 2012).*

Table 3.5 *Mix proportion of the ULCC (kg/m³).*

Water	OPC	SF	SRA	Cenosphere	Fiber	SP (l/m³)	ρw (kg/m³)
262	741	65	20	335	12.3	11.25	1440

Note: OPC: ordinary Portland cement; SF: silica fume; SRA: shrinkage reducing admixture; SP: superplasticizer; ρ_w = density

Expanded clay lightweight aggregate (LWA) with a size ranging between 4 mm and 8 mm was used as coarse aggregate to produce high-strength LWC. The average particle density of this LWA (Liapor F7.0) was approximately 1000 kg/m³. The particle shape and internal pores of the lightweight aggregate are shown in Figure 3.6(b). For strength enhancement, 8% silica fume (undensified) by mass of total cementitious material was used in both the ULCC and LWC. In addition to these materials, ordinary Portland cement, natural sand and superplasticizer were used to produce the high-strength LWC. Before mixing, the LWA was soaked in water for one hour. The mix proportion of the LWC is given in Table 3.6.

3.3 Details of specimens

All beam specimens for both flexural and fatigue tests were 100 × 76 × 406 mm in size with an effective span of 300 mm. The experimental study for the fatigue tests was conducted on 54 ULCC specimens and 42 high-strength LWC specimens. In addition to the fatigue specimens, 12 beams for each mix were prepared for flexural testing. The beams with ULCC and LWC were cured in a fog room for 28 days before testing. The density of all the ULCC and LWC specimens was determined by the water displacement method just after demoulding. The average densities of the ULCC and LWC were 1440 kg/m³ and 1870 kg/m³, respectively, as shown in Tables 3.5 and 3.6.

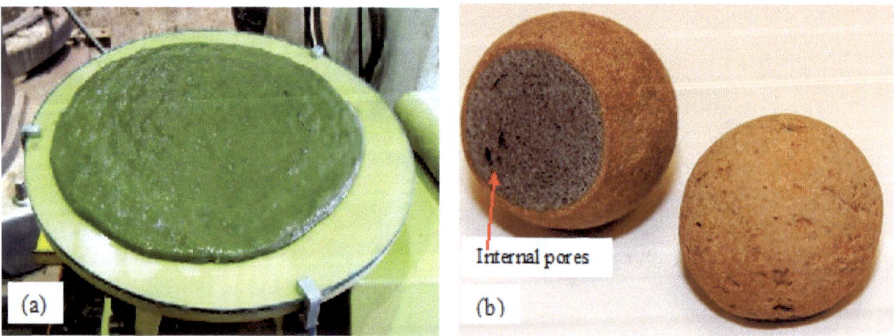

Figure 3.6 *(a) Flow test for the ULCC; (b) Particle shape and internal pores of the lightweight aggregate (Liapor F7.0).*

Table 3.6 *Mix proportion of the LWC (kg/m³).*

Water	OPC	SF	Sand	LWA F7.0	SP (l/m³)	ρw (kg/m³)
175	460	40	793	409	2.5	1870

Note: OPC: ordinary Portland cement; SF: silica fume; SP: superplasticizer; ρ_w = density

Table 3.7 *Mechanical properties of ULCC, LWC and NWC.*

Mix type	Density (kg/m³)	Compressive strength (MPa)	Flexural strength (MPa)	Elastic modulus (GPa)
ULCC	1440	62	6.4	15.0
LWC	1870	63	6.0	20.1
NWC*	2350	58	5.4	–

*The properties are taken from Mohammadi and Kaushik (2005).

The compressive strength of each batch of concrete was determined at 28 days using φ100 × 200 mm² cylinders. At 28 days, the ULCC and LWC had average compressive strengths of 62 MPa and 63 MPa, respectively (Table 3.7). Typically, a lightweight aggregate concrete having a compressive strength of greater than 40 MPa is considered high-strength lightweight concrete (ACI 213R-14, 2014). Therefore, the ULCC can be considered a high-strength lightweight cementitious composite. It has a high specific strength (strength-to-density ratio), greater than 41 kPa/kg.m⁻³, whereas typical normal-weight concrete (NWC) with a strength of 60 MPa and a density of 2400 kg/m³ has a specific strength of 25 kPa/kg.m⁻³. For lightweight aggregate concrete, the specific strength lies in the range 23–39 kPa/kg.m⁻³ for strength grades 30 to 65, according to the strength vs density trend, as shown in Figure 3.7. The dashed

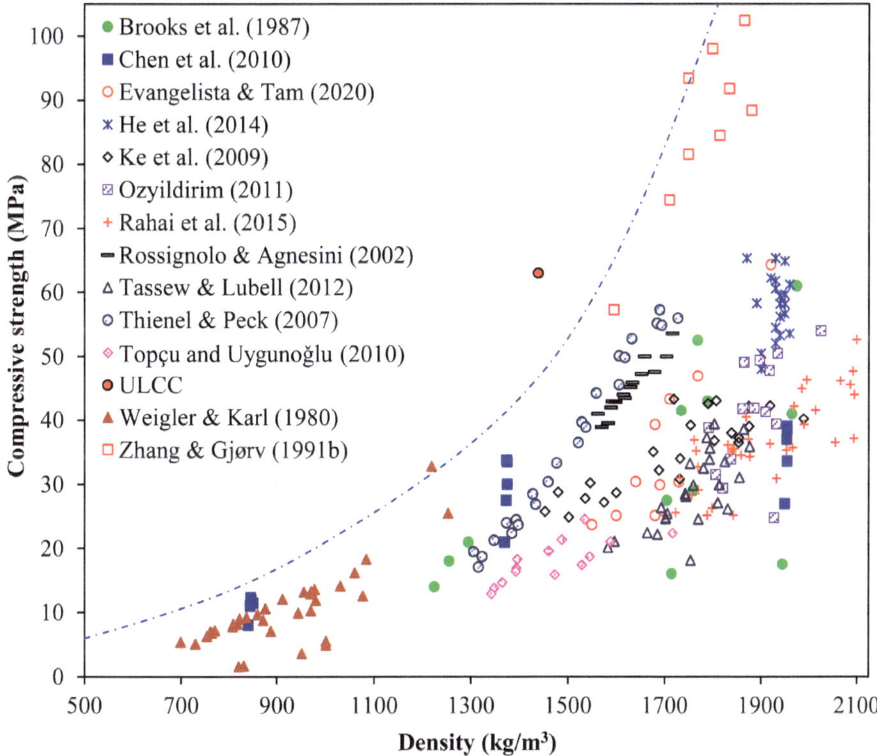

Figure 3.7 *Compressive strength vs density of ULCC compared with other LWC.*

line in Figure 3.7 may represent the current state-of-art of the development of LWC, including high-strength LWC (Weigler and Karl, 1980; Brooks *et al.*, 1987; Zhang and Gjørv, 1991b; Rossignolo and Agnesini, 2002; Thienel and Peck, 2007; Ke *et al.*, 2009; Chen *et al.*, 2010; Topçu and Uygunoğlu, 2010; Ozyildirim, 2011; Tassew and Lubell, 2012; He *et al.*, 2014; Rahai *et al.*, 2015; Evangelista and Tam, 2020). Besides a 40% weight reduction compared with conventional concrete, the ULCC exhibited ultimate tensile and flexural strengths comparable with conventional high-strength LWC of similar compressive strength (Table 3.7).

3.4 Static flexural and fatigue tests

The static flexural tests were carried out just before the fatigue tests for each batch of ULCC and LWC. The flexural strength was determined according to ASTM C78. The third-point loading method was used to determine the flexural strength of the concrete. A similar loading system was also used for the fatigue test. The schematic test set-up is shown in Figure 3.8 according to ASTM C78 standard, and the actual laboratory test set-up is shown in Figure 3.9. A spreader beam was placed above the central mid-span

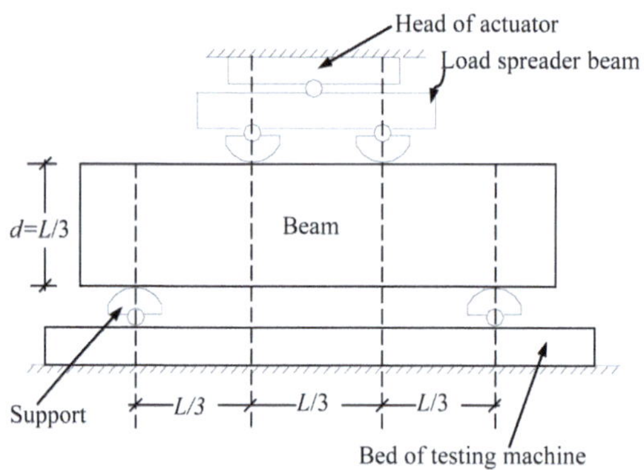

Figure 3. 8 *Schematic of the flexural test arrangement.*

Figure 3.9 *Actual test set-up for the flexural test.*

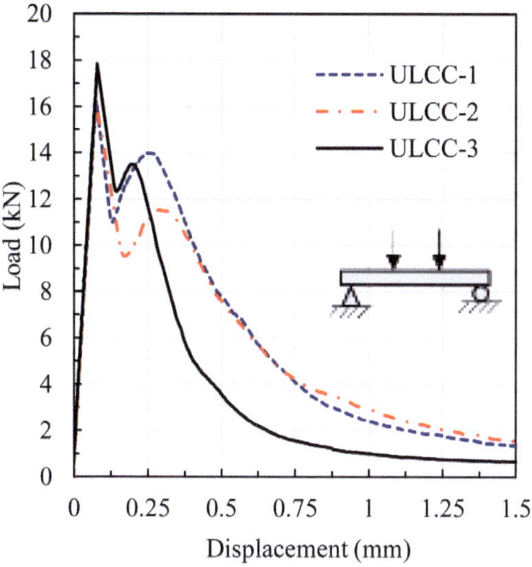

Figure 3.10 *Load–displacement curves of ULCC prisms (100 × 76 × 406 mm) subjected to third-point loading (Sohel et al., 2021).*

of the beam specimen. The load cell was placed at the middle of the spreader beam. A frame was fabricated to hold displacement transducers (LVDT) at the mid-span of the prism (see Figure 3.9). The frame was attached to the specimens and fixed with the specimen above the supports so that the actual displacement at the mid-span with respect to the support could be captured. The load–displacement curves for ULCC are shown in Figure 3.10 (Sohel *et al.*, 2021). The mechanical properties of the ULCC and LWC are given in Table 3.7. All values in the Table 3.7 are averages of three samples. The flexural strength f_r, which was derived for third-point loading, was calculated using Equation (11):

$$f_r = \frac{PL}{bd^2} \tag{11}$$

where P is the peak load, L is the beam span, b is the width of the beam, and d is the depth of the beam.

The flexural fatigue and static tests on the beam specimens were carried out using an Instron servo-controlled hydraulic testing machine (model Instron 8803) of 50 kN static capacity. Constant amplitude sinusoidal loading was applied for the flexural fatigue test, as shown schematically in Figure 3.3. The relationships between the cyclic stress parameters are summarized in Equations (3) to (7) (Gray, 1960; Zhang *et al.*, 1996).

Figure 3.11 shows the fatigue test set-up. The loading position and support conditions for the fatigue tests were kept the same as those for the static flexural tests. Prior to the test, the loading range (maximum and minimum load) and loading frequency were set via the software that controlled the testing machine. The maximum stress levels S (f_{max}/f_r) for the fatigue tests varied between 0.90 and 0.60. At the same

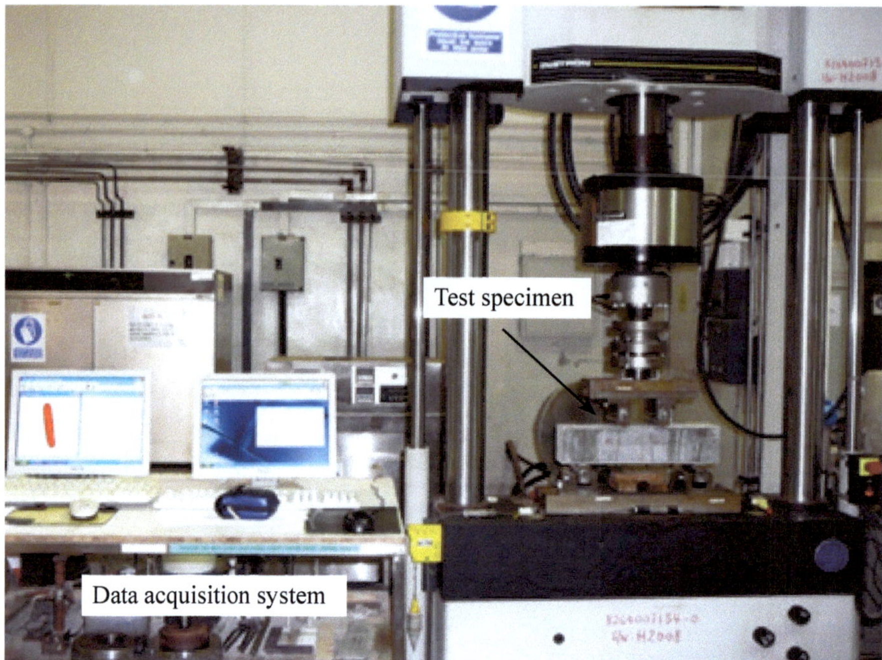

Figure 3.11 *Fatigue test set-up.*

time, the minimum stress level or fatigue stress ratio R (f_{min}/f_{max}) was maintained at a constant level of 0.1 throughout the test in accordance with previous fatigue studies on plain and fiber-reinforced concrete (Tepfers and Kutti, 1979; Singh and Kaushik, 2000; Mohammadi and Kaushik, 2005). The number of loading cycles for the failure of the beam specimen was recorded as the fatigue life, N.

4. Flexural and fatigue performance of ULCC and high-strength LWC

In this section, the flexural strength and fatigue strength of ULCC and high-strength LWC are discussed. Generally, the fatigue performance of cementitious materials is evaluated using experimental results. In the pervious section, the fatigue test procedure was discussed in detail. Using the test procedure described, fatigue test results for ULCC and LWC were obtained by Sohel *et al.* (2018). The fatigue performance of ULCC and high-strength LWC are evaluated using the fatigue test results.

4.1 Failure mode under flexural fatigue load

The fatigue tests were performed under load control mode at a frequency of 5 Hz. When the load dropped sharply due to the formation of a large crack, the specimen was considered to have failed. The failure of the specimen under fatigue load was caused by the initiation of a single crack at the bottom fiber in the middle-third span of the beam, which resembled a flexural test. With an increase in the number of loading cycles, the crack propagated toward the compression zone of the specimen, resulting in complete failure. All specimens split simultaneously into two parts about the visible crack at failure. The failure patterns of the ULCC and LWC specimens under flexural fatigue load are shown in Figure 3.12.

Some of the ULCC specimens showed the initiation of multiple cracks, whereby a crack spread as the number of loading cycles increased, resulting in the failure of the specimen. At higher stress levels, the failure occurred almost immediately after the onset of the first visible crack (Figure 3.12a). However, at lower stress levels ($< S = 0.7$), the specimen could sustain an adequate number of loading cycles (0.5 to 2 million cycles in some cases), despite the onset of the first visible crack. This is due to the fact that the small PVA fibers in the ULCC mix were bridged through the cracks and delayed their further expansion. The load–deflection curves under flexural load on the ULCC prism (Figure 3.10) also supports this explanation.

In case of high-strength LWC, it was observed that aggregates were fractured in all cases, indicating that there was no interfacial bond failure (Figure 3.12b). In other

(a)

(b)

Figure 3.12 *Fatigue failure of (a) the ULCC specimen and (b) the LWC specimen.*

words, the fatigue failure pattern of the LWC specimens is always due to the growth of micro-cracks in the concrete through its weakest plane (in this case through the weak aggregates). A similar observation was reported by Chakrabarti and Ledbetter (1967) in their research report for lightweight concrete with expanded shales coarse aggregates under flexural fatigue load. The fatigue fracture was always accompanied by the fracture of almost all coarse aggregates along the fracture surface. On the contrary, this was not found to be so in the case of the concrete with natural aggregate: none of the coarse aggregates fractured; the failure surface was generally between the mortar and the aggregates (Chakrabarti, 1967). On the other hand, Thomas (1979) observed from experimental observation that concrete made with crushed limestone seems to fail through the aggregate while concrete made with gravel tends to fail around the aggregate.

4.2 S–N curves and Wöhler fatigue equations

The fatigue test results (Sohel *et al.*, 2018) at different stress levels for the ULCC and LWC specimens are summarized in Tables 3.8 and 3.9. The results indicate that the fatigue lives of the ULCC and LWC increase with decreasing stress level S. Scattering of the test data for ULCC is larger than that for plain LWC, which implies that the dispersion of fine PVA fibers in ULCC is important in ensuring satisfactory performance of the hardened concrete.

Table 3.8 *Laboratory fatigue test results for the ULCC (R = S_{min}/S_{max} = 0.10).*

Type of material	Fatigue life data 'N'			
	$S = 0.90$	$S = 0.80$	$S = 0.70$	$S = 0.65$
	265	6450	22659	186790
	378	8748	24331	199025
	428	9989	52816	237522
	555	10068	57397	243059
	751	13980	64165	359484
	1066	17524	82264	373633
	1335	23256	100400	390867
ULCC	1416	25444	134771	405440
	2063	28169	137733	452690
	2783	29295	143110	577820
	2839	30915	162980	582292
	2840	45065	289272	752156
	3334	53098	348716	957314
	3569	–	348788	–

Table 3.9 *Laboratory fatigue test results for the LWC (R = S_{min}/S_{max} = 0.10).*

Type of material	Fatigue life data 'N'			
	S = 0.90	S = 0.80	S = 0.70	S = 0.60
	165	2001	10916	99860
	282	2734	23060	134853
	393	3000	24694	152024
	434	3386	41407	170625
	545	4706	46941	219960
LWC	687	5286	51783	235023
	916	6626	61616	255693
	1201	9859	64588	357401
	1450	15191	85583	477682
	1572	16696	96310	568921
	–	–	98686	781932

The *S–N* curves for ULCC and LWC are shown in Figure 3.13, which represents the relationship between the applied maximum stresses level *S* and number of load cycles *N* before the specimen fails. The relationship can be described by Equation (7) or Equation (8). For comparison, test results for NWC reported by Mohammadi and Kaushik (2005) are also plotted (Figure 3.13c). Based on the least squares method, linear regression analyses were carried out to determine the values of the coefficients *a* and *b* for Equation (7). The values of these coefficients for each type of concrete are listed in Table 3.10.

The material coefficient β in Equation (8) can be obtained from the test data to make the equation applicable to ULCC, LWC and NWC. Using the values of the stress level *S* (i.e. f_{max}/f_r), $\log_{10}$ (*N*) and R, the value of the coefficient β was determined for each test of ULCC, LWC and NWC. The average value of β for the whole population is 0.058503 with a standard deviation of 0.0166 for ULCC with 0.9% PVA fiber content; 0.064209 with a standard deviation of 0.016341 for LWC; and 0.058716 with a standard deviation of 0.00716 for NWC. By incorporating the corresponding values of the coefficients *a*, *b* and β in Equations (7) and (8), the flexural fatigue lives of ULCC, LWC and NWC can be predicted at any stress level.

Figure 3.13d compares the *S–N* curves for ULCC, LWC and NWC under flexural fatigue load. In the stress ranges tested, the flexural fatigue performance of ULCC is better than that of LWC. This is true when the performance of ULCC is compared with that of NWC. This may be attributed to the PVA fibers used in the ULCC, since the fibers would be able to bridge the cracks and extend the fatigue life. When the *S–N* curves are extended, it can be seen that the fatigue performance of ULCC and NWC is similar at stress level *S* < 0.6 (see Figure 3.13d).

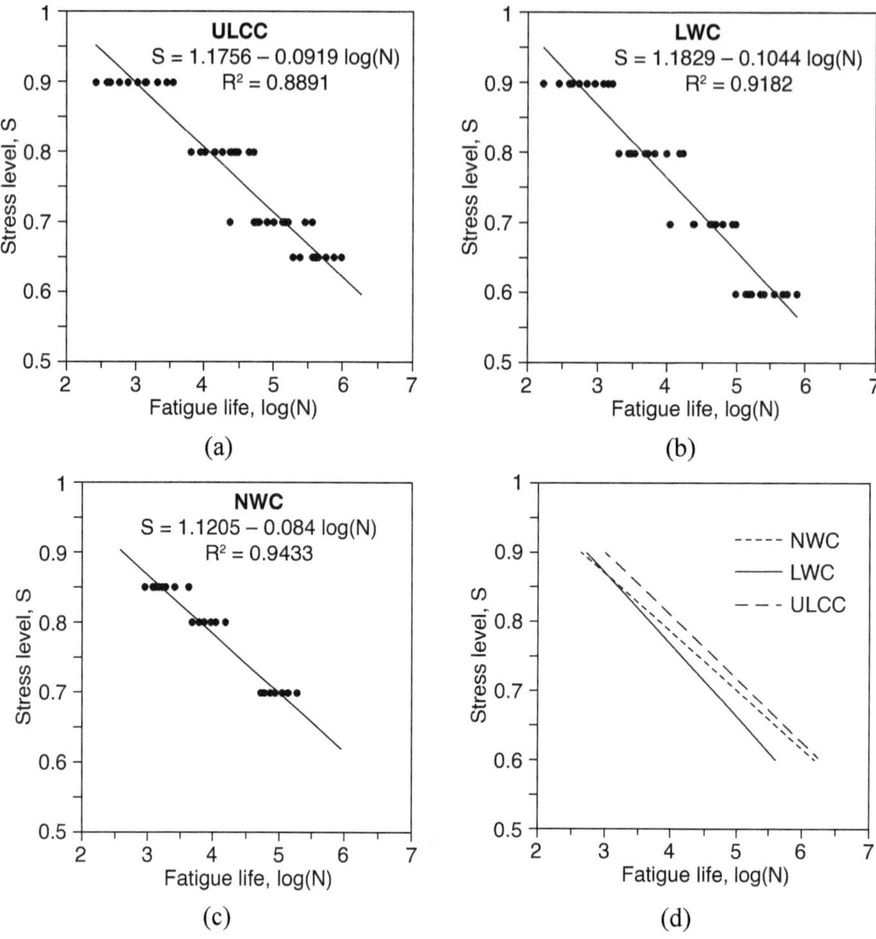

Figure 3.13 *S–N curves for (a) ULCC; (b) LWC from experimental study; (c) NWC; and (d) a comparison of ULCC, LWC and NWC.*

Table 3.10 *Coefficients a and b of Equation (7) and coefficient β of Equation (8) for ULCC, LWC and NWC.*

Concrete type	Based on Eq. (7)		Based on Eq. (8)
	Coefficient a	Coefficient b	Coefficient β
ULCC	1.1756	−0.0919	0.058503
LWC	1.183	−0.104	0.064209
NWC	1.121	−0.084	0.058716

The Wöhler fatigue equation and the modified Wöhler fatigue equation for the ULCC, LWC and NWC are given in Table 3.11. Using these equations, the endurance limit for the ULCC, LWC and NWC can be determined. The endurance limit of concrete is generally defined as the maximum flexural fatigue strength at which a

Table 3.11 *Fatigue equations and endurance limit for ULCC, LWC and NWC.*

Concrete type	Wöhler fatigue equation (Eq. 7)	Modified Wöhler fatigue equation (Eq. 8)	Stress level (S) for 2×10^6 cycles	
			Eq. 7	Eq. 8*
ULCC	$S = 1.1756 - 0.0919 \log_{10} N$	$S = 1 - 0.058503(1 - R) \log_{10} N$	0.60	0.67
LWC	$S = 1.183 - 0.104 \log_{10} N$	$S = 1 - 0.064209 (1 - R) \log_{10} N$	0.53	0.64
NWC	$S = 1.1205 - 0.084 \log_{10} N$	$S = 1 - 0.058716 (1 - R) \log_{10} N$	0.59	0.67

* Value of $R = 0.1$

concrete beam can withstand 2 million cycles (2×10^6 cycles) of non-reversed fatigue loading, expressed as a percentage of its flexural strength (static modulus of rupture) (Ramakrishnan *et al.*, 1989). The endurance limits obtained using Equation (7) for the ULCC, LWC and NWC are 60%, 53% and 59% of the static flexural strength, respectively. The endurance limit of the ULCC is higher than that of both plain NWC and LWC, as shown in Table 3.11. For all type of concrete mixes, it is shown that the Wöhler fatigue equation (Equation (7)) is more conservative than the modified Wöhler fatigue equation (Equation (8)). The result for the high-strength LWC is in agreement with the findings of other studies in that endurance limit of LWC with expanded shale aggregates is 55% of the static flexural strength (Ramakrishnan *et al.*, 1993).

5. Probabilistic fatigue life of ULCC and LWC

The statistical nature of the fatigue test data exhibits a larger scatter than that of the static test data. The statistical variability may arise from the variation of many design factors, such as the applied load and the heterogenetic nature of the material, which lead to increased uncertainties in the design. Therefore, applying probabilistic analysis to the fatigue test data makes it more realistic and provides adequate resistance against fatigue failure of concrete structures. As with all other cementitious materials, the fatigue test data obtained for the ULCC and the high-strength LWC showed considerable variability at the same applied stress level (S). It is accordingly necessary to apply the probabilistic concept in the design to secure a sufficiently reliable fatigue resistance value for the concrete structures. Numerous mathematical probability models have been developed to represent the probabilistic distribution of concrete fatigue-life data. Many studies (Oh, 1991; Singh and Kaushik, 2001) have shown that the 2-parameter Weibull distribution can be used to describe the distribution of the fatigue-life data of cementitious materials, since it provides safer and better reliability, as proved statistically and experimentally. Therefore, the 2-parameter Weibull distribution was used to describe the probabilistic distribution of the flexural fatigue life of the ULCC and LWC. The failure probability function (cumulative distribution function, CDF) for the 2-parameter Weibull distribution is defined by Equation (12):

$$P_f(N) = 1 - \exp\left[-\left(\frac{n}{u}\right)\right]; \quad n > 0 \tag{12}$$

where λ is the shape parameter, u is the scale parameter or characteristic life at stress level S, n is the specific value of the random variable or the fatigue life of concrete at the stress level S, and $P_f(N)$ is the failure probability.

The confidence probability function or survival probability function (L_n) for the 2-parameter Weibull distribution is defined by Equation (13):

$$L_n = \exp\left[-\left(\frac{n}{u}\right)^{\lambda}\right]; \quad n > 0 \qquad (13)$$

Many researchers (Oh, 1986, 1991; Singh and Kaushik, 2000, 2001; Mohammadi and Kaushik, 2005; Singh *et al.*, 2005; Arora and Singh, 2016) have employed the graphical method, the moments method, and maximum-likelihood estimation method to estimate the Weibull distribution parameters (λ and u).

5.1 Estimation of the Weibull parameters by the graphical method

The graphical method is the simplest method to determine the Weibull distribution parameters and to check whether the test data follow the distribution. Taking the natural logarithm twice on both sides of Equation (13) yields

$$\ln\left[\ln\left(\frac{1}{L_n}\right)\right] = \ln(n) - \ln(u) \qquad (14)$$

Equation (14) represents a linear relationship between $\ln(n)$ and $\ln[\ln(1/L_n)]$. If all the data points fall roughly along a straight line, it is assumed that the test data follow the Weibull distribution at a given fatigue stress level. The distribution parameters λ and u can be obtained directly from the linear trend line equation at a particular stress level S. To plot a graph using Equation (14), the test data of a particular stress level need to be arranged in descending order. Using the arranged data, and the empirical survival function L_n, all test data can be calculated from the expression given in Equation (15) (Oh, 1991; Singh and Kaushik, 2001):

$$L_n = 1 - \frac{i}{(1+k)} \qquad (15)$$

where k is the sample size and i is the order number of the fatigue test data at a particular stress level S. A detail calculation procedure is given elsewhere (Oh, 1991). Figure 3.14 shows the graphical analyses of the fatigue-life data for ULCC and LWC at different stress levels. The approximate straight lines of the plots indicate that the two-parameter Weibull distribution is a reasonable assumption for the fatigue life of the ULCC and the LWC at all tested stress levels. The distribution parameters λ and u can be determined from the straight line equations or from the regression analyses. In view of Equation (14), the slope of the straight line in Figure 3.14 is the value of the shape parameter λ for the corresponding fatigue stress level. The scale parameter u can also be obtained from Equation (14) and Figure 3.14. The distribution parameters obtained from the graphical method are given in Tables 3.12–3.14.

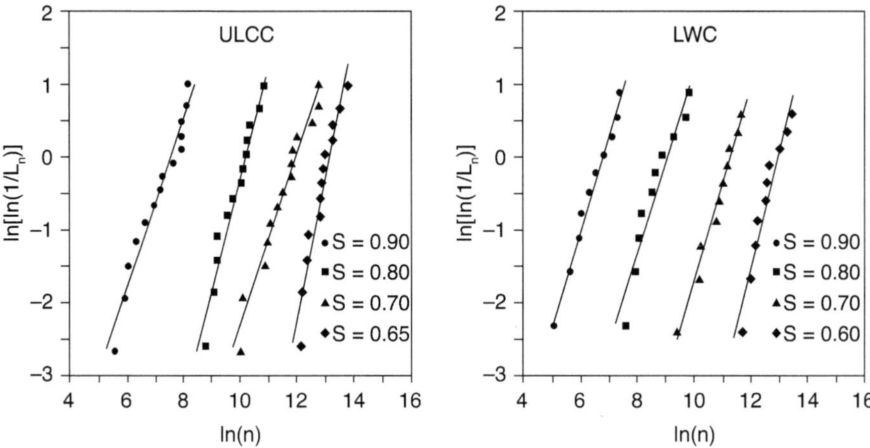

Figure 3.14 *Graphical analysis of fatigue-life data for ULCC and LWC at different stress levels S.*

Table 3.12 *Values of the Weibull distribution parameters λ and u for ULCC at all tested stress levels, using the three methods.*

Method	S = 0.9		S = 0.80		S = 0.70		S = 0.65	
	λ	u	λ	u	λ	u	λ	u
Graphical method	1.156	1925	1.546	26745	1.175	157628	1.991	504398
Method of moments	1.462	1863	1.690	26025	1.283	151892	2.038	496471
Maximum likelihood	1.455	1863	1.795	26256	1.352	154120	2.160	499268
Average	1.358	1884	1.677	26342	1.270	154547	2.063	500046

Table 3.13 *Values of the Weibull distribution parameters λ and u for LWC at all tested stress levels, using the three methods.*

Method	S = 0.9		S = 0.80		S = 0.70		S = 0.60	
	λ	u	λ	u	λ	u	λ	u
Graphical method	1.346	885	1.315	7932	1.464	64939	1.522	359797
Method of moments	1.590	852	1.3483	7576	1.947	62084	1.517	348284
Maximum likelihood	1.689	860	1.496	7763	2.039	62146	1.671	354553
Average	1.541	866	1.386	7757	1.817	63056	1.570	354211

Table 3.14 *Values of the Weibull distribution parameters λ and u for NWC at different stress levels, using the three methods.*

Method	S = 0.85		S = 0.80		S = 0.70	
	λ	u	λ	u	λ	u
Graphical method	3.387	1757	2.703	9282	2.160	105913
Method of moments	3.674	1746	2.794	9203	2.213	103780
Maximum likelihood	3.577	1744	2.838	9203	2.345	104226
Average	3.546	1749	2.778	9229	2.239	104640

Along with the graphical method, other two methods – namely, the moments method and the maximum likelihood estimation (Oh, 1991; Singh and Kaushik, 2001) – were also used to obtain the Weibull distribution parameters for the fatigue life of the ULCC, LWC and NWC. Detailed procedure of these two methods can be found in Mohammadi and Kaushik (2005), Oh (1991) and Singh and Kaushik (2001). Tables 3.12–3.14 give the distribution parameters obtained by these two methods as well as with the graphical method.

It can be seen from Tables 3.12–3.14 that the distribution parameters obtained from the three different methods are almost identical for a given stress level. The average values of these three methods are considered to be the Weibull distribution parameters for ULCC, LWC and NWC at different stress levels of the fatigue-test data. It can also be seen that the average values of the shape parameters (λ) of ULCC and LWC are lower than those of NWC. This indicates that the variability in the fatigue-life distribution of the ULCC and the LWC is much greater than that for NWC.

5.2 Goodness-of-fit test

As shown in the previous section, the 2-parameter Weibull distribution function can be used to model the fatigue life of ULCC, LWC and NWC at a given stress level. To make sure that the Weibull distribution assumption is correct, a goodness-of-fit test is one of the reliable approaches (Jia *et al.*, 2011; Oh, 1991). The Kolmogorov–Smirnov (K–S) goodness-of-fit test, the χ^2-test, and the Anderson–Darling (A–D) test are commonly used as a goodness-of-fit test for concrete fatigue-test data in probabilistic analysis (Oh, 1991). Here, the Anderson–Darling tests are explained elaborately.

The Anderson–Darling test (Anderson and Darling, 1954) was applied as a goodness-of-fit test in this study. The A–D test is sensitive to inconsistencies in the tail areas and is widely used for material test data (Jia *et al.*, 2011; Rust *et al.*, 1989). The A–D goodness-of-fit test statistics can be written in the following form (Equation (16)) for hand calculation (Stephens, 1977; Ang and Tang, 2007):

$$AD = -k - \frac{1}{k}\sum_{i=1}^{k}(2i-1)[\ln F_n(n_i) + \ln(1 - F_n(n_{k+1-i}))] \tag{16}$$

where k is the sample size, i is the order number of the data point, and $F_n = P_f(N)$ is the proposed distribution (CDF). For the 2-parameter Weibull distribution, the cumulative distribution function is given by Equation (17):

$$F_n = 1 - \exp\left[-\left(\frac{n_i}{u}\right)^{\lambda}\right] \tag{17}$$

The adjusted A–D test statistics for the Weibull distribution are given by Equation (18):

$$AD^* = AD\left(1 + \frac{0.2}{\sqrt{k}}\right) \tag{18}$$

Table 3.15 *Anderson–Darling goodness-of-fit test for a stress level $S = 0.8$ for ULCC.*

i	n_i	$F_n(n_i)$	$F_x(n_{k+1-i})$	ith term[*]
1	6450	0.090163	0.960804	0.434256
2	8748	0.145728	0.914592	1.012229
3	9989	0.178597	0.729605	1.165574
4	10068	0.180743	0.697299	1.5646
5	13980	0.292259	0.673394	1.626311
6	17524	0.396429	0.610738	1.58126
7	23256	0.555791	0.555791	1.398823
8	25444	0.610738	0.396429	1.151514
9	28169	0.673394	0.292259	0.969133
10	29295	0.697299	0.180743	0.818314
11	30915	0.729605	0.178597	0.827067
12	45065	0.914592	0.145728	0.436616
13	53098	0.960804	0.090163	0.258605
				$\Sigma = 13.24430$

AD = 13.24430 – 13 = 0.24430 and AD[*]= 0.2440(1 + 0.2/√13) = 0.257854

$* - \dfrac{1}{k}(2i-1)[\ln F_n(n_i) + \ln(1 - F_n(x_{k+1-i}))]$

If AD[*] is less than 0.757 (critical value at 5% significance level for small samples), Weibull distribution is acceptable at 5% significance level (Stephens, 1977; Lockhart and Stephens, 1994; Ang and Tang, 2007). Table 3.15 shows the results obtained by the A–D test for a stress level $S = 0.8$. For the sample size $k = 13$, the A–D statistic is AD = 13.2440 –13 = 0.2440; the corresponding adjusted A–D statistic becomes AD[*] = 0.2440(1 + 0.2/√13) = 0.257854 (using Equation (18)); this is less than critical value $c_\alpha = 0.757$ at the significance level 5%. Therefore, the Weibull distribution is acceptable or a valid distribution at 5% significance level.

The A–D test was also applied to the fatigue-life data of the ULCC and the LWC at all other stress levels tested in this study. It was found that the 2-parameter Weibull distribution function was acceptable at 5% significance level.

5.3 Failure probability and design fatigue life of ULCC and LWC

Since fatigue life is an important factor in the design of concrete structures subjected to such loading, it becomes essential to include the failure probability in the $S–N$ relationship in order to provide better predictability of the fatigue strength of cementitious materials. However, the widely used Wöhler fatigue equation does not contain the failure probability P_f, which is an important parameter in fatigue analysis. As seen in the preceding sections, the flexural fatigue test data of the ULCC, LWC and NWC were scattered in nature and they followed the 2-parameter Weibull distribution at all tested stress levels. Therefore, this distribution function can be used further to calculate the design fatigue life N_D, with different failure probabilities P_f. The design fatigue life N_D should be chosen such that the probability of fatigue failure is very low (Oh, 1986). The design fatigue life with a permissible failure probability P_f can be obtained from Equation (12), as follows:

$$N_D = u \left[\ln\left(\frac{1}{1-P_f} \right) \right]^{\frac{1}{\lambda}} \tag{19}$$

N_D can be calculated for a particular failure probability using Equation (19) with the average values of the Weibull distribution parameters at a particular stress level. Using the average values of scale parameter λ and shape parameter u, presented in Tables 3.12–3.14, the design fatigue lives at different stress levels for the ULCC, LWC and NWC were calculated, corresponding to permissible failure probabilities of 1%, 5% 10% and 25%, or P_f = 0.01, 0.05, 0.1 and 0.25. The calculated fatigue lives are plotted in Figure 3.15

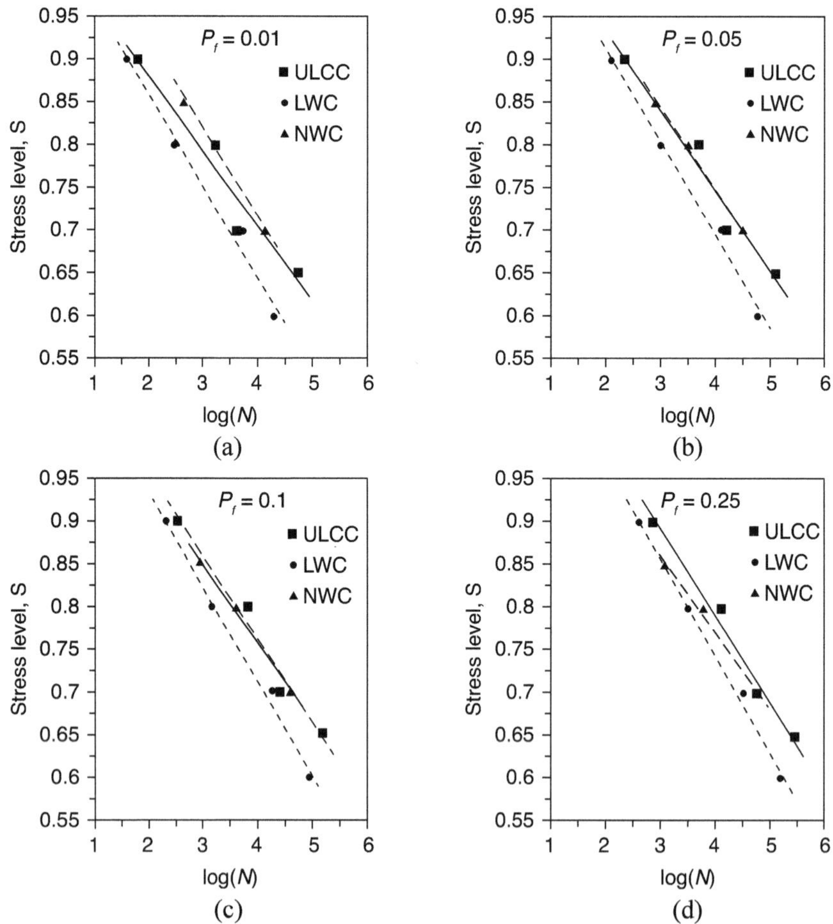

Figure 3.15 *Single-log fatigue curves of ULCC, LWC and NWC corresponding to a failure probability of (a) 1% (P_f = 0.01); (b) 5% (P_f = 0.05); (c) 10% (P_f = 0.1); and (d) 25% (P_f = 0.25).*

Table 3.16 *Single-log fatigue equations for the ULCC, LWC and NWC at failure probabilities of 25%, 10%, 5% and 1%.*

	Concrete type	Fatigue equation	Correlation coefficient (C_o)	Stress level (S) for 2×10^6 cycles
$P_f = 0.25$	ULCC	$S = 1.1952 - 0.1008 \log_{10}(N)$	0.992	0.56
	LWC	$S = 1.1937 - 0.1124 \log_{10}(N)$	0.998	0.49
	NWC	$S = 1.1312 - 0.0897 \log_{10}(N)$	0.997	0.57
$P_f = 0.1$	ULCC	$S = 1.1526 - 0.0973 \log_{10}(N)$	0.986	0.54
	LWC	$S = 1.1571 - 0.1109 \log_{10}(N)$	0.996	0.46
	NWC	$S = 1.1318 - 0.0936 \log_{10}(N)$	0.997	0.54
$P_f = 0.05$	ULCC	$S = 1.123 - 0.0947 \log_{10}(N)$	0.981	0.53
	LWC	$S = 1.1311 - 0.1099 \log_{10}(N)$	0.995	0.44
	NWC	$S = 1.1322 - 0.0967 \log_{10}(N)$	0.997	0.52
$P_f = 0.01$	ULCC	$S = 1.0595 - 0.0888 \log_{10}(N)$	0.968	0.50
	LWC	$S = 1.0736 - 0.1073 \log_{10}(N)$	0.991	0.40
	NWC	$S = 1.1333 - 0.1044 \log_{10}(N)$	0.998	0.48

to obtain single-log fatigue curves for the ULCC, LWC and NWC for $P_f = 0.01$, 0.05, 0.1 and 0.25 or a failure probability of 1%, 5%, 10% and 25%, respectively. Many researchers used the single-log fatigue curve to describe the relation between stress level S and fatigue life N for plain and fiber-reinforced concrete (Hsu, 1981; Singh and Kaushik, 2001; Li *et al.*, 2007). The equation of the curve can be obtained by regression analysis. The fatigue equations obtained corresponding to different failure probabilities are given in Table 3.16. These equations can be used by designers to obtain the flexural fatigue strength (i.e. stress level) of ULCC, LWC and NWC at a desired level of failure probability.

The stress levels S for the two million cycles are 0.56, 0.54, 0.53 and 0.50 at a failure probability P_f of 0.25, 0.1, 0.5 and 0.01, respectively, for the ULCC (see Table 3.16). This indicates that the stress level decreases when the failure probability decreases for the same fatigue life (number of cycles). Similar trends can also be observed for the LWC and NWC. The endurance limits for 2×10^6 cycles of fatigue load are 54%, 46% and 54% of the static flexural strength for the ULCC, LWC and NWC, respectively, with a failure probability of 10%. On the other hand, using the Wöhler fatigue equation the endurance limits for 2×10^6 cycles of fatigue load are 60%, 53% and 59% of the static flexural strength for the ULCC, LWC and NWC, respectively (Table 3.16). From this analysis, it is seen that the S–N relationships incorporating the failure probabilities (1% to 25%) are more conservative than those calculated using the Wöhler fatigue equation.

The experimental and theoretical results obtained for the ULCC and the high-strength LWC are valid for the type and proportion of the materials used and the size of the specimen tested in this investigation. Therefore, more general characterization may not be valid at this stage. More research is necessary to generate additional fatigue data for the newly developed material ULCC with different proportions of materials and mineral admixtures.

6. Remarks

In this chapter, the background to the development of ultra lightweight cement composite (ULCC) and high-strength lightweight concrete (LWC) and their flexural fatigue performance is discussed. Based on the experimental investigation and theoretical analysis, the flexural fatigue performance of the ULCC and LWC may now be summarized.

Under flexural fatigue load, the ULCC showed better performance in terms of load cycles and crack pattern than plain LWC of similar strength. Fatigue test results and their regression analyses showed a quite consistent tendency for both the ULCC and plain high-strength LWC, i.e. a lower stress level indicated a longer fatigue life. Based on the Wöhler fatigue equation, the flexural fatigue strengths of ULCC and high-strength LWC for 2×10^6 cycles were 60% and 53% of the static flexural strength, respectively. With the presence of fine PVA fiber, ULCC displayed a higher fatigue life than plain high-strength LWC. The probabilistic distribution of fatigue lives of the ULCC and high-strength LWC at different stress levels can be modeled by the 2-parameter Weibull distribution with a statistical correlation coefficient greater than 0.90. This behavior was the same as normal weight concrete with the same strength. Weibull distribution was employed to incorporate the failure probability P_f into the S–N relationship of the ULCC, LWC and NWC. The single-log equation obtained by regression analysis was used to examine the flexural fatigue performance of the ULCC, LWC and NWC. When the failure probability was smaller, the fatigue life was shorter at the same stress level. In other words, the stress level reduces as the failure probability decreases for the same fatigue life (number of cycles). S–N relationships that incorporated failure probabilities ($P_f = 0.01 - 0.25$) gave more conservative fatigue lives than the Wöhler fatigue equation. Therefore, establishing relationships would be useful in allowing designers to obtain the trend for the flexural fatigue strength of the ULCC and high-strength LWC at the desired level of the failure probability.

References

Aas-Jakobsen, K. (1970) *Fatigue of Concrete Beams and Columns, Bulletin No. 70-1, NTH Institute for Betonkonstruksjoner. Trondheim*: NTH Institute for Betonkonstruksjoner.

ACI 213R-14. (2014) *Guide for Structural Lightweight-Aggregate Concrete*. Report by ACI Committee 213. Farmington Hills, MI: American Concrete Institute.

ACI 215R-74. (1997) *Considerations for Design of Concrete Structures Subjected to Fatigue Loading (Reapproved 1997)*. ACI Committee 215. Farmington Hills, MI: American Concrete Institute.

Anderson, T.W. and Darling, D.A. (1954) A test of goodness of fit. *Journal of the American Statistical Association* **49**(268), 765.

Ang, A.H.S. and Tang, W.H. (2007) *Probability Concepts in Engineering: Emphasis on Applications to Civil and Environmental Engineering. Second Edition.* New York, NY: John Wiley & and Sons.

Arora, S. and Singh, S.P. (2016) Analysis of flexural fatigue failure of concrete made with 100% coarse recycled concrete sggregates. *Construction and Building Materials* **102**(1), 782–791.

Ballinger, C.A. (1971) Cumulative fatigue damage characteristics of plain concrete. *Highway Research Record No. 370*, 48–60.

Brooks, J.J., Bennett, E.W. and Owens, P.L. (1987) Influence of lightweight aggregates on thermal strain capacity on concrete. *Magazine of Concrete Research* **39**(139), 60–72.

Chakrabarti, J.C. and Ledbetter, W.B. (1967) *Flexural Fatigue Durability of Selected Unreinforced Structural Lightweight Concretes.* Arlington, TX: Texas Transportation Institute, Texas A&M University, Research Report 81-4.

Chen, B., Liu, J. and Chen, L.Z. (2010) Experimental study of lightweight expanded polystyrene aggregate concrete containing silica fume and polypropylene fibers. *Journal of Shanghai Jiaotong University (Science)* **15**(2), 129–137.

Chia, K.S., Zhang, M.H. and Liew, J.Y.R. (2011) High-strength ultra lightweight cement composite–material properties. *Proceedings of 9th International Symposium on High Performance Concrete-Design, Verification & Utilization, 9-11 August, Rotorua. New Zealand.* New Zealand Concrete Society, pp. 9–11.

Elfgren, L. (ed.). (1989) *Fracture Mechanics of Concrete Structures: From Theory to Applications: Report of the Technical Committee 90-FMA Fracture Mechanics to Concrete/Applications.* London: Chapman & Hall.

ESCI. (1971) *Lightweight Concrete. History, Application and Economics.* Salt Lake City, UT: Expanded Shale Clay and Slate Institute.

Evangelista, A.C.J. and Tam, V.W.Y. (2020) Properties of high-strength lightweight concrete using manufactured aggregate. *Proceedings of Institution of Civil Engineers: Construction Materials* **173**(4), 157–169.

Ganesan, N., Bharati Raj, J. and Shashikala, A.P. (2013) Flexural fatigue behavior of self-compacting rubberized concrete. *Construction and Building Materials* **44**, 7–14.

Gray, W.H. (1960) *A Study of the Fatigue Properties of Lightweight Aggregate Concrete: Technical Paper.* Publication FHWA/IN/JHRP-60/20. Lafayette, IN: Joint Highway Research Project, Indiana Department of Transportation and Purdue University. Available at https://doi.org/10.5703/1288284314411.

He, Y., Zhang, X., Zhang, Y. and Zhou, Y. (2014) Effects of particle characteristics of lightweight aggregate on mechanical properties of lightweight aggregate concrete. *Construction and Building Materials* **72**, 270–282.

Hoff, G.C. (1993) High strength lightweight aggregate concrete for Arctic applications. Part 1: Unhardened concrete properties. In Thomas A. Hom (ed.) *Structural Lightweight Aggregate Performance. ACI Special Publication 136.* Farmington Hills, MI: American Concrete Institute, pp. 1–66.

Hoff, G.C. (1994) Observations on the fatigue behavior of high-strength lightweight concrete. *American Concrete Institute, ACI Special Publication* 149. Farmington Hills, MI: American Concrete Institute, pp. 785–822.

Holm, T.A. and Bremner, T.W. (2000) *State-of-the-Art Report on High-Strength, High-Durability Structural Low-Density Concrete for Applications in Severe Marine Environments.* ERDC/SL TR-00-3. Vicksburg, MS and Washington, DC: Structures Laboratory (U.S.) and Engineer and Research Development Center (U.S).

Hsu, T.T.C. (1981) Fatigue of plain concrete. *Journal of the American Concrete Institute* **78**(4), 292–305.

Hsu, T.T.C. (1984) Fatigue and microcracking of concrete. *Matériaux et Constructions* **17**(1), 51–54.

Jia, Y.U., Marquis, G. and Björk, T. (2011) Development of data sheets for statistical evaluation of fatigue data. *Journal of Iron and Steel Research International* **18**(5), 70–78.

Ke, Y., Beaucour, A.L., Ortola, S., Dumontet, H. and Cabrillac, R. (2009) Influence of volume fraction and characteristics of lightweight aggregates on the mechanical properties of concrete. *Construction and Building Materials* **23**(8), 2821–2828.

Kesler, C.E. (1953) Effect of speed of testing on flexural fatigue strength of plain concrete. In F. Burggraf and J. W. Miller (eds) *Proceedings of the Thirty-Second Annual Meeting of the Highway Research Board, Washington, D.C., January 13-16, 1953.* Highway Research Board Proceedings, vol. 32, pp. 251–258. Washington, DC: Highway Research Board.

Kolay, P.K. and Singh, D.N. (2001) Physical, chemical, mineralogical, and thermal properties of cenospheres from an ash lagoon. *Cement and Concrete Research* **31**(4), 539–542.

Koltsida, I.S., Tomor, A.K. and Booth, C.A. (2018) Probability of fatigue failure in brick masonry under compressive loading. *International Journal of Fatigue* **112**, 233–239.

Lee, M.K. and Barr, B.I.G. (2004) An overview of the fatigue behaviour of plain and fibre reinforced concrete. *Cement and Concrete Composites* **26**(4), 299–305.

Li, H., Zhang, M. and Ou, J. ping. (2007) Flexural fatigue performance of concrete containing nano-particles for pavement. *International Journal of Fatigue* **29**(7), 1292–1301.

Lockhart, R.A. and Stephens, M.A. (1994) Estimation and tests of fit for the three-parameter Weibull distribution. *Journal of the Royal Statistical Society: Series B (Methodological)* **56**(3), 491–500.

Marshall, P.W., Sohel, K.M.A., Liew, J.Y.R., Yan, J.-B., Palmer, A. and Choo, Y.S. (2012) Development of SCS sandwich composite shell for arctic caissons. *Society of Petroleum Engineers – Arctic Technology Conference 2012*, Vol. 2.

Mohammadi, Y. and Kaushik, S.K. (2005) Flexural fatigue-life distributions of plain and fibrous concrete at various stress levels. *Journal of Materials in Civil Engineering* **17**(6), 650–658.

Naaman, A.E. and Hammoud, H. (1998) Fatigue characteristics of high performance fiber-reinforced concrete. *Cement and Concrete Composites* **20**(5), 353–363.

Nieto, A.J., Chicharro, J.M. and Pintado, P. (2006) An approximated methodology for fatigue tests and fatigue monitoring of concrete specimens. *International Journal of Fatigue* **28**(8), 835–842.

Oh, B.H. (1986) Fatigue analysis of plain concrete in flexure. *Journal of Structural Engineering* **112**(2), 273–288.

Oh, B.H. (1991) Fatigue-life distributions of concrete for various stress levels. *ACI Materials Journal* **88**(2), 122–128.

Ozyildirim, H.C. (2011) *Laboratory Investigation of Lightweight Concrete Properties.* FHWA/VCTIR 11-R17. Charlottesville, VA and Richmond, VA: Virginia Center for Tansportation Innovation& Research and US Deparment of Transportation Federal Higway Administration. Available at: http://www.virginiadot.org/vtrc/main/online_reports/pdf/11-r17.pdf (accessed 21 August 2022).

Petkovic, G. (1991) Properties of concrete related to fatigue damage with emphasis on high strength concrete. PhD thesis, Norwegian Institute of Technology, University of Trondheim, Norway.

Rahai, A., Dolati, A., Kamel, M.E. and Babaizadeh, H. (2015) Studying the effect of various parameters on mechanical properties of lightweight aggregate concrete using MANOVA. *Materials and Structures/Materiaux et Constructions* **48**(8), 2353–2365.

Raithby, K.D. (1979) Flexural fatigue behaviour of plain concrete. *Fatigue & Fracture of Engineering Materials & Structures* **2**(3), 269–278.

Ramakrishnan, V. and Panchalan, R.K. (2003) Probabilistic modeling of the flexural fatigue performance of lightweight concrete. In J.P. Ries and T.A. Holm (eds) *Theodore Bremner Symposium on High-Performance Lightweight Concrete*. Ottowa, Canada and Farmington Hill, MI: CANMET/ACI, pp. 205–224.

Ramakrishnan, V., Wu, G.Y. and Hosalli, G. (1989) Flexural fatigue strength, endurance limit, and impact strength of fiber reinforced concretes. *Transportation Research Record* **1226**, 17–24.

Ramakrishnan, V., Bremner, T.W. and Malhotra, V.M. (1993) Fatigue strength and endurance limit of lightweight concrete. In Thomas A. Hom (ed.) *Structural Lightweight Aggregate Performance. ACI Special Publication 136.* Farmington Hills, MI: American Concrete Institute, pp. 397–420.

Ries, J.P. and Holm, T.A. (2004) *A Holistic Approach to Sustainability for the Concrete Community. Lightweight Concrete – Two Millennia of Proven Performance.* Cicago, IL: Expanded Shale, Clay and Slate Institute (ESCSI). Information Sheet 7700.1.

Rossignolo, J.A. and Agnesini, M.V.C. (2002) Mechanical properties of polymer-modified lightweight aggregate concrete. *Cement and Concrete Research* **32**(3), 329–334.

Rust, S.W., Todt, F.R., Harris, B., Neal, D. and Vangel, M. (1989) Statistical methods for calculating material allowables for MIL-HDBK-17. In C.C. Chamis (ed.) *Test Methods and Design Allowables for Fibrous Composites. Volume 2.* West Conshohocken, PA: ASTM International, pp. 136–149.

Singh, S.R. and Kaushik, S.K. (2000) Flexural fatigue life distributions and failure probability of steel fibrous concrete. *ACI Structural Journal* **97**(6), 658–667.

Singh, S.P. and Kaushik, S.K. (2001) Flexural fatigue analysis of steel fiber-reinforced concrete. *ACI Materials Journal* **98**(4), 306–312.

Singh, S.P., Mohammadi, Y. and Kaushik, S.K. (2005) Flexural fatigue analysis of steel fibrous concrete containing mixed fibers. *ACI Materials Journal* **102**(6), 438–444.

Sohel, K.M.A., Liew, J.Y.R. and Fares, A.I. (2021) Shear bond behavior of composite slabs with ultra-lightweight cementitious composite. *Journal of Building Engineering* **44**(12), 103284.

Sohel, K.M.A., Al-Jabri, K., Zhang, M.H. and Liew, J.Y.R. (2018) Flexural fatigue behavior of ultra-lightweight cement composite and high-strength lightweight aggregate concrete. *Construction and Building Materials* **173**, 90–100.

Sohel, K.M.A., Liew, J.Y.R., Yan, J.B., Zhang, M.H. and Chia, K.S. (2012) Behavior of steel-concrete-steel sandwich structures with lightweight cement composite and novel shear connectors. *Composite Structures* **94**(12), 3500–3509.

Sohel, K.M.A., Al-Hinai, M.H.S., Alnuaimi, A., Al-Shahri, M. and El-Gamal, S. (2022) Prediction of flexural fatigue life and failure probability of normal weight concrete. *Materiales de Construcción* **72**(347), e291.

Stephens, M.A. (1977) Goodness of fit for the extreme value distribution. *Biometrika*, Oxford Academic **64**(3), 583–588.

Tassew, S.T. and Lubell, A.S. (2012) Mechanical properties of lightweight ceramic concrete. *Materials and Structures/Materiaux et Constructions* **45**(4), 561–574.

Tepfers, R. and Kutti, T. (1979) Fatigue strength of plain, ordinary, and lightweight concrete. *Journal of the American Concrete Institute* **76** (5), 635–652.

Thienel, K. and Peck, M. (2007) Die Renaissance leichter Betone in der Architektur. *DZement + Beton* **47**(5), 522–534.

Thomas, T.L. (1979) The effects of air content, water-cement ratio, and aggregate type on the flexural fatigue strength of plain concrete. Master's thesis, Department of Civil Engineering, Iowa State University. Available at https://core.ac.uk/download/pdf/38943176.pdf.

Topçu, I.B. and Uygunoğlu, T. (2010) Effect of aggregate type on properties of hardened self-consolidating lightweight concrete (SCLC). *Construction and Building Materials* **24**(7), 1286–1295.

Treybig, H.J., Smith, P. and VonQuintus, H. (1977) *Overlay Design and Reflection Cracking Analysis for Rigid Pavements – Vol. 1. Development of New Design Criteria.* Report prepared for the Department of Transportation, Federal Highway Administration, Research and Development, Washington, DC. Austin, TX: Austin Research Engineers Incorporated.

Wang, H.L. and Song, Y.P. (2011) Fatigue capacity of plain concrete under fatigue loading with constant confined stress. *Materials and Structures/Materiaux et Constructions* **44**(1), 253–262.

Wang, J.Y., Zhang, M.H., Li, W., Chia, K.S. and Liew, R.J.Y. (2012) Stability of cenospheres in lightweight cement composites in terms of alkali–silica reaction. *Cement and Concrete Research* **42**(5), 721–727.

Wang, Y., Qian, X., Liew, J.Y.R. and Zhang, M.H. (2015) Impact of cement composite filled steel tubes: An experimental, numerical and theoretical treatise. *Thin-Walled Structures* **87**, 76–88.

Wee, T.H. (2005) Recent developments in high strength lightweight concrete with and without aggregates. In N. Banthia, T. Uomoto, A. Bentur and S.P. Shah (eds) *Proceedings of the Third International Conference on Construction Materials: Performance, Innovations and Structural Implications and Mindess Symposium, 22–24 August, Vancouver, British Columbia, Canada*. Vancouver, BC: University of British Columbia, p. 97.

Weigler, H. and Karl, S. (1980) Structural lightweight aggregate concrete with reduced density lightweight aggregate foamed concrete. *International Journal of Cement Composites and Lightweight Concrete* **2**(2), 101–104.

Wirsching, P.H. and Yao, J.T.P. (1982) Fatigue reliability: Introduction. *Journal of the Structural Division* **108**(1), 3–23.

Wu, Y., Wang, J.Y., Monteiro, P.J.M. and Zhang, M.H. (2015) Development of ultra-lightweight cement composites with low thermal conductivity and high specific strength for energy efficient buildings. *Construction and Building Materials* **87**, 100–112.

Yan, J.B., Wang, J.Y., Liew, J.Y.R., Qian, X. and Zhang, W. (2016) Reinforced ultra-lightweight cement composite flat slabs: Experiments and analysis. *Materials and Design* **95**, 148–158.

Zhang, B., Phillips, D. V. and Wu, K. (1996) Effects of loading frequency and stress reversal on fatigue life of plain concrete. *Magazine of Concrete Research* **48**(4), 361–375.

Zhang, M.H. and Gjørv, O.E. (1991a) Permeability of high-strength lightweight concrete. *ACI Materials Journal* **88**(5), 463–469.

Zhang, M.H. and Gjørv, O.E. (1991b) Mechanical properties of high-strength lightweight concrete. *ACI Materials Journal* **88**(3), 240–247.

Chapter 4

The use of fiber-reinforced lightweight concrete to improve the performance of a high-speed railway bridge approach

Pengpeng Ni, Kaiwen Liu and Qian Su

1. Introduction

By 2018, China had the longest high-speed railway network: more than 29,000 km, accounting for two-thirds of the world's high-speed railways. Bridges are heavily used to cross obstacles, such as rivers and valleys. In soft-soil regions, differential settlement at bridge approaches can occur over time, leading to 'unevenness' in the rail track (Zaman *et al.*, 1991). Thus, the levelness of the railway line is influenced, which is associated with increased dynamic responses in the railway and trains. Therefore, the safe operation of high-speed railways can be jeopardized.

To mitigate this unevenness at the bridge approach (Puppala *et al.*, 2009) techniques have been developed as presented in Figure 4.1. The reinforced concrete slab (approach slab) is often casted *in situ*, with one end of the slab supported on the abutment. The success of the design is directly determined by the thickness of the approach slab, which is related to the flexural resistance of the track. Ground treatment methods are commonly implemented to improve the stability of the soft-soil zone near the bridge approach, including geosynthetics reinforcement, gravel columns and deep columns or piles. Recently, researchers have proposed backfilling the soft-soil zone near the bridge approach, using lightweight materials such as expanded polystyrene (EPS) geofoams and lightweight concrete (LWC), to minimize the consolidation settlement due to overburden pressures.

The approach slab technique has, in the past, been used extensively due to its simplicity of construction. The field performance of bridge approach slab expansion joints from 18 sites in the Houston District has been carefully compiled in a dataset by Ha *et al.* (2002). Similar work was conducted on 63 pile-supported and 21 soil-supported approach slabs in southeastern Louisiana by Bakeer *et al.* (2005), in which they found significant differences in design and practice. White *et al.* (2007) indicated that the settlement of poorly graded sandy backfills could exceed 5–18% of the sand thickness, based on the data from 74 bridges in Iowa. The use of backfills was also found to be of significance where adverse performance of bridge approach slabs could be seen (Cai *et al.*, 2005; Khodair and Nassif, 2005; Roy and Thiagarajan, 2007).

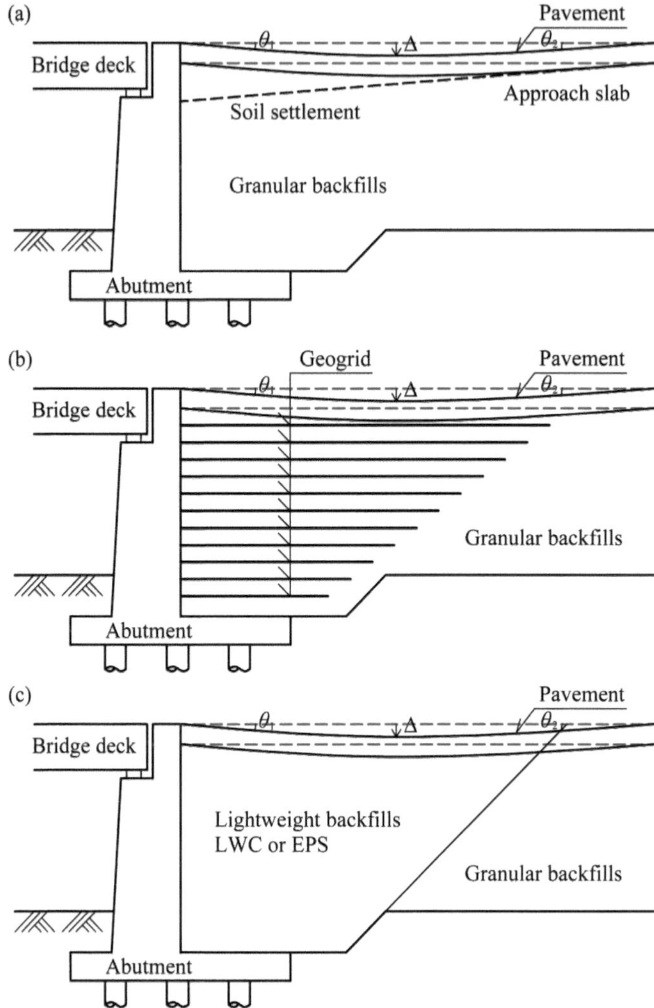

Figure 4.1 *Treatments for bridge approach: (a) approach slab; (b) soil reinforcement; and (c) lightweight backfills.*

However, ground improvement techniques have been developed to improve the resistance of soft soils, including the use of geogrids (Monley and Wu, 1993; Feng *et al.*, 2007; Liu *et al.*, 2007), geosynthetic-wrapped face embankments (Helwany *et al.*, 2003; Rowe and Liu, 2015; Zhang *et al.*, 2016), grouted gravel columns (Liu *et al.*, 2015), reinforced floating columns (Liu and Rowe, 2015) and deep cement mixing columns or piles (Liu and Rowe, 2016; Zhuang and Wang, 2017; Yi and Ni, 2018; Yi *et al.*, 2018; Li *et al.*, 2019; Ni *et al.*, 2020). It should be emphasized that high-speed railway bridge approaches are subjected to cyclic loads throughout their service life, leading to distress in the reinforced zone over time. Geosynthetics are often included in the soft-soil zone and the nonlinearity and stress relaxation behavior can be problematic in the long term.

It is noteworthy that the use of an approach slab or soil reinforcement is simply to increase the capacity of the system. The alternative method of lightweight backfills is to minimize the capacity demand in the system. Jamnongpipatkul *et al.* (2009) incorporated air foam stabilized soils to improve the short-term settlement response of a bridge approach in Thailand. The benefits of backfilling lightweight treated soils under bridge approaches were also reported by de Paiva and Trentin (2013), where the measured differential settlement was reduced. Similar observations were found by Satoh *et al.* (2001), Otani *et al.* (2002), Watabe and Noguchi (2011) and Kikuchi *et al.* (2011). It should be emphasized that the performance of the ground treatment could be influenced by harsh environments (Yi *et al.*, 2014). Lightweight concrete usually has better durability and can be used for high-speed railway networks along coastlines.

Researchers have frequently conducted investigations on the mechanical behavior of lightweight concrete (Holm and Ries, 2007; Nambiar and Ramamurthy, 2008; Just and Middendorf, 2009; Ramamurthy *et al.*, 2009; Zhang *et al.*, 2011; Ranjani and Ramamurthy, 2012; Lim *et al.*, 2015). The successful application of foamed concrete for the runway of an aircraft arresting system was reported by Zhang and Yang (2015), where the aircraft deceleration behavior due to the LWC was satisfactory. Fiber is commonly mixed with foam during the casting of lightweight concrete, as suggested by Gray and Ohashi (1983). The simplified beam-spring mass model is the normal method for evaluating the dynamic response of bridge approaches (Gupta and Traill-Nash, 1980; Zhai and Sun, 1994; Zhai and True, 2000; Zhang and Hu, 2007; Shi *et al.*, 2008; Zhai *et al.*, 2010; Liu *et al.*, 2011; Naeimi *et al.*, 2015). Alternatively, Yang *et al.* (2013) proposed analyzing the vehicle-track-subgrade system for bridge approaches using graded gravel mixed with 5% of cement validated using the three-dimensional (3D) finite element analysis technique. Both the static and dynamic responses of bridge approaches with fiber-reinforced LWC backfills have been demonstrated to be effective (Liu *et al.*, 2018; Liu *et al.*, 2020).

In this chapter, a full 3D numerical calculation is presented regarding the response of bridge approaches with fiber-reinforced lightweight concrete backfills. Uniaxial compressive strength tests, split tensile tests and flexural tests were carried out to measure the optimum fiber content. The static response of a bridge approach is studied in terms of the lateral displacement of the abutments, pavement rotation, vertical stress and settlement. The dynamic response of a bridge approach is analyzed considering the live loads from a CRH380A electric high-speed train.

2. Numerical simulation

2.1 Overview of the bridge approach

The case study for a double-track electric railway is presented, where a U-shaped abutment is connected with an embankment through a bridge approach. The cross-section of the bridge approach is shown schematically in Figure 4.2. The U-shaped abutment is 7 m in height and has a bearing seat of 3 m along the railway. The low-cap pile foundation is 9 m in length along the railway, 1.5 m in height with a pile group (i.e. 3 piles in the longitudinal direction by 4 piles in the transverse direction) and is designed to provide pile end resistance. The diameter of the pile is 1.2 m. There is different spacing between the piles: 3.5 m and 4.0 m in the longitudinal and transverse directions, respectively. The design

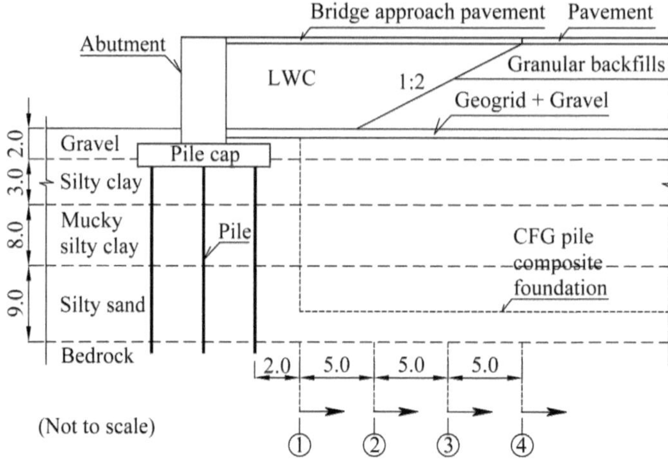

Figure 4.2 *Cross-section of bridge approach backfilled with high-performance fiber-reinforced lightweight concrete (LWC) (unit: m).*

contains no wing walls, given that LWC is poured in placed to form the embankment. The behavior of the embankment follows the principle of a cantilever retaining wall.

A slope angle of 1:2 is employed for the embankment in the transverse direction. At the bridge approach, LWC backfills are used to form an inverse trapezoidal-shaped zone with a full height of 6 m as illustrated in Figure 4.2. The transition zone between the LWC and granular backfills is 20 m long, and the inverse trapezoidal-shaped zone inclines at an angle of 1:2. In the LWC treatment zone, above the subbase layer, graded gravel mixed with 5% of cement is poured to form a base layer with a thickness of 0.4 m. Outside the transition zone, the subgrade also contains a base layer of 0.4 m. Below the subgrade, granular fill type A is used to form a layer 2.3 m thick, and granular fill type B is used below the type A fill to form a layer 3.3 m thick. The foundation are soils depicted in Figure 4.2, shown as gravel, silty clay, mucky silty clay, and silty sand. A cement fly-ash gravel (CFG) pile composite foundation is used below the embankment, to improve the stability of the soft-soil zone. The diameter and length of the CFG pile is 0.5 m and 16 m, respectively. The spacing between the grid pattern CFG piles is 2 m. Geogrid and gravel are mixed between the embankment and the soft-soil zone and the mixture layer has a thickness of 0.6 m.

For the CRH380A electric high-speed train, CRTSIII type non-ballasted track is commonly used. The dimensions of the slab are 5.35 m long, 2.5 m wide, and 0.19 m thick. The spacing between crossties is 0.63 m, where a gasket with dimensions of 0.15 m by 0.194 m (longitudinal by transverse) is also used.

2.2 Element discretization and boundary conditions

The static and dynamic responses of bridge approaches are investigated numerically. The general-purpose finite element code ABAQUS is used in the standard/implicit mode. A half model is only needed, given the symmetry of the problem. Eight-node brick elements (C3D8) are used to discretize all components, such as the track, the

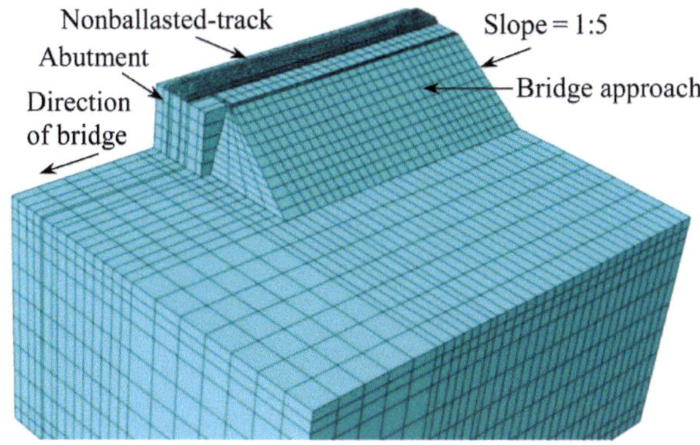

Figure 4.3 *Mesh discretization of bridge approach.*

soil and the pile, etc. Near the embankment, finer meshes are used. The mesh becomes coarser with the distance from the embankment, as shown in Figure 4.3.

The simulation is heavily influenced by the model size, especially for dynamic problems. In general, the model size should be greater than 10 times the structure size, to minimize the boundary effect for soil-structure interaction problems (Liu and Rowe, 2015; Rowe and Liu, 2015; Liu and Rowe, 2016; Ni *et al.*, 2018). The infinite element approach is an alternative to simulate the far field soil boundary (Ni *et al.*, 2017). At boundaries, stress waves can reflect, even when a large model is used. The waveform due to the interference of reflected waves can change, which influences the accuracy of the analysis. The technique of an artificial boundary is normally used at lateral boundaries to simulate the dynamic behavior of the system as realistically as possible. Liu *et al.* (2006) introduced a viscous-spring artificial boundary as follows (Equation (1)):

$$\begin{cases} K_b = a\dfrac{G}{R} \\ c_b = \rho \cdot c \end{cases} \tag{1}$$

where K_b and c_b are the spring stiffness and the damping coefficients, respectively; G represents the shear modulus; ρ is the density; R characterizes the distance between the physical origin of the shear waves and the artificial boundary; c is the propagation velocity (for a P-wave, $c = c_p$ and for an S-wave, $c = c_s$); and a takes the value 4.0 and 2.0 in the normal and tangential directions, respectively.

At boundaries, physical constrains are applied; the normal direction is constrained. The base of the model is fully fixed, and the ground surface is unconstrained. The symmetric boundary is used at the plane of symmetry along the midline of the track. The interface between different components is defined through the contact technique (Nassif *et al.*, 2002).

2.3 Materials

A number of fiber-reinforced LWC specimens were cast to determine the uniaxial compressive strength (UCS), the split tensile strength and the flexural strength. The measured results against the values of LWC specimens with no fiber are shown in Figure 4.4. Comparing with the Portland cement concrete unit weight of 25 kN/m³, the density of LWC by mixing foam, aggregates, and cement paste can be as low as 5–15 kN/m³. The fiber length is 6 mm, the diameter is 14 μm, the tensile strength is 0.3 N/tex, and the elastic modulus is 63.5 GPa, which will increase the durability

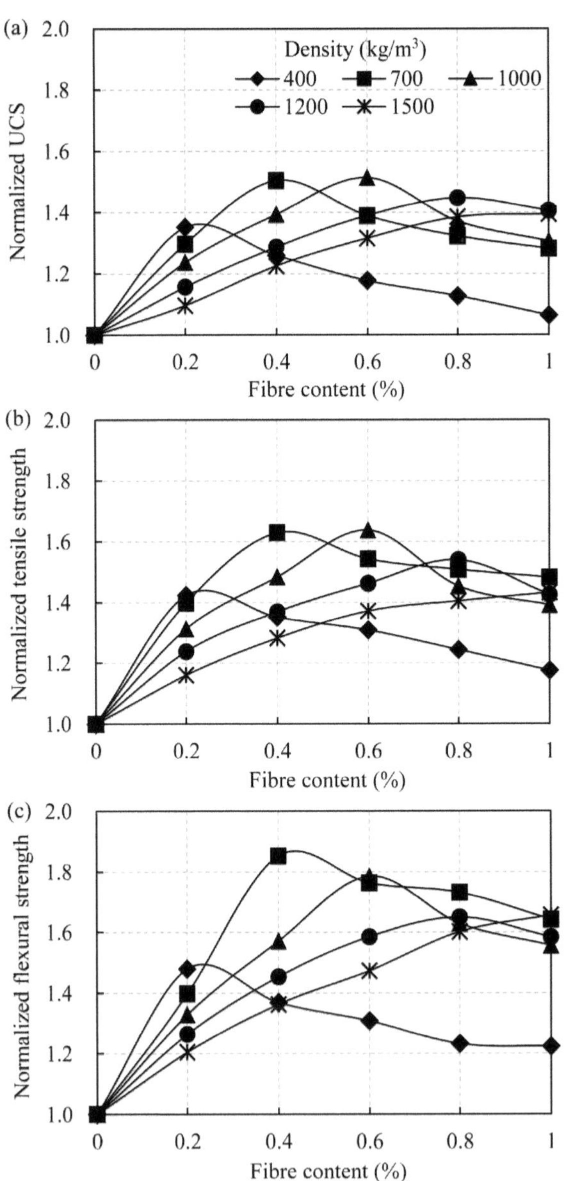

Figure 4.4 *Variation of normalized strength parameters with fiber content: (a) normalized UCS; (b) normalized tensile strength; and (c) normalized flexural strength.*

Table 4.1 *Concrete parameters in the numerical model.*

Type	Model	ρ (kg/m³)	E (MPa)	v (–)
Concrete (abutment)	Linear	2600	30000	0.2
Concrete (track slab)	Linear	2600	35000	0.167
CA mortar	Linear	1800	240	0.4
Concrete (support layer)	Linear	2500	24000	0.2
LWC-D7	Linear	700	360	0.1
LWC-D8	Linear	800	568	0.1
LWC-D9	Linear	900	696	0.1
LWC-D10	Linear	1000	810	0.1

of the LWC. Five concrete densities and six fiber contents are used to form the testing matrix. In the following discussion, normalization is conducted, where the measured strength of fiber-reinforced LWC is divided by the strength of the unreinforced LWC. It is found that the optimal fiber content varies from 0.4% to 0.6%, giving a concrete density of 700–1000 kg/m³. The fiber-reinforced LWC is considered a linear elastic material with properties as listed in Table 4.1.

The standard Mohr–Coulomb model is employed to simulate the behavior of granular materials and the Modified Cam Clay model is adopted to characterize the response of soft soils. Results of similar studies (Liu *et al.*, 2007; Liu and Rowe, 2015; Rowe and Liu, 2015; Liu and Rowe, 2016) are followed to obtain all input parameters for soil models, as tabulated in Table 4.2.

Table 4.2 *Soil parameters in the numerical model (Liu et al., 2007).*

Material	Model	c (kPa)	φ (°)	ψ (°)	E (MPa)	v (–)	λ	κ	M	e_1	$k_w \times 10^{-4}$ (m/day)
Graded gravel + 5% cement	MC	75	35	10	580	0.24					
Graded gravel + 5% cement	MC	75	30	10	350	0.24					
Graded gravel	MC	75	27	8	150	0.30					
Fills A & B	MC	55	20	5	70	0.33					
Granular backfills	MC	38	20	1	50	0.35					
Gravel	MC	15	20	0	15	0.33					
Silty clay	MCC					0.35	0.06	0.012	1.2	0.87	8.64
Mucky silty clay	MCC					0.35	0.15	0.03	0.95	1.79	4.32
Silty sand	MCC					0.35	0.03	0.005	1.28	0.97	4.32

Note: c is the cohesion, ϕ is the internal friction angle, ψ is the dilation angle, E is the elastic modulus, v is the Poisson's ratio, λ is the slope of the normal compression (virgin consolidation) line, κ is the slope of the swelling line, M is the slope of the critical state line (CSL) passing through the origin, e_1 is the void ratio, and k_w is the hydraulic conductivity.

For dynamic analysis, the damping term is defined using the Rayleigh damping model, in which the damping matrix C is correlated with the mass matrix M and the stiffness matrix K (Equation (2)):

$$C = \alpha M + \beta K \tag{2}$$

where α and β are functions of the frequency and the damping ratio.

2.4 Loads and modeling program

For static analysis, four cases of soft-soil treatment are investigated. The zone of the CFG pile composite foundation is shown in Figure 4.2. The boundary in case 1 is at a distance of 2 m from the abutment; the value is 7 m, 12 m, and 17 m for cases 2, 3, and 4, respectively. The self-weight of the CRTSIII type non-ballasted track is 13.7 kN/m², and the dead load of the CRH380A electric high-speed train is 40.4 kN/m², leading to the total applied load of 54.1 kN/m².

For dynamic analysis, the live load from the CRH380A electric high-speed train must be taken into account. Figure 4.5 shows a schematic illustration of a CRH380A electric high-speed train with 8 coaches (not all coaches shown) running at 350 km/h. The dead load of each coach train is 100 kN, and the time history of the live load acting on the rail fastening system is shown in Figure 4.6.

Different configurations of the transition zone are summarized in Table 4.3. The CRH380A high-speed train is designed to operate at a cruise speed of 350 km/h. The designed cruise speed of the train is 350 km/h, and four additional speeds of 300, 250,

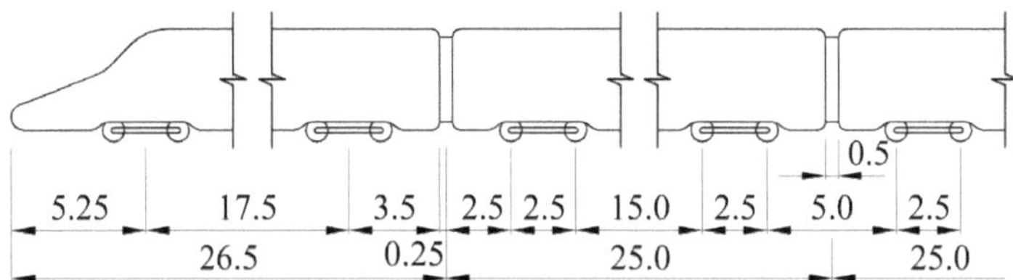

Figure 4.5 *Distribution of wheels and axles for the CRH380A electric high-speed train.*

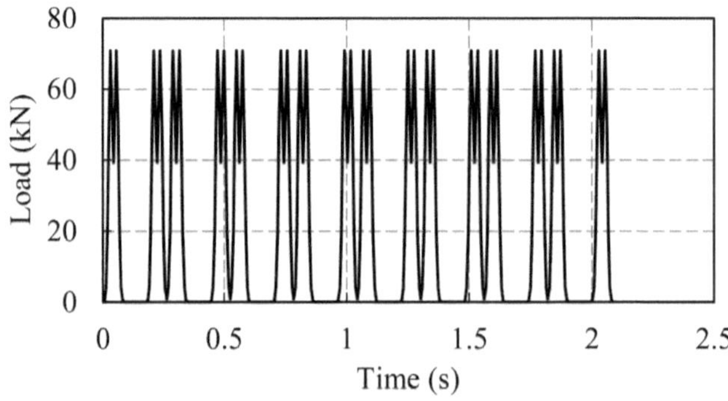

Figure 4.6 *Time history of load acting on the rail fastening system.*

Table 4.3 *Configurations of the transition zone.*

Materials for bridge approach	Inverse trapezoid		Trapezoid	
	No.	*Schematics*	*No.*	*Schematics*
LWC backfills under the base layer	C1	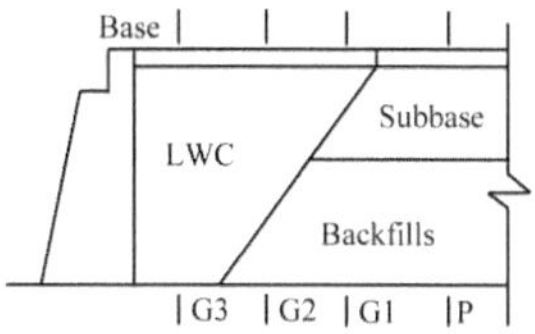 Base / Subbase / LWC / Backfills / G3 G2 G1 P	C2	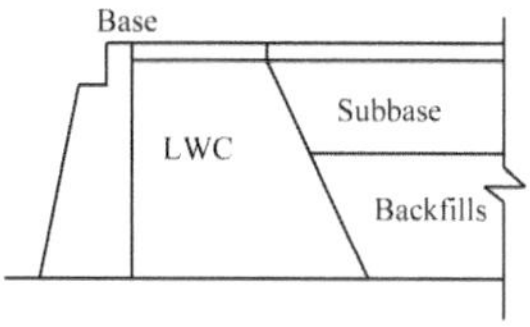 Base / Subbase / LWC / Backfills
Graded gravel + 3% cement for subbase and LWC for backfills	C3	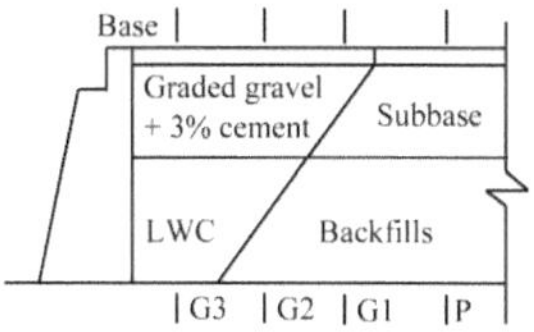 Base / Graded gravel + 3% cement / Subbase / LWC / Backfills / G3 G2 G1 P	C4	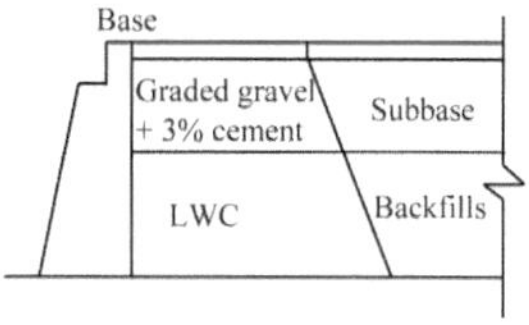 Base / Graded gravel + 3% cement / Subbase / LWC / Backfills
Graded gravel + 3% cement under the base layer	C5	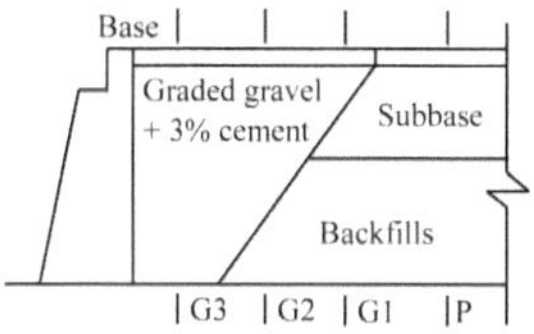 Base / Graded gravel + 3% cement / Subbase / Backfills / G3 G2 G1 P	C6	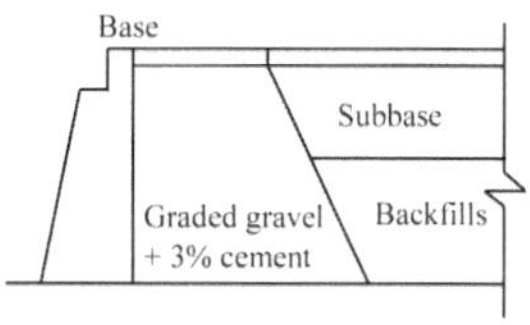 Base / Subbase / Graded gravel + 3% cement / Backfills

200, and 160 km/h are also considered. In total, the testing matrix contains a total of 90 analyses ($\{4 \times 4 \times 5\} + \{2 \times 5\}$). For further analysis, four sections along the longitudinal direction of embankment are defined as given in Table 4.3.

3. Results of the static analysis

3.1 Influence of soft-soil treatment

The size of the soft-soil treatment zone has a direct impact on the performance of the bridge approach. For high-speed rails, the lateral displacement of the abutment and the rotation of the pavement must be strictly controlled. In the design guidelines (TB 10002.1, 2005; TB 10020, 2009), the lateral displacement Δ of the abutment is defined as $\Delta \le 5\sqrt{L}$, where L is the bridge span (L should be taken as 24 m when $L < 24$ m). The rotation of the pavement in the proximity of the abutment should be a maximum of 1.5%.

In Figure 4.7, the influence, in the transition zone, of the backfill materials on the mobilized lateral displacement at the top of abutment can be seen for three construction cases (case 1 of LWC backfills under the base layer, case 2 of Graded gravel + 3% cement for subbase and LWC for backfills, and case 3 of Graded gravel + 3% cement under the base layer, as shown in Table 5.3). In general, the lateral displacement at the top of abutment increases as a linear function of the embankment height. In case 1, where the soft-soil treatment zone is the largest, the lateral displacement is the smallest, as expected. With the decrease in the size of the soft-soil treatment zone, lateral displacement increases, as shown in cases 1 to case 3. For the graded gravel backfilled

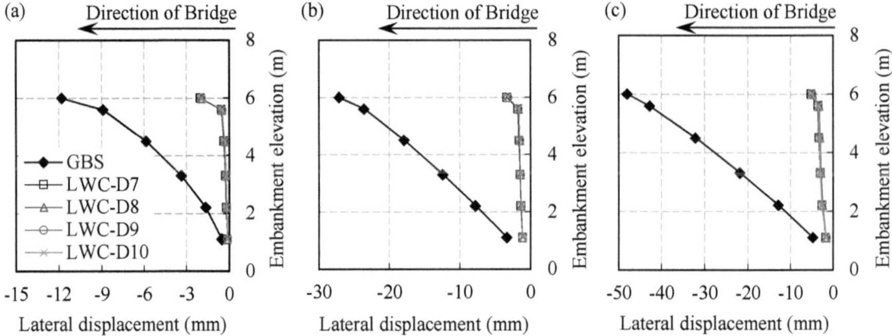

Figure 4.7 *Influence of backfill materials on the mobilized lateral displacement on top of the abutment during the backfilling process: (a) case 1; (b) case 2; and (c) case 3.*

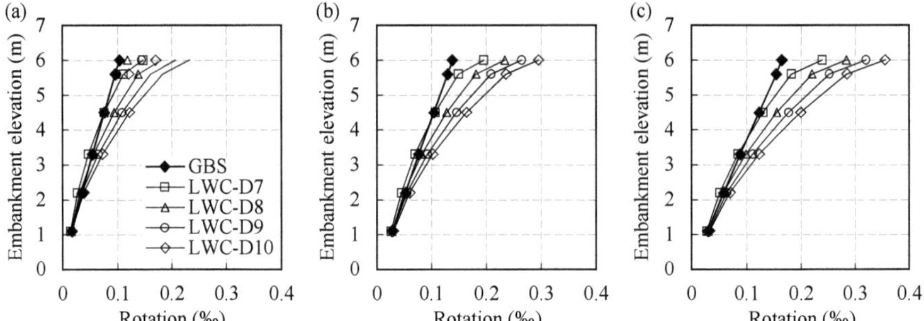

Figure 4.8 *Influence of backfill materials on the mobilized rotation of the pavement on top of the abutment during the backfilling process: (a) case 1; (b) case 2; and (c) case 3.*

system (GBS), the lateral displacement of the abutment is excessive, and can exceed the design specification of $\Delta = 28.3$ mm ($L = 32$ m) in some cases. When the transition zone of the bridge approach is backfilled with LWC, the magnitude of the lateral displacement reduces significantly.

The results of the rotation of the pavement on the top of the abutment during the backfilling process are plotted in Figure 4.8. When the bridge approach is backfilled with LWC, the rotation of the abutment is higher than with GBS. The difference between LWC and GBS backfills becomes more apparent with the increase of the embankment height. It is clear that rotation increases with the density of the backfill material (from LWC-D7 to LWC-D10). At the end of construction, the magnitude of rotation calculated for LWC-D10 is approximately twice that for GBS backfill. However, the maximum rotation calculated for case 3 is only 0.356%, which is still much less than the suggested value of 1.5%.

3.2 Vertical stress in the soil

With the increase in embankment elevation, the additional vertical stress between the piles under the pile cap increases. For soft soils, the increase in additional stress can result in consolidation settlement. The development of negative skin friction at the

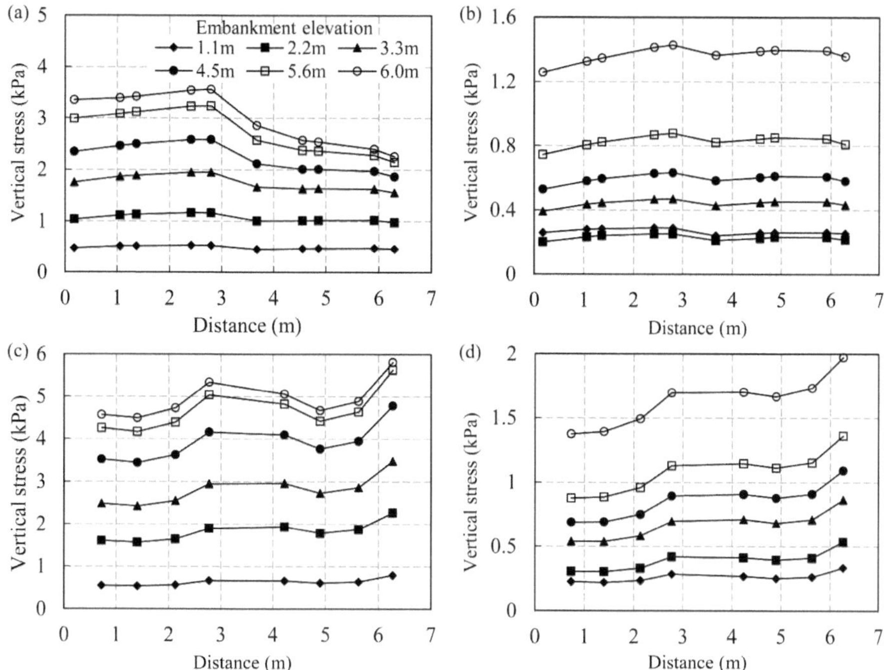

Figure 4.9 *Distribution of vertical stress in the foundation soils between piles: (a) at a depth of 7.5 m under the abutment with GBS backfills; (b) at a depth of 7.5 m under the abutment with LWC-D7 backfills; (c) at a depth of 10 m under the abutment with GBS backfills; and (d) at a depth of 10 m under the abutment with LWC-D7 backfills.*

soil–pile interface is detrimental to the stability of the abutment. In this case, all piles are designed as end-bearing piles, and the deployment of negative skin friction could damage the piles.

For case 1, with the maximum area of soft-soil treatment, the distribution of vertical stress between the piles at two depths as a function of the embankment height is plotted in Figure 4.9. At a depth of 7.5 m below the pile cap, the vertical stress is high on the left side, and decreases with the distance from left to the right. At a depth of 10 m, a lower vertical stress is found on the left, and a higher value is found on the right. As expected, the magnitude of the vertical stress increases with depth. The difference in the pattern of vertical stress could be caused by the depth of the soft-soil treatment zone. It should be noted that LWC backfills produce lower levels of vertical stress in the soil than do GBS backfills. The beneficial effect of LWC backfills is attributed to its lower density. The density is 2000 kg/m³ and 700 kg/m³ for GBS and LWC, receptively. The 3-fold difference in density between the two types of backfill explains the difference in the vertical stress of 2–3 times.

3.3 Ground settlement

With the increase in embankment height, the soil in the bridge approach settles. The numerical model can calculate the profile of ground surface settlement, based on which

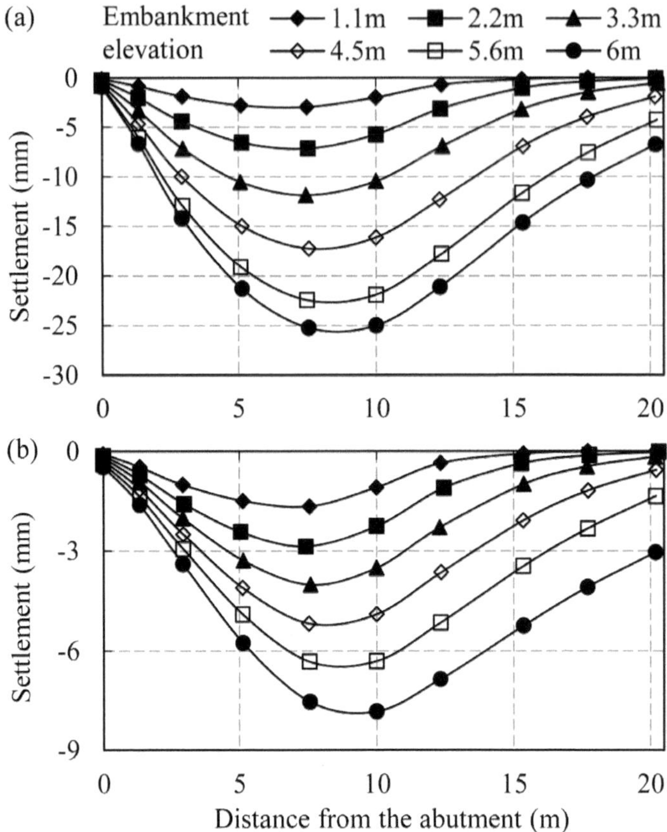

Figure 4.10 *Profile of ground settlement: (a) case 1 with GBS backfills; and (b) case 1 with LWC-D7 backfills.*

the embankment elevation can be increased to counteract the settlement. Figure 4.10 shows the calculated ground surface settlement as a function of embankment elevation for case 1 with GBS and LWC-D7 backfills in the transition zone. One can see that the profile of ground surface settlement has a concave upward shape. The patterns of settlement calculated for the two backfills are similar. The use of LWC can effectively reduce the maximum settlement.

For the three cases of the soft-soil treatment, correlations between maximum settlement and embankment elevation are plotted in Figure 4.11. It may be observed that when the size of the soft-soil treatment zone reduces, the maximum settlement increases, especially for GBS backfills. For LWC backfills, the magnitude of the maximum settlement is not significantly influenced by the zone of the soft-soil treatment. Overall, the maximum settlement with LWC backfills is approximately 3–4 times less than the value with the GBS backfills. It should be emphasized that the difference in density for LWC does not alter the results too much.

After the completion of construction, the dead load of the CRH380A electric high-speed train is applied to the embankment. The profile of ground settlement induced

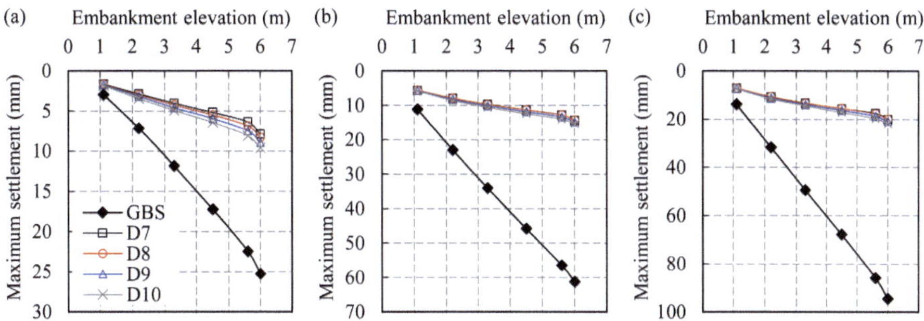

Figure 4.11 *Variation of maximum settlement with embankment elevation: (a) case 1; (b) case 2; and (c) case 3.*

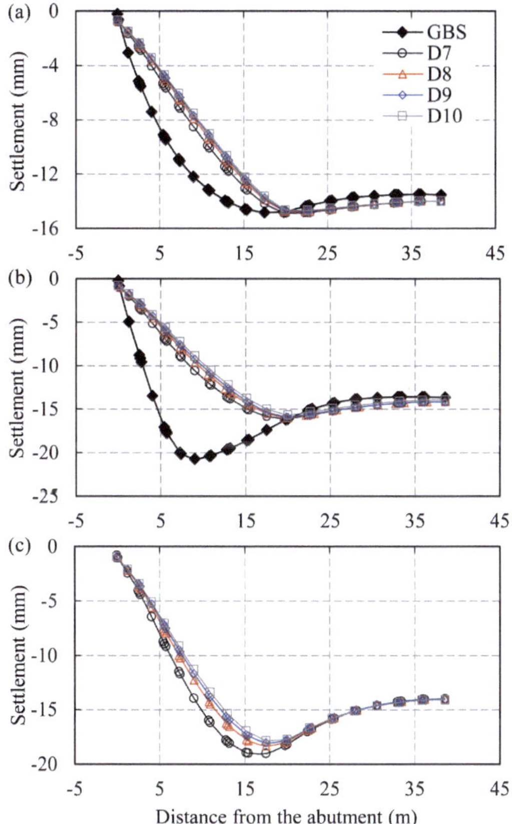

Figure 4.12 *Profile of ground settlement induced by dead loads from the train: (a) case 1; (b) case 2; and (c) case 3 (excessive settlement occurs with GBS backfills).*

by dead loads from the train is illustrated in Figure 4.12. For all three cases, the settlement of the embankment outside the transition zone is not influenced by the backfill material of the bridge approach. A maximum settlement of 13.96 mm is obtained, which is less than the design specification of 15 mm (TB 10020-2009, 2009).

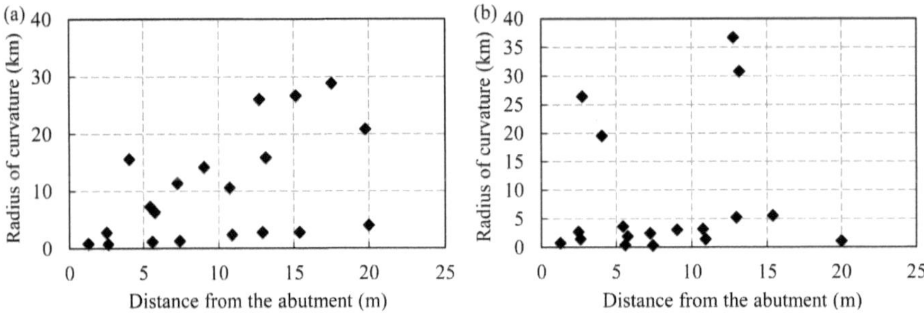

Figure 4.13 *Radius of curvature of bridge approach pavement with GBS backfills: (a) case 1; and (b) case 2.*

The primary source of settlement is induced by the deformation of the soft-soil zone rather than the embankment itself. In Figure 4.12c, the profile of ground settlement with GBS backfills is not included. This is because as the size of the soft-soil treatment zone decreases, the ground settlement increases, eventually resulting in convergence problems. The performance of the LWC backfilled bridge approach is better, as the settlement curve is smoother.

The radius of curvature can be calculated by double differentiation of the settlement curve, and the results are presented in Figure 4.13. The design guidelines (TB 10020-2009, 2009) suggest that the minimum radius of curvature (R_{sh}) should be calculated using Equation (3):

$$R_{sh} \geq 0.4v_e^2 \tag{3}$$

The rail speed v_e of 350 km/h gives a R_{sh} value of 49 km. From Figure 4.13, it can be seen that the calculated radius of curvature for GBS backfills is always smaller than 40 km. Hence, using GBS as backfills in the transition zone of the bridge approach cannot produce satisfactory results. For all cases with LWC backfills, the calculated radius of curvature fulfills the requirement of the design guidelines.

4. Results of dynamic analysis

4.1 Acceleration

Taking two cross-sections as an example, the profile of acceleration with depth is presented in Figure 4.14. Greater acceleration in the soil is associated with higher rail speeds, and the magnitude of the acceleration reduces with depth. In the LWC zone (section G3), the calculated maximum acceleration is much smaller than that obtained in the gravel backfilled zone (section P). When the rail speed is 350 km/h, the magnitude of maximum acceleration is 1.84 m/s² for gravel backfills, and the corresponding value is 0.72 m/s² for LWC backfills. The benefits of adding LWC to absorb vibration energies can be clearly seen.

The influence of backfills on the magnitude of acceleration is shown in Figure 4.15. It should be emphasized that for a specific height, the calculated acceleration with LWC backfills is always lower in value. When the LWC density increases, the

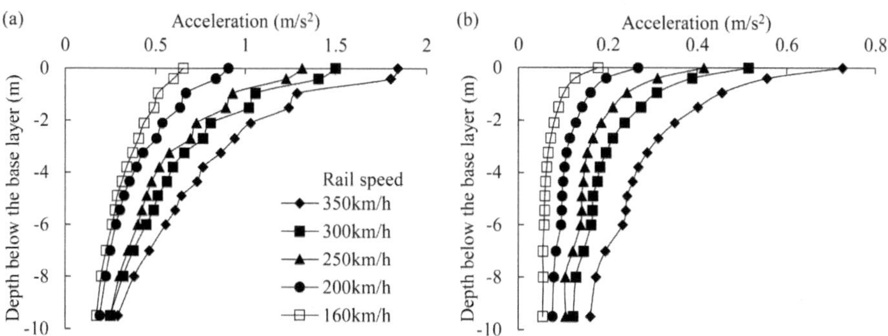

Figure 4.14 *Degradation of acceleration with depth as a function of rail speed: (a) section P; and (b) section G3.*

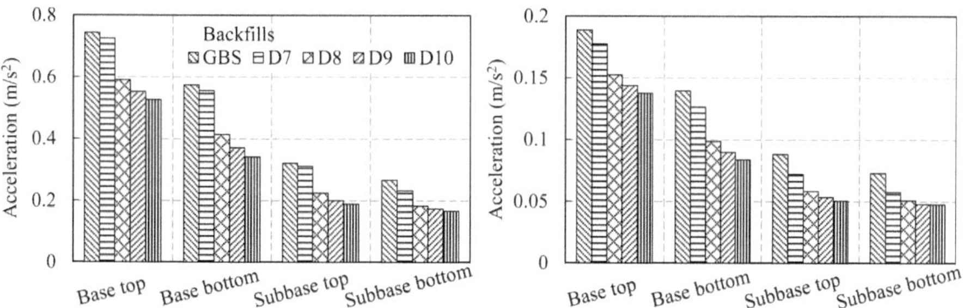

Figure 4.15 *Influence of backfill materials on the magnitude of acceleration at different depths: (a) rail speed of 350 km/h; and (b) rail speed of 160 km/h.*

acceleration reduces more dramatically. The elastic modulus of LWC is a function that increases with density. The density of LWC-D7 is closer to that for GBS, and the density of LWC-D10 becomes much higher. It is demonstrated that the elastic modulus of backfills can apparently change the measured response more than the density can.

Different structural types of bridge approach can change the dynamic response of the system. It is interesting that the configuration in the transition zone has minimal influence on the calculated response in sections P (pure gravel backfills) and G3 (pure LWC backfills). Comparisons of the acceleration profile in sections G1 and G2 are presented in Figure 4.16. Regardless of the configuration in the transition zone (either trapezoid or inverse trapezoid), the profile of acceleration with depth does not change appreciably when different backfills are used. Among all cases, the calculated acceleration for C5 is the minimum. The case with an inverse trapezoidal-shaped transition zone often results in a smaller acceleration compared to the case with a trapezoidal-shaped transition zone and should be adopted in practice.

It is commonly believed that, at the bridge approach, the flexural rigidity is altered quite suddenly, causing changes in acceleration. The vibration affects the operation of the train, causing bumps in the vertical plane, and influences the ride quality of the passengers. The variations of acceleration with distance from the abutment are

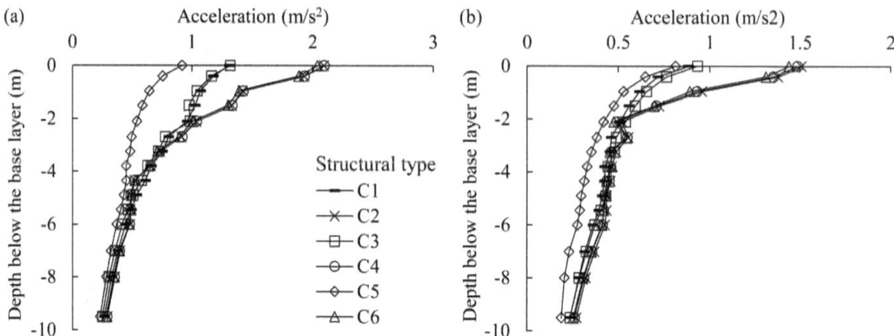

Figure 4.16 *Influence of the structural type of bridge approach on the acceleration versus depth curve: (a) section G1, rail speed of 350 km/h, LWC-D7 backfills; and (b) section G2, rail speed of 350 km/h, LWC-D7 backfills.*

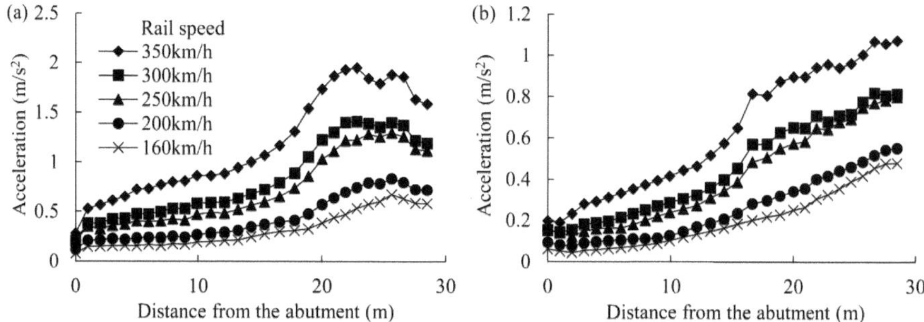

Figure 4.17 *Profile of acceleration versus distance from the abutment (configuration C1 and LWC-D7 backfills): (a) base top; and (b) subbase bottom.*

calculated at two elevations, as illustrated in Figure 4.17. It is seen that the magnitude of acceleration increases with rail speed. Similar patterns are observed for all curves, demonstrating that the propagation of vibration in the longitudinal direction is not significantly affected by the variations in rail speed. The acceleration in the transition zone with LWC backfills is generally lower and increases outside the transition zone with gravel backfills.

The profile of acceleration in the longitudinal direction is influenced by the backfill materials used in the transition zone, as shown in Figure 4.18. It should be emphasized that the type of backfill at the bridge approach affects the profile of the acceleration only in the transition zone, and the acceleration outside the transition zone is almost unaffected. In the scenario with LWC backfills, the acceleration is lower, and the beneficial effect of LWC is more significant when the modulus of elasticity is higher (LWC-D10).

The structural type of bridge approach can also change the acceleration at distances from the abutment, as depicted in Figure 4.19. Three inverse trapezoidal-shaped configurations, C1, C3 and C5 result in similar outcomes, whereas the cases for three trapezoidal-shaped configurations of C2, C4 and C6 are almost identical. Overall, the

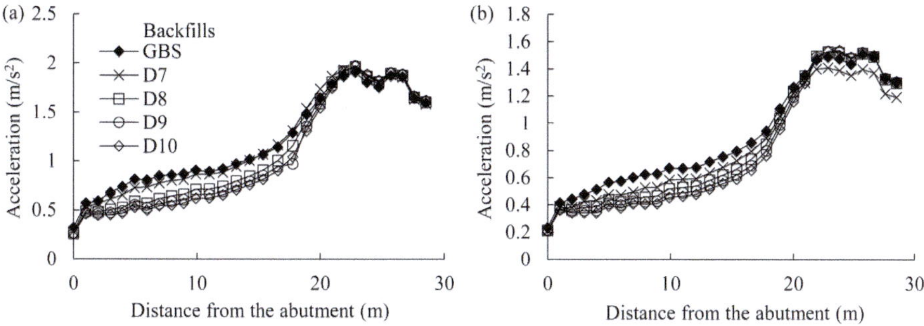

Figure 4.18 *Influence of backfill materials on the profile of acceleration versus distance from the abutment (configuration C1 and base top): (a) rail speed of 350 km/h; and (b) rail speed of 300 km/h.*

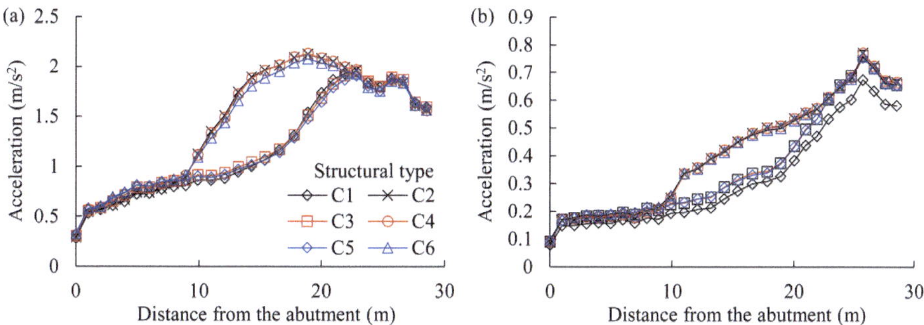

Figure 4.19 *Influence of the structural type of bridge approach on the profile of acceleration versus distance from the abutment (LWC-D7 backfills and base top): (a) rail speed of 350 km/h; and (b) rail speed of 160 km/h.*

inverse trapezoidal-shaped configuration can improve the performance of the bridge approach. When the transition zone changes at a distance of 10–20 m from the abutment, the calculated acceleration can be reduced. The inverse trapezoidal-shaped configuration in the transition zone is suggested for use in practice.

4.2 Displacement

Elastic deformation in the embankment can be due to the vibration of the train. If the embankment experiences excessive displacement, the structural integrity of the track is affected, jeopardizing the safe operation of the train. The permissible elastic deformation under live loads is suggested to be 3.5 mm following different design guidelines (TB 10020-2009, 2009). Figure 4.20 plots the displacement against the distance from the abutment. As expected, a higher rail speed can lead to an increased displacement. Backfilling the transition zone with LWC is beneficial in lowering the displacement and is more apparent when the LWC has a higher elastic modulus (LWC-D10). It is suggested that the density of LWC is greater than 800 kg/m^3, with a maximum displacement of less than 1.4 mm, and the design requirement is fulfilled.

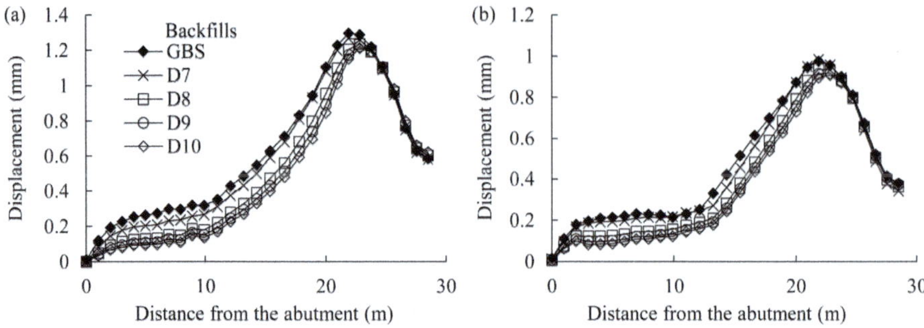

Figure 4.20 *Influence of backfill materials on the profile of displacement versus distance from the abutment (configuration C1 and base top): (a) rail speed of 350 km/h; and (b) rail speed of 160 km/h.*

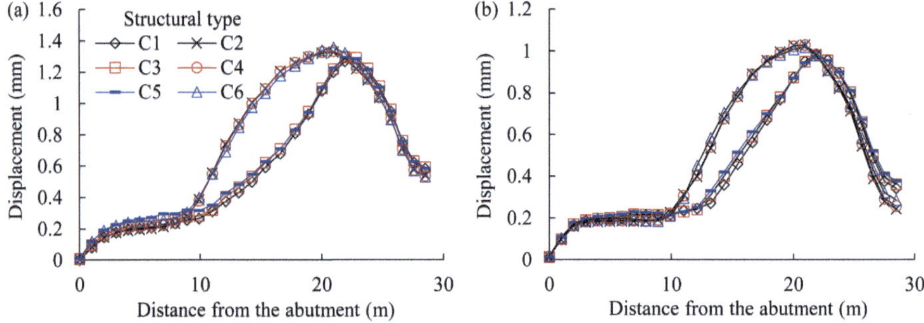

Figure 4.21 *Influence of the structural type of bridge approach on the profile of displacement versus distance from the abutment (LWC-D7 backfills and base top): (a) rail speed of 350 km/h; and (b) rail speed of 160 km/h.*

The displacement peaks at about 20–25 m from the abutment and is outside the transition zone of the bridge approach. The flexural stiffness alters from the edge of the transition zone, leading to differences in the elastic deformation.

The profiles for the displacement against distance from the abutment are calculated for different structural types of bridge approach and are shown in Figure 4.21. The inverse trapezoidal-shaped transition zones of C1, C3 and C5 are more beneficial, leading to smaller displacements than those calculated for the trapezoidal-shaped transition zones of C2, C4 and C6. However, the type of backfill does not result in huge differences in displacement. Again, use of the inverse trapezoidal-shaped transition zone is recommended to improve the dynamic response of the system.

4.3 Time-frequency analysis

The fast Fourier transform (FFT) is used to conduct time–frequency analysis, in which the dynamic response is sampled over time and is converted into frequency components. Hence, the dominant frequencies of the system can be identified. In Figure 4.22, the time histories of the acceleration for sections P (pure gravel backfills) and G3 (pure LWC backfills) are presented, based on which the FFT spectrum can be calculated.

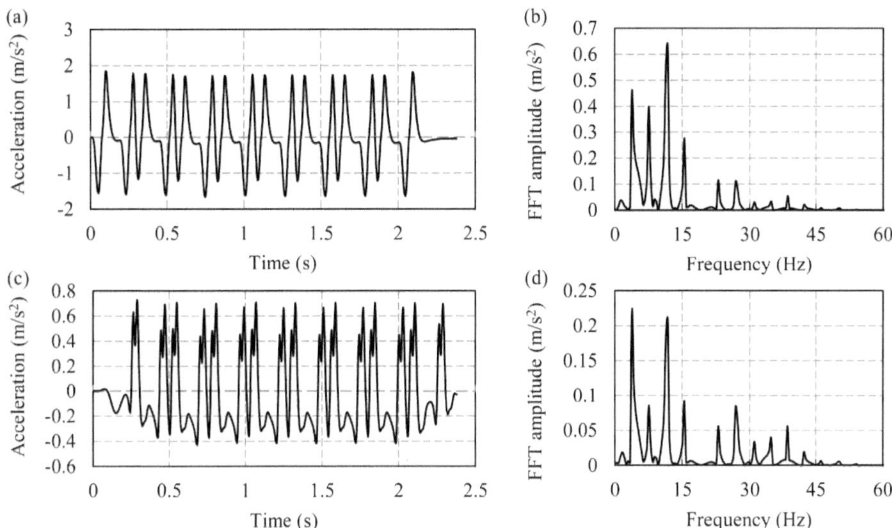

Figure 4.22 *Results of time–frequency analysis (configuration C1, LWC-D7 backfills and base top): (a) time history of acceleration at section P with a rail speed of 350 km/h; (b) Fast Fourier Transform (FFT) amplitude at section P with a rail speed of 350 km/h; (c) time history of acceleration at section G3 with a rail speed of 350 km/h; and (d) Fast Fourier Transform (FFT) amplitude at section G3 with a rail speed of 350 km/h.*

The time histories of the acceleration in different zones are different due to the changes in flexural stiffness. One can see that the magnitude of acceleration is more than 50% lower when gravel is replaced by LWC. Low-frequency components are found to be dominant within the frequency range of 0–45 Hz, being coincident with the field test observations of Nie *et al.* (2005). Outside the transition zone, the frequency at peak FFT amplitude occurs within the range of 8–12 Hz, while the dominant frequency falls within 2.5–4 Hz in the transition zone.

For sections P and G3, the time histories of stress in the embankment under a high-speed train running at 350 km/h are plotted in Figure 4.23, together with the calculated FFT spectrum. It should be noticed that the stress response hardly changes for the gravel or the LWC. The dominant frequency occurs at 3.77 Hz. Considering the distribution of wheels and axles of the CRH380A electric high-speed train, the resonant frequency can be computed as 3.88 Hz. In fact, the excited frequency of 3.77 Hz obtained from FFT is very close to the calculated resonant frequency of 3.88 Hz.

5. Conclusions

In this chapter, the static and dynamic responses of the bridge approach backfilled with fiber-reinforced lightweight concrete (LWC) are studied using three-dimensional numerical analyses. The following conclusions can be drawn:

- The mechanical properties of LWC are improved by adding fibers at a fiber content of 0.4–0.6%, leading to the optimal performance of fiber-reinforced LWC with a density of 700–1000 kg/m³.

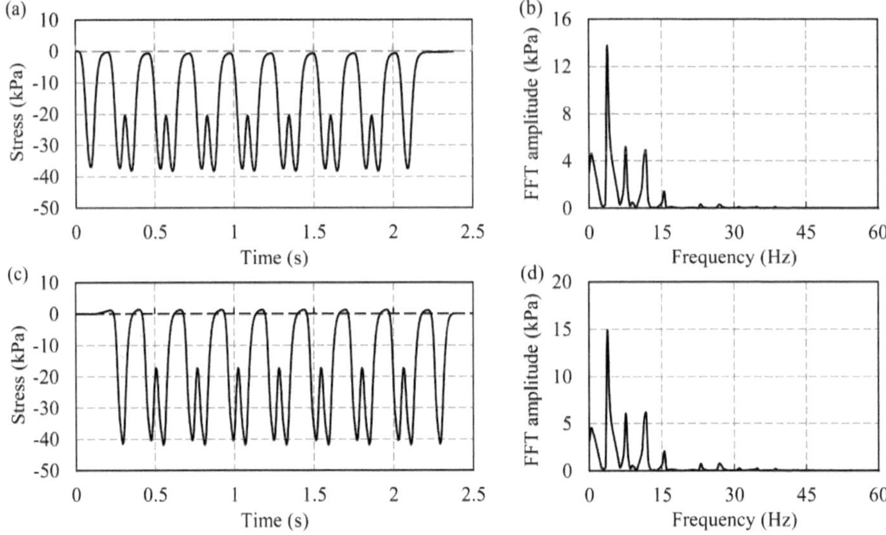

Figure 4.23 *Results of time–frequency analysis (configuration C1, LWC-D7 backfills and base top): (a) time history of stress at section P with a rail speed of 350 km/h; (b) Fast Fourier Transform (FFT) amplitude at section P with a rail speed of 350 km/h; (c) time history of stress at section G3 with a rail speed of 350 km/h; and (d) Fast Fourier Transform (FFT) amplitude at section G3 with a rail speed of 350 km/h.*

- The use of fiber-reinforced LWC in the transition zone can effectively reduce the lateral displacement at the top of the abutment and can result in a gentle increase in the rotation of the pavement, but this rotation will still be within the design specifications.
- LWC backfills can reduce the surface settlement greatly. Gravel backfills in the transition zone lead to a radius decrease in the ground curvature compared to LWC.
- The acceleration in the embankment is higher outside the transition zone. With an increase in rail speed, the magnitude of acceleration increases. For LWC with a higher density, the elastic modulus increases, such that the level of acceleration decreases. Similar observations can be found for the magnitude of displacement in the embankment.
- The configuration of LWC backfills in the transition zone can influence significantly the performance of the bridge approach. It is suggested that the inverse trapezoidal-shaped zone with LWC backfill is used to reduce the dynamic response of the bridge approach.
- For a bridge approach backfilled with LWC, the dominant frequency falls within the range 0–45 Hz based on time–frequency analysis.

Acknowledgement

This chapter incorporates material from Liu *et al.* (2018) and Liu *et al.* (2020), for which copyright was retained by the authors.

References

Bakeer, R.M., Shutt, M.A., Zhong, J., Das, S.C. and Morvant, M. (2005) Performance of pile-supported bridge approach slabs. *Journal of Bridge Engineering* **10**(2), 228–237.

Cai, C.S., Shi, X.M., Voyiadjis, G.Z. and Zhang, Z.J. (2005) Structural performance of bridge approach slabs under given embankment settlement. *Journal of Bridge Engineering* **10**(4), 482–489.

de Paiva, C.E.L. and Trentin, L.C. (2013) A finite element method for bridge approach. *International Journal of Engineering Research and Applications* **3**(1), 729–732.

Feng, G., Luo, R. and Chen, C. (2007) Mechanism of geogrid reinforced soil at bridge approach. *Journal of Wuhan University of Technology. Materials Science Edition* **22**(2), 376–379.

Gray, D.H. and Ohashi, H. (1983) Mechanics of fiber reinforcement in sand. *Journal of Geotechnical Engineering* **109**(3), 335–353.

Gupta, R. and Traill-Nash, R. (1980) Bridge dynamic loading due to road surface irregularities and braking of vehicle. *Earthquake Engineering & Structural Dynamics* **8**(2), 83–96.

Ha, H., Seo, J. and Briaud, J.-L. (2002) *Investigation of Settlement at Bridge Approach Slab Expansion Joint: Survey and Site Investigations.* College Station, TX and Washington, DC: Texas Department of Transportation, U.S. Department of Transportation and Federal Highway Administration. Report No. 4147-1.

Helwany, S.M., Wu, J.T. and Froessl, B. (2003) GRS bridge abutments—an effective means to alleviate bridge approach settlement. *Geotextiles and Geomembranes* **21**(3), 177–196.

Holm, T.A. and Ries, J.P. (2007) High-performance lightweight concrete. In Y.A. Holm and J.P. Ries (eds) *Reference Manual for the Properties and Applications of Expanded Shale, Clay and Slate Lightweight Aggregate.* Chicago, IL: Expanded Shale, Clay, and Slate Institute (ESCSI), ch.9.

Jamnongpipatkul, P., Dechasakulsom, M. and Sukolrat, J. (2009) Application of air foam stabilized soil for bridge-embankment transition zone in Thailand. In L.F. Walubita, L. du Plessis, S.-C. Huang, G.S. Simate and Z. Liu (eds) *Asphalt Material Characterization, Accelerated Testing, and Highway Management: Selected Papers from the 2009 GeoHunan International Conference.* Reston, VA: American Society of Civil Engineers (ASCE), 181–193.

Just, A. and Middendorf, B. (2009) Microstructure of high-strength foam concrete. *Materials Characterization* **60**(7), 741–748.

Khodair, Y. and Nassif, H. (2005) Finite element analysis of bridge approach slabs considering soil–structure interaction. *Bridge Structures: Assessment, Design and Construction* **1**(3), 245–256.

Kikuchi, Y., Nagatome, T., Mizutani, T.-A. and Yoshino, H. (2011) The effect of air foam inclusion on the permeability and absorption properties of light weight soil. *Soils and Foundations* **51**(1), 151–165.

Li, W., Ni, P. and Yi, Y. (2019) Comparison of reactive magnesia, quick lime, and ordinary Portland cement for stabilization/solidification of heavy metal-contaminated soils. *Science of the Total Environment* **671**, 741–753.

Lim, S.K., Tan, C.S., Zhao, X. and Ling, T.C. (2015) Strength and toughness of lightweight foamed concrete with different sand grading. *KSCE Journal of Civil Engineering* **19**(7), 2191–2197.

Liu, H., Kong, G., Chu, J. and Ding, X. (2015) Grouted gravel column-supported highway embankment over soft clay: *Case study. Canadian Geotechnical Journal* **52**(11), 1725–1733.

Liu, H.L., Ng, C.W.W. and Fei, K. (2007) Performance of a geogrid-reinforced and pile-supported highway embankment over soft clay: Case study. *Journal of Geotechnical and Geoenvironmental Engineering* **133**(12), 1483–1493.

Liu, J.B., Gu, Y. and Du, Y.X. (2006) Consistent viscous-spring artificial boundaries and viscous-spring boundary elements. *Chinese Journal of Geotechnical Engineering* **28**(9), 1070–1075.

Liu, K., Su, Q., Ni, P., Zhou, C., Zhao, W. and Yue, F. (2018) Evaluation on the dynamic performance of bridge approach backfilled with fibre reinforced lightweight concrete under high-speed train loading. *Computers and Geotechnics* **104**, 42–53.

Liu, K.-W. and Rowe, R.K. (2015) Numerical modelling of prefabricated vertical drains and surcharge on reinforced floating column-supported embankment behaviour. *Geotextiles and Geomembranes* **43**(6), 493–505.

Liu, K.-W. and Rowe, R.K. (2016) Performance of reinforced, DMM column-supported embankment considering reinforcement viscosity and subsoil's decreasing hydraulic conductivity. *Computers and Geotechnics* **71**, 147–158.

Liu, K.-W., Yue, F., Su, Q., Zhou, C., Xiong, Z. and He, Y. (2020) Assessment of the use of fiberglass reinforced foam concrete in high-speed railway bridge approach involving foundation cost comparison. *Advances in Structural Engineering* **23**(2), 388–396.

Liu, W., Luna, R., Stephenson, R. and Wang, S. (2011) Earthquake-induced deformation analysis of a bridge approach embankment in Missouri. *Geotechnical and Geological Engineering* **29**(5), 845–854.

Monley, G.J. and Wu, J.T.H. (1993) Tensile reinforcement effects on bridge-approach settlement. *Journal of Geotechnical Engineering* **119**(4), 749–762.

Naeimi, M., Zakeri, J.A., Esmaeili, M. and Mehrali, M. (2015) Dynamic response of sleepers in a track with uneven rail irregularities using a 3D vehicle–track model with sleeper beams. *Archive of Applied Mechanics* **85**(11), 1679–1699.

Nambiar, E.K.K. and Ramamurthy, K. (2008) Models for strength prediction of foam concrete. *Materials and Structures* **41**(2), 247.

Nassif, H., Abu-Amra, T. and Shah, N. (2002) *Finite Element Modeling of Bridge Approach and Transition Slabs.* Trenton, NJ and Washington, DC: New Jersey Department of Transportation Division of Research and Technology and U.S. Department of Transportation Federal Highway Administration. Final Report No. FHWA NJ 2002-007.

Ni, P., Mangalathu, S., Mei, G. and Zhao, Y. (2017) Permeable piles: An alternative to improve the performance of driven piles. *Computers and Geotechnics* **84**, 78–87.

Ni, P., Moore, I.D. and Take, W.A. (2018) Numerical modeling of normal fault-pipeline interaction and comparison with centrifuge tests. *Soil Dynamics and Earthquake Engineering* **105**, 127–138.

Ni, P., Yi, Y. and Liu, S. (2020) Bearing capacity of composite ground with soil-cement columns under earth fills: Physical and numerical modeling. *Soils and Foundations* **59**(6), 2206–2219.

Nie, Z.H., Li, L., Liu, B.C. and Bo, R. (2005) Testing and analysis on vibration of subgrade for Qinhuangdao-Shenyang Railway. *Chinese Journal of Geotechnical Engineering* **28**(9), 1067–1071.

Otani, J., Mukunoki, T. and Kikuchi, Y. (2002) Visualization for engineering property of in-situ light weight soils with air foams. *Soils and Foundations* **42**(3), 93–105.

Puppala, A.J., Saride, S., Archeewa, E., Hoyos, L.R. and Nazarian, S. (2009) *Recommendations for Design, Construction, and Maintenance of Bridge Approach Slabs: Synthesis Report.* College Station, TX and Washington, DC: Texas Department of Transportation and U.S. Department of Transportation and Federal Highway Administration Report No. 0-6022-1.

Ramamurthy, K., Nambiar, E.K. and Ranjani, G.I.S. (2009) A classification of studies on properties of foam concrete. *Cement and Concrete Composites* **31**(6), 388–396.

Ranjani, G.I.S. and Ramamurthy, K. (2012) Behaviour of foam concrete under sulphate environments. *Cement and Concrete Composites* **34**(7), 825–834.

Rowe, R.K. and Liu, K.-W. (2015) Three-dimensional finite element modelling of a full-scale geosynthetic-reinforced, pile-supported embankment. *Canadian Geotechnical Journal* **52**(12), 2041–2054.

Roy, S. and Thiagarajan, G. (2007) Nonlinear finite-element analysis of reinforced concrete bridge approach slab. *Journal of Bridge Engineering* **12**(6), 801–806.

Satoh, T., Tsuchida, T., Mitsukuri, K. and Hong, Z. (2001) Field placing test of lightweight treated soil under seawater in Kumamoto Port. *Soils and Foundations* **41**(5), 145–154.

Shi, X., Cai, C.S. and Chen, S. (2008) Vehicle induced dynamic behavior of short-span slab bridges considering effect of approach slab condition. *Journal of Bridge Engineering* **13**(1), 83–92.

TB 10002.1-2005. (2005) *Fundamental Code for Design on Railway Bridge and Culvert.* Beijing, China: China Railway Culture Media Co.

TB 10020-2009. (2009) *Code for Design of High Speed Railway.* Beijing: People's Republic of China Ministry of Railways (MOR).

Watabe, Y. and Noguchi, T. (2011) Site-investigation and geotechnical design of D-runway construction in Tokyo Haneda Airport. *Soils and Foundations* **51**(6), 1003–1018.

White, D.J., Mekkawy, M.M., Sritharan, S. and Suleiman, M.T. (2007) 'Underlying' causes for settlement of bridge approach pavement systems. *Journal of Performance of Constructed Facilities* **21**(4), 273–282.

Yang, C.W., Sun, H.i., Zhang, J.J., Zhu, C.B. and Yan, L.P. (2013) Dynamic responses of bridge-approach embankment transition section of high-speed rail. *Journal of Central South University* **20**(10), 2830–2839.

Yi, Y. and Ni, P. (2018) Stabilization of marine soft clay with two industry by-products. In L. Li, B. Cetin and X. Yang (eds) *Proceedings of GeoShanghai 2018 International Conference:Ground Improvement and Geosynthetics.* Singapore: Springer, 121–128.

Yi, Y., Ni, P. and Liu, S. (2018) Numerical investigation of T-shaped soil-cement column supported embankment over soft ground. In W. Wu and H.-S. Yu (eds) *Proceedings of China–Europe Conference on Geotechnical Engineering. Volume 2.* Cham, Switzerland: Springer, 1068–1071.

Yi, Y., Li, C., Liu, S. and Al-Tabbaa, A. (2014) Resistance of MgO-GGBS and CS-GGBS stabilised marine soft clays to sodium sulfate attack. *Geotechnique* **64**(8), 673–679.

Zaman, M., Gopalasingam, A. and Laguros, J.G. (1991) Consolidation settlement of bridge approach foundation. *Journal of Geotechnical Engineering* **117**(2), 219–240.

Zhai, W. and Sun, X. (1994) A detailed model for investigating vertical interaction between railway vehicle and track. *Vehicle System Dynamics* **23**(S1), 603–615.

Zhai, W. and True, H. (2000) Vehicle-track dynamics on a ramp and on the bridge: Simulation and measurements. *Vehicle System Dynamics* **33**(S1): 604–615.

Zhai, W., He, Z. and Song, X. (2010) Prediction of high-speed train induced ground vibration based on train-track-ground system model. *Earthquake Engineering and Engineering Vibration* **9**(4), 545–554.

Zhang, H.L. and Hu, C.S. (2007) Determination of allowable differential settlement in bridge approach due to vehicle vibrations. *Journal of Bridge Engineering* **12**(2), 154–163.

Zhang, J., Zheng, J., Zhao, D. and Chen, S. (2016) Field study on performance of new technique of geosynthetic-reinforced and pile-supported embankment at bridge approach. *Science China Technological Sciences* **59**(1), 162–174.

Zhang, M.-H., Liu, X. and Chia, K.-S. (2011) High-strength high-performance lightweight concrete: A review. In M. Khrapko (ed.) *Proceedings of 9th International Symposium on High Performance Concrete*. New Zealand: New Zealand Concrete Society.

Zhang, Z.Q. and Yang, J.L. (2015) Improving safety of runway overrun through foamed concrete aircraft arresting system: An experimental study. *International Journal of Crashworthiness* **20**(5), 448–463.

Zhuang, Y. and Wang, K. (2017) Numerical simulation of high-speed railway foundation improved by PVD-DCM method and compared with field measurements. *European Journal of Environmental and Civil Engineering* **21**(11), 1363–1383.

Chapter 5
Self-compacting lightweight aggregate concrete

Payam Shafigh and Muhammad Aslam

1. Introduction

Normal-weight concrete (NWC) is considered the most widely used construction material in civil engineering. This is due to its numerous superior properties (Shafigh *et al.*, 2010). The concrete industry uses huge quantities of natural resources: annual consumption amounts to 1.5 billion tonnes of cement, 10–12 billion tonnes of aggregates and 1 billion tonnes of mixing water, respectively (Shafigh *et al.*, 2013). Given the enormous daily production of concrete, even a slight reduction in production would result in considerable benefits to the environment (Altwair and Kabir, 2010; Silva *et al.*, 2016). The use of sustainable materials in construction has received growing attention and the increasing demand for sustainable development in the construction industry has motivated researchers to study the use of waste or recycled materials as possible alternative construction materials (Mo *et al.*, 2015; Aslam, Shafigh, Jummaat and Lachemi., 2016; Aslam *et al.*, 2016b;).

Self-compacting concrete (SCC) is a special and relatively recent type of concrete that was developed in Japan. The concept was introduced by Okamura in 1986. It was then modified and a prototype was presented by Ozawa in 1988 (Ozawa, 1989; Okamura, 1997). SCC has a number of advantages over NWC. For example, it is capable of filling forms without the need for vibration, even for narrow structural members with congested or heavy rebar, which reduces the construction time and cost as well as noise pollution. And SCC has significantly higher deformability and excellent cohesive properties, which avoids bleeding or segregation. It also has a superior interfacial transition zone (ITZ), which improves its durability and reduces permeability and ultimately ensures excellent structural performance (Shi and Wu, 2005; Shi, 2005).

SCC was successfully produced by Kanadasan and Razak. (2014a, 2014b, 2015). In the case of SCC, the strength, rheological properties, shrinkage and durability are governed by several factors, including aggregates, the composition of materials, the addition of mineral and chemical admixtures, particle packing density, water-to-binder ratio (w/b ratio), and design methods (Siddique *et al.*, 2012; Esmaeilkhanian *et al.*, 2014; Han *et al.*, 2014; Wang *et al.*, 2014). And, as one might expect, the use of waste materials in the production of concrete leads to a sustainable form of the product that reduces the environmental impact (Bravo *et al.*, 2015; Shi, 2005; Aslam Shafigh, Jumaat and Lachemi, 2016; Aslam *et al.*, 2016b).

Lightweight concrete (LWC) has been used since ancient times. It outperforms NWC on several fronts: greater resistance to fire, heat and frost; enhanced sound absorption;

more effective seismic damping; and greater anti-condensation properties (Shafigh *et al.*, 2010; Aslam, Shafigh, Jumaat and Lachemi, 2016; Aslam *et al.*, 2016b). With the growing interest in lightweight concrete as a sustainable building material, more research is being conducted regarding the utilization of lightweight aggregates (LWA) in the production of self-compacting lightweight concrete. The use of LWA is the most popular method of producing lightweight concrete (Polat *et al.*, 2010). LWAs are generally classified as either natural or artificial. Naturally-arising materials – high pressure and temperature from earth processes – are expanded clay, shale, slate, perlite and vermiculite. Artificial aggregates are further classified into modified naturally-arising materials and industrial by-products. Examples of industrial by-products normally used as construction materials are sintered pulverized fuel ash, sintered slate, foamed blast-furnace slag and colliery waste (Shafigh *et al.*, 2010). Common naturally-occurring LWAs are diatomite, pumice, volcanic cinders, tuff and scoria (Neville and Brooks, 2008).

Several aggregates can be used successfully to produce lightweight self-compacting concrete. Natural sources include pumice, expanded shale, scoria, volcanic cinders, lightweight expanded clay aggregate (LECA), Lytag, tuff, diatomite, perlite and vermiculite. Industrial by-products include oil-palm boiler clinker (OPBC) or palm oil clinker (POC), foamed slag, slag aggregate, and coarse bottom ash (CBA). The use of these materials promotes the utilization of sustainable materials in construction whilst ensuring economic viability and adequate structural performance (CEB/FIP, 1983); Neville and Brooks, 2008; Emdadi *et al.*, 2014; Aslam, Shafigh, Jumaat and Lachemi, 2016; Aslam *et al.*, 2016b).

OPBC is an alternative aggregate found in tropical regions where palm oil industries are common/abundant. It has a higher specific gravity and density compared to other LWAs. Several studies have successfully developed LWC using OPBC as coarse aggregate (Aslam, Shafigh, Jumaat and Lachemi, 2016; Aslam *et al.*, 2016b; Javed *et al.*, 2018).

This chapter reviews and explores the potential of various alternative materials for producing sustainable and environmentally friendly SCC. The physical, mechanical and chemical properties of the suitable waste materials are evaluated for this purpose and a comparison with conventional materials is provided. In addition, the use of different types of aggregates in SCLWC is examined for different concrete mix designs. Their effects on the fresh and hardened properties of SCLWC are discussed. By summarizing and analyzing the properties of the alternative materials we hope to highlight the important successes that have been, and may be, accomplished.

2. Aggregates: Origin and properties

The mechanical behavior of concrete is determined by the physical properties of its constituents. Accordingly, properties such as the specific gravity, surface texture and shape, thickness, loose and compacted bulk densities and the water absorption have been thoroughly investigated for the alternative materials used to produce SCLWC.

2.1 Origin

Oil-palm boiler clinker (OPBC): In the palm oil industry, the fibers and shells are used as fuel for heating the boiler during the extraction process. After approximately 3–4 hours of burning at a temperature of around 800 °C, the ash from the shells and fibers, together with a few other pollutants, combine to create a by-product that is known locally as 'boiler stone' or 'oil-palm boiler clinker (OPBC)' (Soleymani, 2012; Aslam Shafigh, Jumaat and Lachemi, 2016; Aslam *et al.*, 2016b).

Oil-palm-shell (OPS) or Palm-Kernel Shell (PKS): OPS is from the oil-palm industry, which is the most important agro-industry in tropical region countries like Malaysia, Indonesia. and Thailand. The oil-palm industry globally produces over 190 million tonnes of waste in the form of fibers, shells, empty fruit bunches, trunk, fronds, and palm-oil mill effluent (Lee and Ofori-Boateng, 2013). It is considered as an organic lightweight aggregate with a color varying from dark grey to black, and with a specific gravity of 1.14 and loose bulk density of 545 kg/m^3 (Okpala, 1990). OPS as a LWA can produce structural lightweight concrete of grade 30 without the addition of any cementitious materials, however, grade 50 high-strength OPSC with 28-day compressive strength of about 53 MPa with oven-dried density in the range of 1790 to 1922 kg/m^3 was successfully developed, which is about 18–24% lower than the conventional concrete with average density of 2350 kg/m^3 (Shafigh *et al.*, 2011).

Pumice aggregate (PA): Pumice is a natural extrusive igneous (pyroclastic) rock formed by the discharge of lava. It has been utilized in lightweight concrete around the world. Its lightweight nature is the result of gases escaping during its formation. The availability of pumice is obviously dependent on location and the transportation cost, and time may be a deterrent. However, the presence of approximately 7.4 billion m^3 of pumice reserves in Turkey (Kotan and Gul, 2010; Kurt *et al.*, 2016) suggests that pumice aggregate could play a significant role in the production of lighter, workable, economical and environmentally-friendly concrete.

Tuff aggregate (TA): Tuff is a soft rock that has been used in the construction industry for a long time. It is composed of volcanic ash ejected during the volcanic eruption. After expulsion and deposition, the ash is compacted into solid rock through a process known as consolidation (Asniar *et al.*, 2019).

Diatomite aggregate (DA): This is an earthy and loosely-cemented porous lightweight sedimentary rock. It is a siliceous material, consisting principally of the fossilized skeletal remains of diatoms, unicellular aquatic plants related to algae during the tertiary and quaternary periods (Paschen, 1986; Arik, 2003).

Lytag aggregate (LA): This material has been used as a lightweight aggregate for around 50 years. The main ingredient used in the manufacture of lightweight aggregates is pulverized fuel ash (fly ash), the waste material produced from electricity production in coal-fired power stations at a temperature of about 1100 °C. The aggregate is called 'sintered pulverized fuel ash lightweight aggregate' but is more commonly known as Lytag (Lytag® Lightweight Solutions, 2017).

Coal bottom ash (CBA): This is a waste product of thermal power plants. The characteristics of CBA make it suitable as an alternative material in construction (Arenas *et al.*, 2013). It generally has good absorption characteristics with the potential to be used commercially rather than being disposed of at ash ponds (Toraldo *et al.*, 2013). It is derived from the unwanted agglomerated ash particles (i.e. those not fine enough or light enough) found in the flue gases. CBA is reported to have angular particles and a porous surface texture (Abdul Talib, 2010).

Lightweight expanded clay aggregate (LECA): LECA aggregates are normally made from expanded clay. They are formed by heating clay in a rotary kiln at a very high temperature of around 1200 ºC. Their main characteristics as an aggregate are that they are very spongy and light and they are highly water absorbent (Mazaheripour *et al.*, 2011).

Expanded shale aggregate (ESA): This is an artificial LWA that is mined, crushed and prepared at high temperature in a rotating kiln, creating a clean, inactive, permeable/porous and lightweight material. ESA has been used as a basic building material for around 100 years. It is included in concrete to deliver lighter buildings and bridges, since its weight is half that of standard crushed stones (Karahan *et al.*, 2012).

Slag aggregate (SA): Slag aggregate is a glass-like industrial by-product and usually comprises a mixture of silicon dioxide and metal oxides; however, the slag can also contain elemental metals and sulfides. The slag aggregates produced from the steel and iron manufacturing processes have been used in the construction industry since the Roman period, when the crushed slag from crude iron production was used to build roads. Currently, slag aggregates are widely used in all kinds of civil works.

2.2 Specific gravity

In producing sustainable SCC, researchers have generally utilized various types of alternative materials in the form of fine and coarse aggregates. Values for the specific gravity of all selected aggregates are shown in Table 5.1. It will be seen that while the specific gravities of the aggregates vary, they never exceed the values of crushed granite. The specific gravity of OPBC/POC generally falls between 1.7 and 2.2 (Kanadasan and Razak, 2014a, 2014b, 2015; Aslam *et al.*, 2016b). The wide range is attributed to the difference in sources (Hemmings *et al.*, 2009).

Compared to all the selected materials, the aggregates gave the lowest specific gravity results, as can be seen in Table 5.1. The specific gravity of OPBC/POC is around 15–35% lower than for conventional NWAs, 47–53% higher than for artificial LWAs and about 24% higher than natural LWAs such as diatomite, volcanic cinders and pumice. The specific gravity of OPS/PKS is found to be in the range 1.2–1.6, which is about 50% lighter than NWAs and about 28% lighter than OPBC. Topçu and Uygunoğlu (2010) reported the specific gravities for pumice, tuff and diatomite aggregates as 1.8–2.3, 2.04 and 2.23, respectively. In comparison with conventional NWAs the specific gravity of pumice aggregate is lighter by around 18–28%, tuff by almost 27%, and diatomite by about 20%. LECA gave a specific gravity in the range 0.8–1.2 (Mazaheripour *et al.*, 2011; Lotfy *et al.*, 2014; Aslam *et al.*, 2016b), whereas for Lytag aggregate the specific gravity lies in the range 0.8–0.9 (Ahmad *et al.*, 2013; Aslam *et al.*, 2016b; Lytag® Lightweight Solutions, 2017).

Karahan *et al.* (2012) reported a specific gravity of 1.43 for ESA, which is around 49%, 36%, 36%, 35%, 30%, and 11%, lighter than NWA, PA, DA, OPBC/POC, TA and OPS, respectively. However, ESA was found to be around 16% and 37% heavier than LECA and Lytag aggregates. The loose density of ESA varies between 776 kg/m^3 and 845 kg/m^3 (Karahan *et al.*, 2012; Chen *et al.*, 2018). Slag aggregate (SA) was found to be around 43% lighter than NWA and gave a specific gravity similar to the organic natural aggregate OPS. In addition, the specific gravity of CBA as fine aggregate was found to be in the range 1.4–2.3 (Cadersa and Auckburally, 2014; Ibrahim *et al.*, 2015; Sadon *et al.*, 2017), whereas conventional sand used as fine aggregate generally yielded a value in the range 2.38–2.64. This means that CBA is significantly

Table 5.1 *Physical properties of LWAs used for SCC (crushed granite included as a reference NWA).*

Type	Size (mm)	Specific gravity	Loose bulk density (kg/m³)	Compacted density (kg/m³)	Water absorption 24 hours (%)	References
Crushed granite	0.2–20	2.4–2.8	1377–1412	1400–2680	0.5–3.34	Dinakar *et al.*, 2008; Kou and Poon, 2009; Carballosa *et al.*, 2015; Kanadasan and Razak, 2015.
OPBC/POC	5.0–14	1.7–2.2	740–790	800–860	3.0±2.0	Kanadasan and Razak, 2014a, 2014b, 2015; Aslam *et al.*, 2016b.
Oil palm shell (OPS) and palm kernel shell (PKS)	2.0–11	1.2–1.6	500–600	600–740	14–33	Alengaram *et al.*, 2013; Aslam, Shafigh, Jumaat and Lachemi., 2016; Aslam *et al.*, 2016a.
Pumice aggregate (PA)	4.0–16	1.8–2.3	–	740–1200	28.75	Uygunoğlu and Topçu, 2009; Topçu and Uygunoğlu, 2010; Kaffetzakis and Papanicolaou, 2016; Kurt *et al.*, 2016.
Tuff aggregate (TA)	4.0–16	2.04	–	762	23.82	Topçu and Uygunoğlu, 2010.
Diatomite aggregate (DA)	4.0–16	2.23	–	435	94.24	
LECA	4.75–12	0.8–1.2	530–622	–	16.2–30.0	Mazaheripour *et al.*, 2011; Lotfy *et al.*, 2014; Aslam *et al.*, 2016b.
Lytag aggregate (LA)	4.75–12	0.8–0.9	750–850	1400	18.0	Ahmad *et al.*, 2013; Aslam *et al.*, 2016b; Lytag® Lightweight Solutions, 2017.
Expanded shale aggregate (ESA)	12.0	1.43	776–845	–	12.9	Karahan *et al.*, 2012; Chen *et al.*, 2018
Limestone powder (LSP)	16.0	1.64	–	840.0	5.02	Khan *et al.*, 2019.
Slag aggregate (SA)	2.5–10	1.61	950.0	–	8.0	Abouhussien *et al.*, 2015.
CBA as fine aggregate	< 5.0	1.4–2.3	1210	–	26.0	Cadersa & Auckburally, 2014; Ibrahim *et al.*, 2015; Sadon *et al.*, 2017.

lighter – by about 13–41% – compared to normal sand, which is responsible for the light weight of structural concrete.

Khan *et al.* (2019) investigated limestone powder (LSP) for SCLWC and reported a specific gravity of 1.64 for LSP, which is very close to that of slag aggregate (1.61). However, it has a rodded density of 840 kg/m³, which is in the range of densities for OPBC, PA and LA. Sadeq (2020) used expanded slate lightweight aggregate for the production of SCLWC. The expanded fine slate aggregate (EFSA) utilized had a specific gravity of 1.80 and a water absorption value of 10%, whereas expanded coarse slate aggregate (ECSA) has a specific gravity of 1.53 and a water absorption value of 7.10%. The specific gravity of ECSA is higher than that of LA by around 41%. Furthermore, the specific gravity of ECSA is found to lie in the range for OPS, PA and CBA fine aggregate. In general, the specific gravity of coarse slate (ECSA) is about 15% lower than its finer particles (EFSA), whereas, it is heavier than lytag (LA), and it has similar specific gravity values to the other types of aggregates such as OPS, PA and CBA.

2.3 Grain size distribution

Particle size distribution plays a vital role for all types of aggregates utilized in concrete and LWA. It must, therefore, comply with concrete design criteria and pertinent regulations. It is common practice to prepare LWA in the required sizes, as specified in Figure 5.1. Generally, lightweight coarse aggregates for use in the production of SCLWC are found to range in particle size from 4 mm to 16 mm, whereas a particle size of less than 4.75 mm is considered a fine aggregate. The gradation of all the selected alternative materials as fine and coarse aggregates is shown in Figure 5.1.

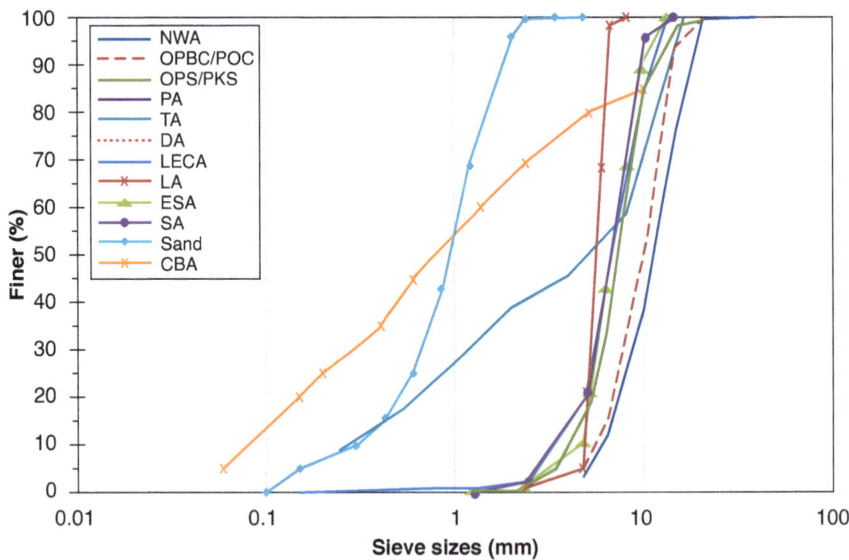

Figure 5.1 *Particle size distribution of some alternative materials used as lightweight aggregates with a comparison with NWAs (Uygunoğlu and Topçu, 2009; Topçu and Uygunoğlu, 2010; Mazaheripour et al., 2011; Karahan et al., 2012; Ahmad et al., 2013; Kanadasan and Razak, 2014a, 2014b; Abouhussien et al., 2015; Carballosa et al., 2015; Ibrahim et al., 2015; Kanadasan and Razak, 2015; Aslam et al., 2016b).*

2.4 Shape and surface texture

The shape and surface texture of the selected aggregates are shown in Figure 5.2. OPBC/POC aggregate is generally found to be greyish in color with the appearance of porous stone, and is irregular and flaky in shape, with spiky and rough broken edges, which generally contributes to its lightweight nature (Zakaria, 1986; Aslam, Shafigh, Jumaat and Lachemi, 2016; Aslam *et al.*, 2016b). OPS/PKS aggregate is an organic natural material. It has a smooth surface on both convex and concave faces, which generally reduces the bond strength between the shells and the cement paste (Aslam *et al.*, 2015). Topçu and Uygunoğlu (2010) reported that the interaction between LWAs and the cement paste matrix is found to be different compared to NWC, which was mainly due to the rough surface texture of the aggregates (PA, TA and DA). It was further observed that pumice has a half-open porous surface texture while tuff is roughly surfaced, which significantly increases the interconnection between materials. Diatomite aggregate, however, has a highly porous structure, which may reduce the interaction between cement paste and aggregate.

LECA is generally made from clay and is characterized as a spongy honeycombed, very light material (Mazaheripour *et al.*, 2011). Lytag aggregate is generally spherical

Figure 5.2 *Shape and surface texture of alternative materials prepared as LWAs for self-compacting concrete (Karahan et al., 2012; Alengaram et al., 2013; Cadersa and Auckburally, 2014; Abouhussien et al., 2015; Aslam et al., 2016b; Sadon et al., 2017; Wang et al., 2018).*

in shape with a rough surface texture. It has significant fire resistance and freeze–thaw properties (Lytag® Lightweight Solutions, 2017). Coal bottom ash (CBA) particles are glassy, porous and grayish. It consists mostly of irregularly shaped particles with very few angular ones, whereas normal fine aggregates are granular and mainly consist of angular particles (Cadersa and Auckburally, 2014; Sadon *et al.*, 2017).

2.5 Bulk density and water absorption

Bulk density provides a measure of the extent to which aggregates can be packed together and it reflects the approximate size distribution and shape of the particles (Ahmad *et al.*, 2007b). The bulk densities (loose and compacted) of materials selected for the alternative LWAs are given in Table 5.1. The loose bulk densities of all the coarse lightweight aggregates were found to be in the range 500–950 kg/m^3, which is significantly lower than the range for NWA. The compacted bulk densities were: 800–860 kg/m^3 for OPBC/POC; 600–740 kg/m^3 for OPS/PKS; 740–1200 kg/m^3 for PA; 762 kg/m^3 for TA; 435 kg/m^3 for DA; and 1400 kg/m^3 for Lytag aggregate (Uygunoğlu and Topçu, 2009; Topçu and Uygunoğlu, 2010; Karahan *et al.*, 2012; Abouhussien *et al.*, 2015; Carballosa *et al.*, 2015; Kanadasan and Razak, 2014a, 2014b; Lotfy *et al.*, 2014; Kanadasan and Razak, 2015; Kaffetzakis and Papanicolaou, 2016).

For structural application requirement, generally, the bulk density of aggregates can be selected within the range of 700 to 1400 kg/m^3 (Aslam *et al.* 2016b). Bohan and Ries (2008) reported that the approximate bulk density of aggregates for conventional concrete ranging from 1200 to 1750 kg/m^3, and for LWAC it varies between 560 to 1120 kg/m^3. Although the density of OPS/PKS is lower but several researchers utilized it as coarse aggregate to produce high-strength LWACs. The DA showed the lowest density among all the LWAs, with about 84% lighter than conventional crushed granite aggregates having a compacted density of 2680 kg/m^3.

Generally, in concrete mix design the degree of water absorption in aggregates plays a very important role because it directly affects the consistency and the workability of the concrete (Aslam *et al.*, 2016b). All the lightweight aggregates showed significantly higher water absorption compared to conventional NWA, as can be seen in Table 5.1. The 24-h water absorption for coarse OPBC aggregate lies in the range 3–5% (Kanadasan and Razak, 2014a, 2014b, 2015), whereas the absorption values for OPS aggregate range from 14% to 33% (Aslam *et al.*, 2015; Aslam *et al.*, 2016b), which is significantly higher (by about 90%) than NWA. Artificial alternative aggregates (PA, TA, LECA and LA) displayed a final absorption in the range 18–30% (Topçu and Uygunoğlu, 2010; Mazaheripour *et al.*, 2011; Lotfy *et al.*, 2014), while for ESA and slag aggregates these values ranged from 8% to 13% (Karahan *et al.*, 2012; Abouhussien *et al.*, 2015). At around 94% the diatomite aggregate displayed the highest absorption value compared to all the selected lightweight and normal-weight materials (Topçu and Uygunoğlu, 2010).

3. Cementitious materials

Palm-oil clinker (POC) or OPBC: The principal source of this material is the palm oil industry. Lumps of coarse material are collected and converted into powder form to be used as an additional cementitious material for the production of self-consolidating concrete (Kanadasan and Razak, 2014a, 2014b, 2015).

Palm oil fuel ash (POFA): Another by-product from the same palm oil industry is POFA, a solid waste material produced by burning palm shells and husk fibres in the plant boiler that generates the energy to be used in the oil palm mill. For the most part, after combustion, almost 5% palm oil fuel ash by weight of the solid wastes is formed (Sata *et al.*, 2010; Abdullah *et al.*, 2015). POFA increase the resistance of concrete to chloride and sulfate penetration, whereas, to increase the durability of concrete in harsh environmental conditions another binder material namely silica fume can be used. Silica fume, an inherent co-product of silicon and ferrosilicon, is generally produced in large electric smelting furnaces at a temperature in excess of 2000 °C via the chemical reaction $SiO_2 + 2C = Si + 2CO$ (Fidjestol and Dastol, 2008).

Rubble powder (RP): Rubble powder has generally been obtained from the reuse and recycling of rubble from building demolition. This comprises brickwork waste and crushing concrete. Materials passing through sieve ASTM no. 100 of 150 μm can be used as an additional binder (Corinaldesi and Moriconi, 2011). Lv *et al.* (2019) performed a detailed study of SCLWC using rubber particles as an additional binder and reported that increasing the rubber particles in the mixture led to a rise in the plastic viscosity and yield stress of the mortar and also improved the filling capacity, flow ability and segregation resistance of SCLWC.

Metakaolin (MK): This is a form of the clay mineral kaolinite and does not contain interlayer water and cations. The temperature of de-hydroxylation hinges on the structural layer stacking arrangement. Ordered kaolinite dehydroxylates between 570 °C and 630 °C, whilst disordered kaolinite dehydroxylates between 530 °C and 570 °C. Dehydroxylated disordered kaolinite displays better pozzolanic activity compared to ordered kaolinite (Kakali *et al.*, 2001). The dehydroxylation of kaolin to metakaolin is an endothermic reaction process due to the higher energy consumption required to remove the chemically-bonded hydroxyl particles. Above the temperature range for dehydroxylation, kaolinite changes into metakaolin, a complex amorphous structure that retains a degree of long-range order due to layer stacking (Bellotto *et al.*, 1995).

Red mud (RM): Red mud is a by-product of the refining of bauxite into alumina in an alumina plant. Chemical analysis of red mud shows that it contains gismondine, goosecrekite and epistilbite. These three mineral compounds, with a three-dimensional structure made up of Si-O tetra-hedroids and Al-O tetra-hedroids, belong to the zeolite group, which may enhance the properties of concrete (Poon *et al.*, 1999; Janotka *et al.*, 2003; Feng and Peng, 2005; Bilim, 2011).

4. Chemical composition of SCLWC materials

The chemical properties of the materials play an important role in ensuring that they behave properly as a binder.

4.1 Alternative cementitious materials

The Portland cement used to make concrete is composed of the primary minerals C_2S, C_3S, C_3A and C_4AF. Although concrete has a number of good properties, it reacts poorly with clay and dust particles, and there is a necessity to use both fine and coarse aggregates if rich concrete is to be attained. These drawbacks are mainly due to the fact that normal Portland cement contains a large amount of CaO (63–67%). During hydration of cement, the calcium-silicate-hydrate (C-S-H) forms along with other chemical compounds such as calcium aluminate hydrate, calcium alumino-ferrite and calcium silicate hydrate etc. These compounds are essential for concrete binding and strength, but are chemically reactive with surrounding environment and may affect the durability of concrete (Chandra and Berntsson, 2002).

With regard to the permeability of concrete, pozzolanic materials are generally used to fill the voids of the blended aggregates. This also strengthens the interface bond between the binder paste and the aggregates, due to pozzolanic reaction (Hwang and Hung, 2005). This led to the development of new binders such as silica fume (SF), fly ash (FA), POC powder, Ground granulated blast-furnace slag (GGBFS), limestone and rubble powders, metakaolin, POFA and red mud, all of which fulfill the requirements for the durability of concrete. These materials contain a significantly lower amount of calcium oxide compared to normal Portland cement, as can be seen in Table 5.2. In addition, these supplementary cementing materials may also contribute to properties such as the strength and the durability of the concrete.

4.2 Alternative materials as LWA

Typically, the lower permeability of the LWAC can be improved by an improvement in the interfacial transition contact zone between the aggregate and binder paste. The improvement in the contact zone may be due to the vesicular nature and the internal curing ability of the aggregate, which permits the cement paste to soak into the LWA and provide a better bond; it also facilitates the chemical bond between the paste and the aggregate (CEB/FIP, 1983). In LWC, the expansive products appear during the chemical reactions, as they appear, it can move into the pores of aggregates and minimize the internal stress (Bremner and Holm, 1995).

Table 5.3 displays the chemical composition of the selected alternative coarse aggregate materials compared to conventional aggregates (NWA) as can be seen, the amount of CaO in crushed granite (limestone) is normally around 55%, whereas in LWAs only slag aggregate (SA) contains around 40% CaO. However, all the alternative materials have significantly higher contents of SiO_2 when compared to the NWA. Generally, these LWAs are produced from natural materials by thermal treatment and are industrial by-products and similar, which have good chemical composition and can be utilized for the development of high-strength lightweight concrete with better properties such as thermal resistance and durability when compared to conventional concrete.

5. Mechanical properties of LWA

Mechanical properties – as measured, for example, by aggregate impact and crushing values and the Los Angeles abrasion test – are important factors in the performance of structural lightweight aggregate concrete (Aslam *et al.*, 2016b). The crushing value of an aggregate measures its resistance to crushing under a compressive load. The value for crushed granite is generally in the range 18–20% (Topçu and Bilir, 2009; Kim *et al.*, 2010; Kanadasan and Razak, 2014a, 2014b, 2015; Yehia *et al.*, 2016), while the crushing value of OPBC/POC is around 6–10% lower than the value for crushed granite aggregate

Table 5.2 *Chemical composition of cementitious materials used in the production of SCLWC. (V. Corinaldesi and Moriconi, 2004; Dinakar et al., 2008; Sukumar et al., 2008; Topçu and Bilir, 2009; Topçu and Uygunoğlu, 2010; Corinaldesi and Moriconi, 2011; Karahan et al., 2012; Zhao et al., 2012; Gesoğlu et al., 2014; Kanadasan and Razak, 2014a, 2014b; Lotfy et al., 2014; Carballosa et al., 2015; Güneyisi et al., 2015; Kanadasan and Razak, 2015; Nagaratnam et al., 2016).*

Compounds		Concentration (%)									
		Cement	POC Powder	Silica Fume	Fly Ash	GGBFS	Limestone powder	Rubble powder	Metakaolin	POFA	Red Mud
Calcium Oxide	CaO	61.4–64.1	6.4	0.23–0.45	0.6–12	43.9	18.6–40.6	2.94	0.36–0.40	3.21	4.98
Silica Dioxide	SiO_2	19.6–21.0	59.9	90.0–95.2	46.2–64.0	33.8	6.4–38.7	84.99	52.0–64.0	52.63	45.76
Ferric Oxide	Fe_2O_3	2.94–5.1	6.9	0.13–1.31	2.69–16.0	0.52	0.06–3.34	3.91	1.1–4.28	1.06	2.85
Sulphur trioxide	SO_3	2.2–3.6	0.39	0.33–0.41	0.08–2.5	0.10	1.20	1.30	0.05–0.22	–	2.15
Aluminum Oxide	Al_2O_3	3.1–5.37	3.89	0.21–0.71	20–40	13.4	0.11–8.0	4.47	30.0–38.3	8.99	40.7
Magnesium Oxide	MgO	2.5–3.13	3.30	–	0.14–15.9	5.40	0.04–2.93	1.10	0.08–0.30	1.45	–
Phosphorous pentoxide	P_2O_5	0.7	3.47	–	–	–	–	–	–	–	1.10
Potassium Oxide	K_2O	0.17–1.0	15.10	1.51	0.7–3.8	0.31	0.7–1.4	0.77	0.37–0.50	3.13	0.45
Titanium Oxide	TiO_2	0.12	0.29	–	0.9–2.0	0.55	–	0.11	–	0.31	2.03
Natrium Oxide (Sodium Monoxide)	Na_2O	0.19–0.7	–	0.45–0.85	0.2–1.8	0.20	0.36–1.0	0.41	–	0.56	–
Manganese trioxide	Mn_2O_3	0.12	–	–	–	0.30	–	–	–	–	–
Manganese Oxide	MnO	–	–	–	–	–	–	–	–	–	–
Loss on ignition	LOI	0.6–2.3	1.89	1.9–3.1	0.3–6.0	1.00	34.2–72.5	26.57	0.68–6.56	27.7	–

Table 5.3 *Chemical composition of alternative LWA materials (Uygunoğlu and Topçu, 2009; Topçu and Uygunoğlu, 2010; Mazaheripour et al., 2011; Karahan et al., 2012; Lotfy et al., 2014; Abouhussien et al., 2015; Ibrahim et al., 2015; Kurt et al., 2016; Aslam et al., 2016b).*

Compounds	Concentration (%)									
	Pumice (PA)	LECA	ESA	OPBC/POC	Lytag (LA)	Lime-Stone	Tuff (TA)	Diatomite (DA)	Slag (SA)	CBA
Calcium Oxide	1.0–5.5	1.5–2.5	2.0	2.3–8.2	3.0–4.0	55.0	0.53	1.36	40.0	1.61
Silica Dioxide	57–75	62–66	64.2	60–82	50–53	0.01	73.0	67.2	37.1	68.9
Ferric Oxide	1.0–7.0	6.5–9.0	4.9	4.6–5.2	5–6	0.05	1.03	2.7	1.9	6.50
Sulphur trioxide	0.14	0.3–16.0	0.7	0.73	0.3	–	–	–	–	–
Aluminum Oxide	11.2–17	0.2–20.6	20.2	3.5–3.7	23–25.0	0.17	14.1	10.0	8.8	18.67
Magnesium Oxide	0.6–2.0	1.0–4.0	3.6	1.2–5.0	2.8–3.0	0.64	0.0	0.63	11.5	0.53
Phosphorous pentoxide	–	0.21	–	0.8–5.3	–	–	–	–	–	–
Potassium Oxide	2.9–8.0	2.0–3.5	–	4.6–11.6	0.2	–	3.6	0.7	–	1.52
Titanium Oxide	–	0.78–0.80	–	0.2	–	–	–	–	–	1.33
Natrium Oxide (Sodium Monoxide)	3.0–5.0	0.7–2.0	3.2	0.1–0.3	0.3	–	1.1	0.3	0.8	0.24
Manganese trioxide	–	–	–	–	–	–	–	–	–	–
Manganese Oxide	–	0.09–0.14	–	–	–	–	–	–	0.6	–
Loss on ignition	1.5–4.6	0.3–0.84	0.3	–	3.1	43.66	6.3	8.0	2.0	2.68

(Kanadasan *et al.*, 2014a, 2014b; Aslam *et al.*, 2015; Kanadasan and Razak, 2015; Aslam, Shafigh, Jumaat and Lachemi, 2016; Aslam *et al.*, 2017; Aslam, Shafigh, Nomeli and Jumaat, 2017). The crushing value is restricted to 30% for roads and for pavement concrete, but 45% is permitted for other structures (Mehta and Monteiro, 2006).

In structural concrete, the toughness of aggregates is generally measured as the impact resistance of the material to failure. IS 283-1970 recommended that aggregates with an impact value of less than 45% by weight can be used in all engineering structures except the wearing surface, whereas an impact value up to 30% by weight can be utilized for the concrete in the wearing surfaces (Mehta and Monteiro, 2006; Neville and Brooks, 2008).

Resistance to wear is an additional critical value for aggregates. Normally, the abrasion resistance of LWA is poorer compared to that of NWA because of the lower stiffness of lighter aggregates (Alengaram *et al.*, 2013). Neville and Brooks (2008) reported that, for the wearing surface, the abrasion value must be less than 30% but that in all other structures it should no greater than 50%. The abrasion value for crushed granite is in the range 17–31% (Bogas *et al.*, 2012; Yehia *et al.*, 2016), for OPBC 20–27% (Kanadasan and Razak, 2014b, 2015; Aslam *et al.*, 2016a; Aslam, Shafigh, Jumaat and Lachemi, 2016), and for pumice, tuff and diatomite aggregates it is 53%, 49% and 64%, respectively (Topçu and Uygunoğlu, 2010). It was observed that the mechanical properties of some of the alternative LWAs were found to be within suitable ranges, and hence may be used in the construction industry as an aggregate.

6. Design methods of SCC

Mixture design is considered a key aspect for the production and application of concrete. Several researchers have investigated the different methods for producing self-compacting concrete (SCC). In this section, the authors have selected and critically reviewed the five most highly used methods, namely the compressive strength method (CSM), the empirical design method (EDM), the rheology of paste model (RPM), the statistical factorial model (SFM), and the close aggregate packing method (CAPM).

6.1 Compressive strength method (CSM)

In this method, the quantity of the ingredients, such as the binder/cement, the mineral admixtures, water and the aggregate contents in the mix proportion, is determined according to the required compressive strength of the concrete. This method was originally proposed by ACI 211.1-91 (ACI, 1991) for SCC with a compressive strength in the range 15–40 MPa. This range was further extended up to 75 MPa, with a maximum water-to-cement ratio (w/c ratio) in the range 0.29–0.8. Based on the ACI 211.1-91 method, Kheder and Al-Jadiri (2010) proposed a mixture design method for SCC. They reported that the coarse aggregate content generally depends on the maximum particle size of the aggregate and the fineness modulus of the fine aggregates, while the water content can be estimated based on concrete strength and maximum aggregate size. The flow chart for the compressive strength method for SCC is presented in Figure 5.3.

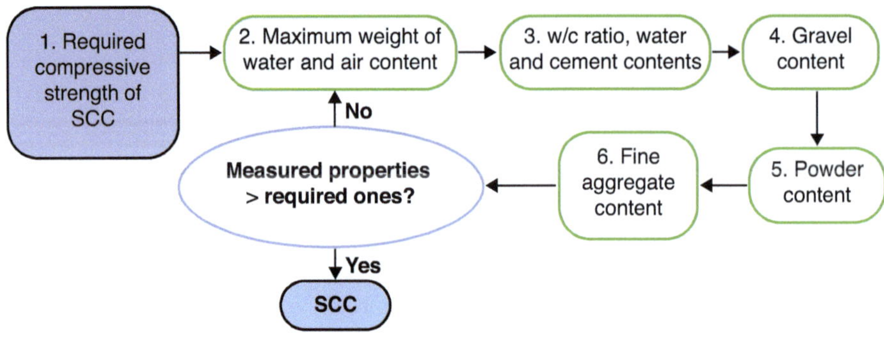

Figure 5.3 *Compressive Strength Method for SCC (Kheder and Al-Jadiri, 2010).*

6.2 Empirical design method (EDM)

This method generally works on the empirical data involving for the contents of fine and coarse aggregates, cementitious materials, water and admixture dosage to estimate the initial mix proportions. However, it requires several trials to arrive at a best estimate of mixture proportions for the required properties of SCC. The method was proposed by Okamura and Ozawa (1995). The procedure included the following: (a) the coarse aggregate amount in the SCC was fixed at 50% of the solid volume; (b) the amount of fine aggregate was fixed as 40% of the mortar volume; (c) the water-to-binder/powder ratio was in the range of 0.9–1.0 by volume, as depending on the binder properties; (d) the dosage of superplasticizer and the final w/c ratio were determined so as to ensure self-compactability. It was further improved by Edamatsu and colleagues (Edamatsu *et al.*, 1999; Edamatsu and Ouchi, 2003) and investigated the interactions between coarse aggregates with the fixed ratios for the fine aggregate, w/b ratio and dosage of superplasticizer. Compared to Okamura, Edamatsu's approach can be utilized for any type of binder and aggregate material. The flow chart for EDM for SCC concrete is shown in Figure 5.4.

6.3 Rheology of paste model (RPM)

The rheology of paste model (RPM) was developed by Saak *et al.* (2001). They reported that the rheology of cement paste is generally responsible for the segregation resistance and workability of the fresh concrete, given a specified particle size distribution and the volume fraction of aggregates. The reliability of this method was tested by considering the flow properties of the concrete. This concept was extended further by Bui *et al.* (2002), who investigated the effects of size distribution, shape of aggregates, fine-to-coarse aggregate ratio, the volumetric ratio of aggregate to cement paste, voids content and the average diameter of solid skeleton particles. The average diameter of solid skeleton particles (d_{av}) and the average aggregate spacing (d_{ss}) can be estimated using Equations (1) and (2):

$$d_{av} = \frac{\sum_i d_i m_i}{\sum_i m_i} \tag{1}$$

$$d_{ss} = d_{av}\left[\sqrt[3]{1 + \frac{V_{paste} - V_{void}}{V_{concrete} - V_{paste}}} - 1\right] \tag{2}$$

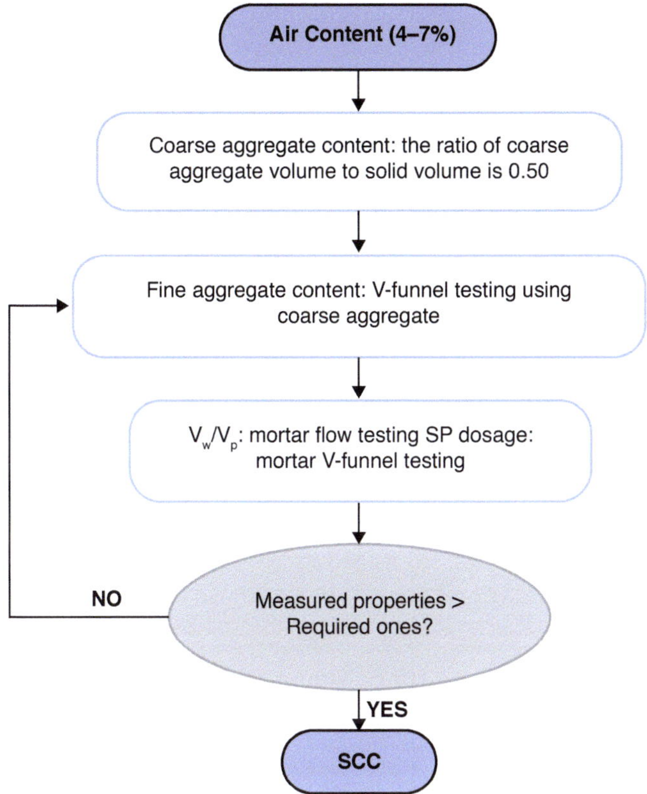

Figure 5.4 *Empirical Design Method for SCC (Edamatsu et al., 1999; Edamatsu and Ouchi, 2003).*

where d_i is the average diameter of the aggregate fraction i and m_i is the mass of that fraction. The rheology of paste model and the criteria related to the average diameter of the aggregate and the average aggregate spacing can be applied for different cement contents, different coarse-to-total aggregate ratios, different w/b ratios, including different types of fly ash. The flow chart for the rheology of paste model for the mix design of SCC is shownn Figure 5.5.

6.4 Statistical factorial model (SFM)

Khayat *et. al.* (1999, 2000) developed the SFM method of mixing design for SCC by considering five key factors: namely, the content of various cementitious materials, the w/b ratios, the concentrations of superplasticizers, the volume of coarse aggregates, and the viscosity-enhancing agent (VEA). This method is generally used to derive design charts correlating input variables and output properties mainly to understand the interaction between the materials. Later, Bouziani (2013) developed a simplex-lattice mixture design method (based on the same concept) to observe the effects of three types of sands – river sand (RS), dune sand (DS) and combinations of sand (CS) – on

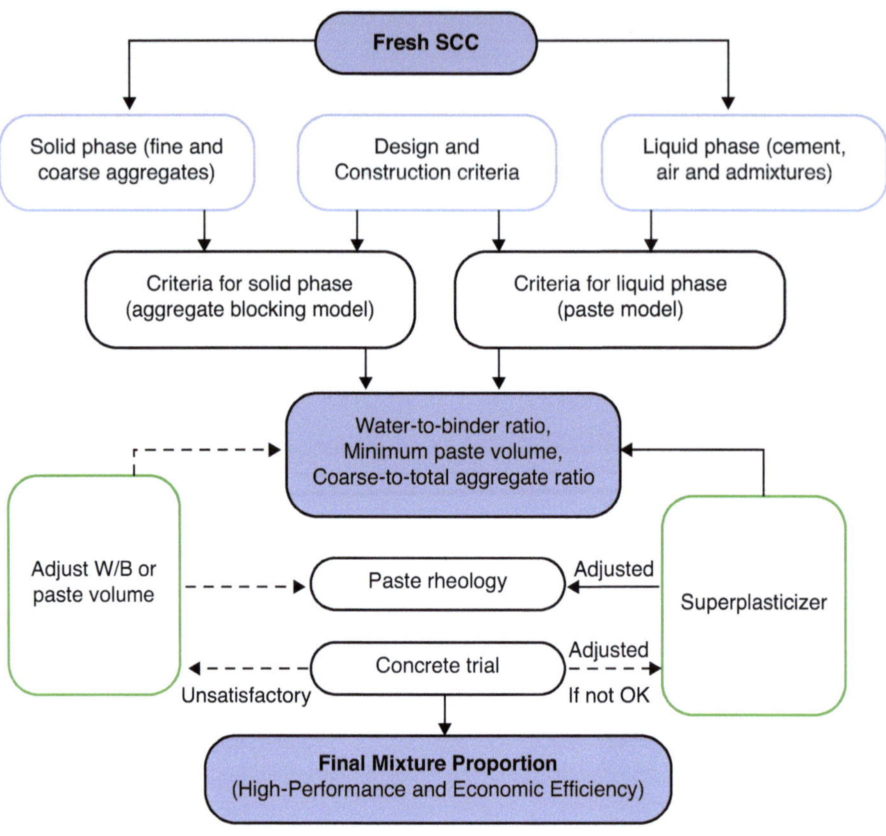

Figure 5.5 *Rheology Paste Model for SCC (Saak et al., 2001; Bui et al., 2002).*

the fresh and hardened properties of SCC. The simplex-lattice is a space-filling design that generally produces a triangular grid of combinations, as shown in Figure 5.6. The number of combinations C can be estimated using Equation (3):

$$C = \frac{(q+m-1)!}{m!(q-1)!} \tag{3}$$

where q is the number of factors and m is the number of levels. The effect of the three sands on the properties of concrete can be computed using Equation (4):

$$Y = (b_1 \times RS) + (b_2 \times CS) + (b_3 \times DS) + (b_4 \times RS \times CS) + (b_5 \times RS \times DS) + (b_6 \times CS \times DS) \tag{4}$$

where the model coefficients (b_i) represent the effect of the involved variables on the response of Y, which can be determined using statistical software.

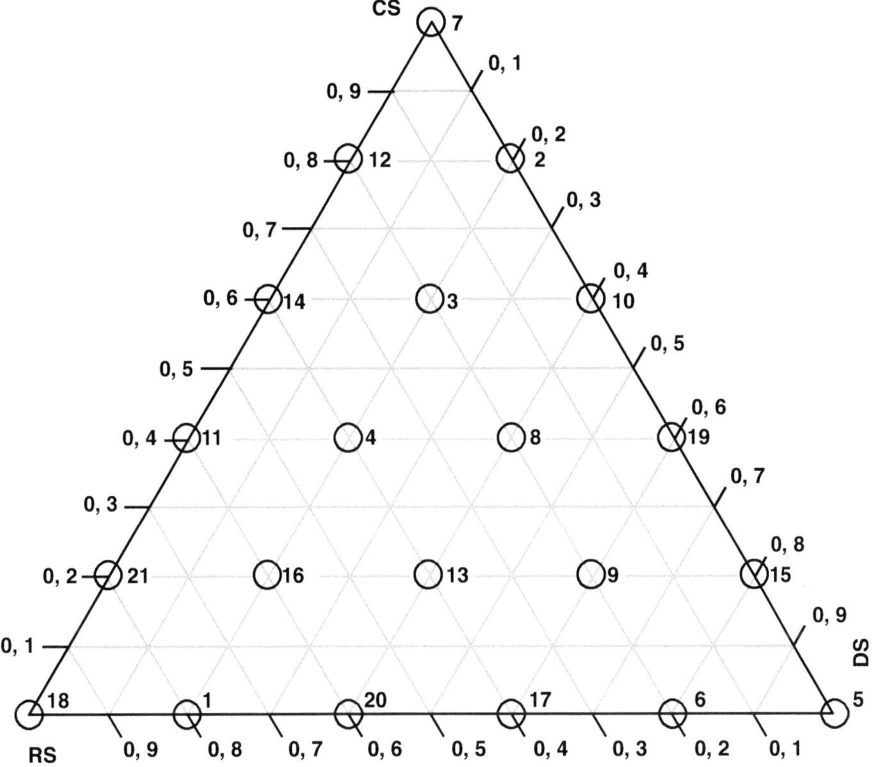

Figure 5.6 *SFM: Simplex-lattice design method with three factors (RS, CS and DS) and five levels for SCC (Bouziani, 2013).*

6.5 Close aggregate packing method (CAPM)

CAPM is a mixture design procedure that determines the mixture constituents by considering the least void (packing) between the aggregates, then applies the pastes to fill the voids between the aggregates. Hwang (2005) proposed the CAPM design mixture model based on the densified mixture design algorithm (DMDA). It generally considers the effects of aggregate packing in three ways: primitive packing, dense packing and gap gradation. In primitive packing, the sand is normally used to fill the voids between the coarse aggregates, and the binder is used to fill the voids between the fine aggregates. In dense packing, standard sieves are used for the gradation of aggregates into various sizes and the remaining fine material is omitted. A flow chart for the DMDA-based model is shown in Figure 5.7. SCC designed by the CAPM based on maximum density theory is cost effective and has higher workability and durability. This method significantly reduces problems due to size and shape, particle distribution, gap gradation of the aggregates, and the large amount of cement paste (Shi *et al.*, 2015).

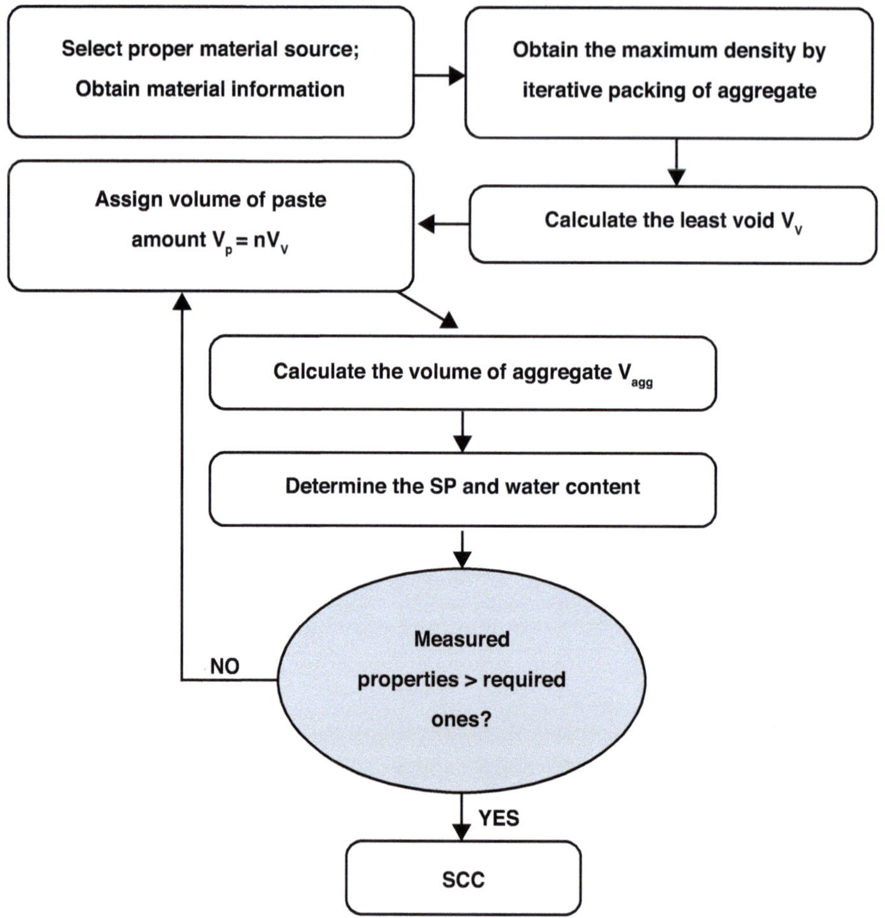

Figure 5.7 *Close Aggregate Packing Method for SCC (Hwang 2005).*

7. Mix proportions for SCLWC

In well-proportioned mixes, the relationship between binder/cement content and strength remains consistent, either within a specific type or when comparing different types. Researchers have undertaken different design methods and trials with varying amounts of binder and with alternative materials as aggregates, including NWAs, to produce concretes with the required range of compressive strengths. In this study, to aid understanding, the authors have classified the mix proportioning for SCC based on the different methods discussed in the previous section.

Particle Packing Mix Design Method: Kanadasan and Razak (2014a; 2014b; 2015) used the particle packing mixing design method to explore the properties of the self-compacting concrete (SCC) by using OPBC/POC as an aggregate. They considered the actual packing level of the paste volume and the aggregate during this method. It was found that the particle packing mixture design could be employed for a variety of combinations of aggregates for the production of SCC. Topçu and Uygunoğlu (2010) investigated the effects of aggregate type on the physical and mechanical properties of hardened SCLWC by using three types of coarse lightweight aggregates: pumice, volcanic tuff and diatomite. They also considered different w/b ratios and superplasticizer dosage levels. The cement content was in the range 443–455 kg/m^3 for self-compacting normal-weight concrete SCNWC and 397–415 kg/m^3 for SCLWCs, while the volume of aggregates was in the range 768–832 kg/m^3 (crushed stone), 390–442 kg/m^3 (pumice), 431–488 kg/m^3 (tuff) and 351–401 kg/m^3 (diatomite), respectively. They concluded that the SCLC containing LWAs (PA, TA and DA) showed lower mechanical properties and higher strain capability in terms of maximum stress when compared with the stress-strain curves of SCC made of NWAs. Further, the substitution of lower strength LWAs with NWAs reduced the modulus of elasticity in comparison to conventional SCC. Consequently, they suggested pumice as a useful aggregate for the production of SCLWC, because of high interlocking in the interfacial transition zone (ITZ), higher compressive and splitting tensile strength, low unit weight and high insulation properties. Iqbal *et al.* (2015) investigated the mechanical properties of SCLWC containing micro-steel fibers and expanded clay (LECA) as lightweight aggregate. They reported that it is possible to produce high-strength SCLWC with a density of about 1700 kg/m^3 and a high proportion of steel fiber (up to 125 kg/m^3) which yet shows a slump value of more than 600 mm. Similarly, Mazaheripour *et al.* (2011) produced SCLWC by using LECA as lightweight aggregate but they used propylene fibers in the concrete instead of steel fibers. In all mixes, the contents were fixed at 500 kg/m^3 cement, 105 kg/m^3 LECA as coarse aggregate, and 160 kg/m^3 of water, which produced SCLWC concrete with a dry density in the range 1800–2000 kg/m^3. In addition, it was reported that the addition of 0.3% volume fraction of propylene fiber in lightweight SCC significantly reduced the slump flow by about 40%.

Sukumar *et al.* (2008) used the particle-packing method to investigate the early-age strength of SCC containing high-volume fly ash. The conventional concrete mixes are generally designed on the assumption that particles of varying sizes fill larger voids, resulting in dense packing with the cement paste. In contrast, self-compacting concrete (SCC) is composed of five main constituents: cement, mineral admixtures, fine aggregates, coarse aggregates, and chemical admixtures, in which first two materials are fillers. Therefore, filler material like fly ash facilitates better flow characteristics in SCC in the fresh state and also has high fluidity, segregation resistance and minimum risk of blocking.

Empirical Design Method: Bogas *et al.* (2012) studied the behavior of SCLWC containing expanded clay as lightweight aggregate. The mixture design procedure was mainly based on Okamura's method, with modifications suggested by Okamura (2003) and Nepomuceno (2005). They produced SCLWCs with self-compactability, better deformation capacity and enough stability within the compressive strength range 37–61 MPa. Furthermore, it was proposed that the use of pre-wetted aggregate had no effect on the stability and self-compactability of the mixtures.

Statistical Factorial Model/Binary and Ternary Combinations: Nagaratnam *et al.* (2016) utilized agro-industrial wastes with binary and ternary combinations to investigate the workability and heat of hydration of SCC. For workability, the replacement of waste was in the range 10–40% of total cementitious material, whereas it was limited to a range of 30–40% for the heat of hydration. They reported that mixes with fly ash (FA) need less superplasticizer (SP) compared to POFA mixes. Furthermore, ternary use FA and POFA mixes showed higher workability and better segregation resistance compared to other POFA mixes.

Other methods: Lo *et al.* (2007) conducted a comparative study on self-compacting LWC and NWC and investigated the workability and mechanical properties of the concrete. The self-compacting lightweight concrete (SCLWC) contained expanded shale as fine and coarse aggregates, a binder range of 500–650 kg/m^3, and a density of 1650 kg/m^3, which was about 25% lighter compared to self-compacting normal-weight concrete (SCNWC). They reported that SCLWC showed excellent workability, lower density and relatively higher compressive strength and elastic modulus compared to SCNWC. Gesoğlu *et al.* (2014) studied the permeation properties of SCC by replacing conventional aggregates with lightweight aggregates. They produced SCC based on a fixed slump flow of 720 ± 20 mm by adjusting the amount of mineral admixture, used a cement content of 440 kg/m^3 and a w/b ratio of 0.32. Karamloo *et al.* (2016) investigated the effects of maximum aggregate size on the fracture behavior of SCLWC. LECA with a nominal maximum size of 9.5, 12.5 and 19 mm, a content of 250 kg/m^3, and a cement content in the range of 400–450 kg/m^3 was utilized for the production of SCLWC. It was reported that, as the maximum size of aggregate increased, so the fracture energy was also increased – by 48% and 87% for w/c ratios of 0.35 and 0.40, respectively.

Ibrahim *et al.* (2015) explored the effect that replacing fine aggregate with CBA had on the splitting tensile strength of SCC. The fine aggregate was partially replaced with a CBA at an interval of 0, 10, 20 and 30% with a w/b ratio in the range 0.35–0.45. It was reported that by incorporating CBA in the SCC the slump flow, L-box ratio and the sieve segregation were reduced, while T500 slump flow time was increased. They achieved a maximum value of 4.25 MPa for the splitting tensile strength value at 28 days. Li and colleagues (Li *et al.*, 2017; Li Tan *et al.*, 2021) investigated SCLWAC considering the strength and rheological properties of the mortar and they reported that the SCLS with w/b ratio of 0.35 using performance design method meets the design requirements of SCC. Furthermore, the SCLC prepared using the mixture design method gives better quality concrete in terms of strength and the self-compacting performance index with a reasonable quantity of raw materials. Furthermore, Li and colleagues (Li, Zhao *et al.*, 2021a) explored steel fiber-reinforced self-compacting lightweight concrete (SFSLC) using paste film thickness theory and investigated various mechanical properties. It was found that this new mixture method can quantitatively determine the quantities of raw materials and that the addition of fiber improves the filling ability of concrete and also reduces segregation in the mixture. Additionally, the micro-steel fiber significantly improves all the mechanical properties compared to long steel fibers.

8. Workability of SCLWC

The slump test is considered one of the standard tests for measuring the consistency of concrete, according to ACI 116R 1990. The workability or slump also reflects the variations in the uniformity of the given mixture proportions (Neville, 2008). The fresh properties, such as slump flow diameter, T500 flow and V-funnel test, of all types of self-compacting lightweight aggregate concretes, are shown in Table 5.4. The slump flow diameter for

Table 5.4 *Fresh properties of self-compacting lightweight concretes.*

Self-compacting lightweight concretes (SCLWC)		Fresh properties of SCLWCs			Hardened density of SCLWCs (kg/m³)	References
Binder type	Lightweight aggregate type	Slump flow diameter (mm)	T500 flow (s)	V-funnel test (s)		
OPC + Fly Ash	Sintering fine sediment excavated from reservoir	550–650	–	–	–	Hwang and Hung, 2005
OPC (Type I)	POC as fine and coarse aggregates	670–730	2–5	7–13	2050–2100	Kanadasan and Razak., 2014a, 2014b, 2015
OPC + Pulverized Fuel Ash (PFA)	Expanded shale aggregate	700–830	–	–	1650	Lo *et al.*, 2007
OPC + FA	Spherical pellets manufactured through cold bonding process	715–730	1–2.9	6.5–9.9	–	Gesoğlu *et al.*, 2014
OPC + FA	Tuff + pumice + diatomite	435–840	–	–	1340–1720	Topçu and Uygunoğlu, 2010
OPC	Coal bottom ash (CBA) as fine aggregate	550–740	2.3– 4.9	–	2275–2380	Ibrahim *et al.*, 2015
OPC + FA + Silica fume (SF) + fibers	LECA	660–710	–	7–12	1230–1600	Corinaldesi and Moriconi (2015)
OPC + limestone filler + SF	Pumice as fine and coarse aggregate	610–880	–	3–7.1	1480–1766	Kaffetzakis and Papanicolaou, 2016
OPC + FA + Steel fibers	LECA as fine and coarse aggregate	600–700	–	16–46	1760–1875	Grabois *et al.*, 2016
OPC + FA	LECA	620–790	5–9	–	1740–1750	Iqbal *et al.*, 2015
OPC + FA	Expanded clay ceramsite + plastic particles	350–750	4.2–7.9	–	1400–1600	Yang *et al.*, 2015

(Continued)

Table 5.4 *(Continued)*

Self-compacting lightweight concretes (SCLWC)		Fresh properties of SCLWCs			Hardened density	References
Binder type	Lightweight aggregate type	Slump flow diameter (mm)	T500 flow (s)	V-funnel test (s)	of SCLWCs (kg/m³)	
OPC + FA + SF	LECA as fine and coarse aggregate	345–755	–	2.3–28.7	1430–1550	Lotfy *et al.*, 2014
OPC + FA	LECA and Arlita AF7	670–730	2–4.8	–	1670–1800	Bogas *et al.*, 2012
OPC + FA + Metakaolin	Coarse expanded shale	690 –710	–	4.5–6.3	1950–1990	Karahan *et al.*, 2012
OPC + SF	Expanded polystyrene as fine and coarse aggregate	615–680	2–3.1	6.2–16.1	–	Madandoust *et al.*, 2011
OPC + SF + Propylene fiber	LECA as fine and coarse aggregate	410–745	5–11	12–23	1800–2000	Mazaheripour *et al.*, 2011
OPC + FA	Expanded shale aggregate	680–780	6.6–9.2	17.8–23.3	1880–1920	Wu *et al.*, 2009
OPC + FA + Blast furnace slag (BFS)	Pumice as coarse aggregate	560–650	5–11	9–24	840–2280	Kurt *et al.*, 2016
OPC	LECA	622–742	–	–	–	Karamloo *et al.*, 2016
OPC + FA + Metakaolin	Lightweight slag aggregate	715–745	2.3–7	7– 43	1950–2050	Abouhussien *et al.*, 2015

self-compacting concrete produced by sintering fine sediment excavated from reservoir as an LWA was reported in the range 550–650 mm (Hwang and Hung, 2005). Kanadasan and Razak (2014a, 2014b, 2015) have developed a SCLWC utilizing OPBC/POC as fine and coarse aggregates. They reported a slump flow diameter of 670–730 mm, T500 flow as 2–5 seconds and a V-funnel test range of 7–13 seconds. SCLWC was also produced using expanded shale as a coarse aggregate. It showed better workability results in terms of a slump flow diameter ranging from 680 mm to 830 mm, and V-funnel ranges of between 4 and 24 seconds compared to SCLWCs made of POC and sintered aggregates (Lo *et al.*, 2007; Wu *et al.*, 2009; Karahan *et al.*, 2012).

Several researchers (Topçu and Uygunoğlu, 2010; Kaffetzakis and Papanicolaou, 2016; Kurt *et al.*, 2016) investigated SCLWCs using natural lightweight aggregates such as pumice, diatomite and tuff. The slump flow diameter and V-funnel test results were found to be in the range 435–880 mm and 3–24 seconds, respectively. LECA used in the production of SCLWCs has also shown better results in terms of the fresh properties of concrete. The European standards and guidelines (SCC-EG, 2005) for self-compacting concrete have classified the slump flow diameters (SF) into three classes: SF1, generally ranging between 550 mm and 650 mm; SF2, ranging between 660 mm and 750 mm; and SF3, ranging between 760 mm and 850 mm. Based on this classification, it was found that almost all the types of alternative material utilized for the production of SCLWCs gave results in the acceptable ranges and can be utilized in structural applications.

9. Physical and mechanical properties of SCLWC

9.1 Density

The hardened density plays an important role in developing the required compressive strength of concrete and is considered one of the most important factors in the design of concrete structures. The density of structural conventional concrete is generally 2400 kg/m^3; however, for lightweight aggregate concrete (SLWAC) it varies typically between 1400 kg/m^3 and 2000 kg/m^3. Neville (2008) reported that the dry density of LWAC was to be found in the range 350–1850 kg/m^3, whereas Clarke (2002) recommended a range between 1200 kg/m^3 and 2000 kg/m^3. Table 5.4 lists the hardened density of all the types of self-compacting lightweight aggregate concretes prepared with alternative materials as an aggregate. Kanadasan and Razak (2014a, 2014b, 2015) reported that replacing conventional aggregates with POC gave a dry density of SCLWC in the range 2050–2100 kg/m^3 and reduction was about 16%. A number of researchers (Lo *et al.*, 2007; Wu *et al.*, 2009; Karahan *et al.*, 2012) prepared SCLWC using expanded shale aggregate and reported a dry density in the range 1650–1990 kg/m^3; even this concrete was found lighter compared to SCLWC containing POC aggregate. The natural lightweight aggregates pumice, diatomite and tuff in SCLWC showed significantly lower dried density results, ranging from 1340 kg/m^3 to 1765 kg/m^3 (Topçu and Uygunoğlu, 2010; Kaffetzakis and Papanicolaou, 2016). However, in some cases, pumice aggregate also produced a density of up to 2280 kg/m^3 (Kurt *et al.*, 2016). Most researchers have explored the use of LECA as lightweight aggregate for the production of SCLWC. Corinaldesi and Moriconi (2015) used it with synthetic fibers and reported a density in the range of 1200–1600 kg/m³, whereas when LECA was utilized as both fine and coarse aggregates the dry density increased and was to be found in the range 1430–2000 kg/m³ (Mazaheripour *et al.*, 2011; Lotfy *et al.*, 2014; Grabois *et al.*, 2016). Trumble and Santizo (1993) reported that, generally, in building or towers, the volume of slabs usually ranged between 70% and 90% and, therefore, LWC can be utilized in floors instead of the compression members (CEB/FIP, 1977).

9.2 Compressive strength

In structural concrete, the compressive strength is generally considered the most germane property for any innovative material used in the construction industry. In general, it affects all the remaining mechanical properties of the concrete, such as the splitting tensile strength and flexural strength and the modulus of elasticity. Neville and Brooks (2008) reported the ACI conditions that the cylindrical 28-day compressive strength of lightweight concrete should be greater than 17 MPa. In this chapter, it was observed that several researchers have investigated the alternative materials as fine and coarse aggregates for the production of self-compacting lightweight concretes (SCLWC) by considering different design methodologies to achieve varying grades of strength (see Table 5.5). Hwang and Hung (2005) investigated the durability performance of SCLWC using sintered fine sediments as an aggregate. They selected a w/b ratio in the range 0.28–0.40 and successfully developed a compressive strength ranging from 32 MPa to 52 MPa. The relationship between compressive strengths and the w/b ratio is shown in Figure 5.8. It was observed that, as the w/c ratio was increased, the compressive strength was significantly reduced.

Kanadasan and Razak (2014a, 2014b, 2015) studied waste by-products from the oil palm industry, namely palm-oil clinker (POC) and oil-palm boiler clinker (OPBC), as an alternative material for the production of SCLWC. They utilized POC/OPBC as an aggregate (both fine and coarse) and as a binder, and they prepared the mix pattern with w/b ratios of 0.28 to 0.29, and managed to achieve better results for the compressive strength of the SCLWC, varying from 35 to 50 MPa. Lo *et al.* (2007) compared the mechanical properties of SCNWC and SCLWCs using conventional and expanded shale (ESA) as LWA. They selected a w/b ratio of 0.30 for SCLWC and achieved a compressive strength in the range 42–60 MPa, almost similar to the compressive strength of conventional concrete (SCC); however, the SCLWC had around 25% lower dried density compared to SCC. Karahan *et al.* (2012) utilized similar material ESA as coarse aggregate with an additional binder of metakaolin up to 15% at an interval of 5% for the production of SCLWC. They chose a w/b ratio of 0.33 and achieved the same compressive strength value of around 40 MPa at all replacement levels (Table 5.5). Generally, it was observed that the utilization of metakaolin with the expanded shale aggregate slightly increased the density of the concrete and even affected the passing and filling ability of the SCLWC but had no positive effect on the strength of the concrete.

The effect of metakaolin with other binders – namely OPC and fly ash – on the properties of semi-lightweight self-compacting concrete made of lightweight slag aggregate (LSA) was investigated by Abouhussien *et al.* (2015). They prepared eleven mixes with w/b ratios of 0.35 and 0.40 and achieved a 28-day compressive strength in the range 11–49 MPa (Table 5.5). They reported that increasing the percentage of LSA in the concrete decreased the unit weight but significantly reduced the passing ability and the flowability of the concrete. In their opinion, this reduction in passing ability and the flowability may be due to the use of metakaolin as a binder.

Some researchers have investigated the mechanical properties, specifically the compressive strength of SCLWCs, by using naturally available lightweight aggregates such as pumice, diatomite and tuff. Kurt *et al.* (2016) studied the effect of pumice powder on SCLWC made of pumice aggregate with an aggregate size in the range 0–16 mm. The range selected for the w/b ratio was between 0.30 and 0.45 and they achieved a very wide range of 28-day compressive strengths: from 10 MPa to 65 MPa. The effect of various natural aggregate types on the hardened properties of SCLWC was also studied by Topçu and Uygunoğlu (2010). They prepared three sets of SCLWCs with similar w/b ratios in the range 0.36–0.48 and reported a 28-day compressive strength in the range 18–27 MPa for pumice aggregate, 16–22 MPa for tuff aggregate, and between 13 MPa and 22 MPa for diatomite SCLWCs. Based on analysis from collected data in Table 6.5, it was observed that, for the same w/b ratios, the pumice aggregate gave better compressive strength results: about 90% higher compared to SCLWCs made of tuff and diatomite aggregates

Several researchers studied the use of lightweight expanded clay aggregate (LECA) as an aggregate for the development of SCLWCs. Corinaldesi and Moriconi (2015) investigated the properties of SCLWC using synthetic fibers and three types of binders as OPC, fly ash and silica fume. They further classified the mix proportions based on micro- and macro-fibers, and LECA with varying sizes and recycled aggregates. The first set contained micro-fibers with a w/b ratio in the range 0.35–0.36 and showed a 28-day compressive strength in the range 44–52 MPa. However, the second set, with macro-fibers and a w/b ratio in the range 0.35–0.39, gave values for compressive strength in the range 38–59 MPa. It was found that the incorporation of macro-fibers in SCLWC gave better compressive strength and post-cracking behavior compared to SCLWC containing micro-fibers. In addition, Iqbal *et al.* (2015) investigated the mechanical properties of high-strength SCLWC using LECA as coarse aggregate and also used steel fibers up to 1.25%. They chose a w/b ratio in the range 0.46–0.48 and achieved a compressive strength in the range 60–68 MPa. It was observed that, as the steel fiber content increased beyond 1%, workability was highly influenced. However, a reduction in compressive strength of about 12% was observed compared to SCLWC containing 0% steel fibers. The same material was considered by Karamloo *et al.* (2016), to observe the effect of maximum aggregate size on the fracture behavior of SCLWC. With a w/b ratio in the range 0.35–0.40 they achieved a 28-day compressive strength of 27–49 MPa. It was observed that SCLWC with a w/b ratio of 0.4 showed 87% higher fracture energy as the maximum aggregate size increased from 9.5 to 19 mm.

Few researchers have used LECA as both fine and coarse aggregates for the production of SCLWCs. Grabois *et al.* (2016) investigated the properties of SCLWC made with LECA as both fine and coarse aggregate, and with steel fibers. With w/b ratios in the range 0.33 to 0.40 they achieved a 28-day compressive strength in the range 31–37 MPa. Similarly, LECA as both fine and coarse was investigated by Mazaheripour *et al.* (2011), but instead of steel fibers they utilized propylene fibers as an additional reinforcement. They prepared a total of eight mixes with varying ranges of fibers and materials and with the same w/b ratio of 0.32 and they achieved a range of compressive strengths between 21 MPa and 27 MPa. By comparing both studies with different fiber ratios, it was found that SCLWC with steel fibers showed better results in terms of compressive strength even at a higher w/b ratio when compared to SCLWC containing propylene fibers (see Table 5.5).

During the investigation of the SCLWCs made of alternative materials, a few new types of material were also found with which the SCLWC was able to be successfully developed. Yang *et al.* (2015) studied the properties of SCLWC made with expanded clay ceramsite and plastic waste particles. Of all the mixes shown in Table 5.5, they chose the lowest w/b ratio – 0.25 – for the development of SCLWC but only managed a highest 28-day compressive strength value of 27 MPa. Gesoğlu *et al.* (2014) investigated the properties of SCLWC made with spherical pellets (manufactured through a cold bonding process) as an aggregate. They prepared ten mixes with the same w/b ratio of 0.32 and achieved a compressive strength in the range 52–70 MPa, which was significantly higher compared to the compressive strength results of all the SCLWCs (Table 5.5 and Figure 5.8). The relationship between the compressive strengths and the w/b ratios of the SCLWCs made of alternative materials is given in Figure 5.8. As expected, increasing the w/c ratio reduces the compressive strength of SCLWCs; however, in the case of the addition of steel fibers to SCLWC made of LECA as coarse aggregate the compressive strength increases with increasing w/b ratios.

Yu *et al.* (2019) investigated the mechanical properties of SCLWC under complex stresses (uni-axial, bi-axial and multi-axial stresses) using a tri-axial testing machine. They obtained various results in terms of stress–strain curve eigenvalues (peak stress and strain) and failure morphology under different lateral stresses. It was found that the uni-axial compression had no significant effect on SCLWC, whereas bi-axial compression completely changed the failure morphology towards the shear morphology for SCLWC in comparison to conventional concrete. Furthermore, the strength and deformantion of SCLWC is strongly affected by lateral stresses higher than the conventional NWC and LWAC and the failure criteria for SCLWC based on Kupfer bi-axial failure criteria and octahedral stress space has positive applicability in real structures.

9.3 Flexural strength

The flexural strength or modulus of rupture (MOR) for the SCLWCs produced by various alternative materials is shown in Table 5.5. Corinaldesi and Moriconi (2015) experimented with SCLWC made of LECA aggregates and synthetic fibers. They classified the mix proportions based on micro- and macro-fibers. With micro-fibers the flexural strength values were found to be in the range 2.0–3.5 MPa, whereas the results for macro-fiber SCLWCs were in the range 2–3 MPa and were approximately 5–6% of the 28-day compressive strength of SCLWC concrete. Iqbal *et al.* (2015) investigated high-strength SCLWC using LECA as coarse aggregate and steel fibers up to 1.25%. The w/b ratio was in the range 0.46–0.48 and they achieved a MOR in the range 3.7–7.6 MPa, approximately 11% of the compressive strength. They reported that the addition of steel fibers to the SCLWC significantly improved the flexural strength of the concrete and also produced strain hardening material.

Table 5.5 *Mechanical properties of self-compacting lightweight concretes.*

Binder type	Lightweight aggregate type	Mix No. / Mix Codes		Water-to-binder ratios		Compressive strength (MPa)		Splitting tensile strength (MPa)		Flexural strength (MPa)	Modulus of elasticity (GPa)		References
OPC + Fly Ash	Sintering fine sediment excavated from reservoir	D28-150 D32-140 D32-150 D32-160	D32-170 D32-180 D32-190 D40-150	0.28 0.32 0.32 0.32	0.32 0.32 0.32 0.40	52 40 45 49	48 47 46 32			–	–		Hwang and Hung, 2005
OPC (Type I)	POC as fine and coarse aggregates	POC25 POC50 POC75 POC100		0.28 0.28 0.28 0.28		50 40.5 48 41				–	–		Kanadasan and Razak, 2014b
OPC + Pulverized Fuel Ash (PFA)	Expanded shale aggregate	1 2 3 4	5 6 7	0.30 0.30 0.30 0.30	0.30 0.30 0.30	42 48 47 52	55 58 60			–	22.5 24.1 24.7 25	26.4 26.9 27.2	Lo *et al.*, 2007
OPC + FA	Spherical pellets manufactured through cold bonding process	MC10 MC20 MC30 MC40 MC50	MCF10 MCF20 MCF30 MCF40 MCF50	0.32 0.32 0.32 0.32 0.32	0.32 0.32 0.32 0.32 0.32	70 65 61 58 55	68 63 59 56 52	3.7 3.35 3.0 2.8 2.6	3.6 3.3 2.9 2.7 2.55	–	–		Gesoğlu *et al.*, 2014

(Continued)

Table 5.5 *(Continued)*

| Self-compacting lightweight concretes (SCLWC) | | Mix No. / Mix Codes | | Water-to-binder ratios | | Mechanical properties of SCLWCs (28 days) | | | | | | | | References |
Binder type	Lightweight aggregate type					Compressive strength (MPa)		Splitting tensile strength (MPa)		Flexural strength (MPa)		Modulus of elasticity (GPa)		
OPC + FA	Pumice + tuff + diatomite	SCLC1		0.36		27		1.8		–		18.0		Topçu and Uygunoğlu, 2010
		SCLC2		0.40		24		1.7				17.0		
		SCLC3	SCLC11	0.43	0.36	23	20	1.6	1.2			13.0	9.0	
		SCLC4	SCLC12	0.46	0.40	20.5	17	1.5	1.1			11.2	8.0	
		SCLC5	SCLC13	0.48	0.43	18	16	1.4	1.0			5.0	6.0	
		SCLC6	SCLC14	0.36	0.46	22	15	1.5	0.95			10.6	4.0	
		SCLC7	SCLC15	0.40	0.48	21	12.5	1.47	0.75			10.5	3.5	
		SCLC8		0.43		19		1.42				9.5		
		SCLC9		0.46		18		1.4				8.0		
		SCLC10		0.48		16		1.3				6.0		
OPC + FA + Silica fume (SF) + fibers	LECA	EC+FA+microF	EC+FA+Sand+microF	0.36	0.36	44.7	54.5	1.66	1.90	2.96	2.64	13.10	18.65	Corinaldesi and Moriconi (2015)
		EC+FA+RCA	EC+FA+Sand+macroF	0.35	0.35	49.0	58.7	2.25	2.19	2.84	3.00	15.98	23.31	
		EC+FA+RCA+macroF	EC+SF+microF	0.35	0.39	46.7	37.8	1.99	1.70	3.47	2.00	18.36	10.19	
		EC+FA+Sand	EC+SF+macroF	0.35	0.37	52.1	40.3	1.76	2.33	2.64	2.11	19.00	13.60	
OPC + FA + Steel fibers	LECA as fine and coarse aggregate	SCLC1000		0.33		36.4		–		–		21.4		Grabois *et al.*, 2016
		SCLC1000SF		0.33		34.8						21.1		
		SCLC7030		0.40		37.0						22.0		
		SCLC7030SF		0.40		31.7						19.4		

(Continued)

Table 5.5 *(Continued)*

Self-compacting lightweight concretes (SCLWC)		Mix No. / Mix Codes		Water-to-binder ratios		Mechanical properties of SCLWCs (28 days)								References
Binder type	Lightweight aggregate type					Compressive strength (MPa)		Splitting tensile strength (MPa)		Flexural strength (MPa)		Modulus of elasticity (GPa)		
OPC + FA	LECA	Mix-0 Mix-0.5 Mix-0.75	Mix-1 Mix-1.25	0.46 0.46 0.46	0.46 0.48	67.8 64.0 61.0	60.9 59.7	4.10 4.76 5.02	5.42 5.64	3.7 4.43 6.13	6.36 7.62	16.58 15.78 15.60	15.67 15.35	Iqbal *et al.*, 2015
OPC + FA	Expanded clay ceramsite + plastic particles	P0 P10 P15	P20 P30	0.25 0.25 0.25	0.25 0.25	25.0 24.6 27.0	26.0 23.0	2.40 2.80 3.00	2.50 2.41	3.00 4.10 4.20	4.20 4.10	17.5 – 16.2	– –	Yang *et al.*, 2015
OPC + FA + Metakaolin	Coarse expanded shale	MK0% MK5% MK10% MK15%		0.33 0.33 0.33 0.33		39.7 39.8 40.1 39.4		1.9 2.0 2.0 2.1		3.7 4.0 3.8 3.6		–		Karahan *et al.*, 2012

(Continued)

Table 5.5 *(Continued)*

Self-compacting lightweight concretes (SCLWC)		Mix No. / Mix Codes		Water-to-binder ratios		Mechanical properties of SCLWCs (28 days)								References
Binder type	Lightweight aggregate type					Compressive strength (MPa)		Splitting tensile strength (MPa)		Flexural strength (MPa)		Modulus of elasticity (GPa)		
OPC + SF + Propylene fiber	LECA as fine and coarse aggregate	GP0.0	IP0.0	0.32	0.32	25.3	21.5	2.77	2.32	4.86	4.36	23.9	22.1	Mazaheripour *et al.*, 2011
		GP0.1	IP0.1	0.32	0.32	24.6	21.7	2.98	2.34	5.10	4.52	24.8	21.6	
		GP0.2	IP0.2	0.32	0.32	26.3	22.7	3.02	2.62	5.28	4.69	24.1	22.0	
		GP0.3	IP0.3	0.32	0.32	24.6	22.8	3.17	2.65	5.38	4.74	24.0	22.4	
OPC + FA + Blast furnace slag (BFS)	Pumice as coarse aggregate	CS1	FANP100	0.35	0.30	13.9	19.9	1.9	1.8					Kurt *et al.*, 2016
		CS2	FANP80	0.40	0.30	12.7	22.2	1.9	2.2					
		CS3	FANP60	0.45	0.30	10.6	26.7	1.7	2.2					
		PP1	FANP40	0.35	0.30	13.2	30.9	1.9	3.1					
		PP2	FANP20	0.40	0.30	11.7	36.1	1.8	3.4					
		PP3	FANP00	0.45	0.30	10.7	53.3	1.7	4.2	—		—		
		PP4	BFSNP100	0.35	0.30	12.5	21.3	1.8	2.1					
		PP5	BFSNP80	0.40	0.30	11.4	25.1	1.8	2.8					
		PP6	BFSNP60	0.45	0.30	10.5	26.8	1.7	3.0					
		PP7	BFSNP40	0.35	0.30	12.3	32.6	1.8	3.3					
		PP8	BFSNP20	0.40	0.30	11.5	37.8	1.8	4.0					
		PP9	BFSNP00	0.45	0.30	10.5	65.0	1.7	7.3					

(Continued)

Table 5.5 *(Continued)*

Binder type	Lightweight aggregate type	Mix No. / Mix Codes	Water-to-binder ratios	Compressive strength (MPa)	Splitting tensile strength (MPa)	Flexural strength (MPa)	Modulus of elasticity (GPa)	References
OPC	LECA	SCLC1	0.35	48.83	3.44		25.12	Karamloo et al., 2016
		SCLC2	0.40	40.45	3.05		23.14	
		SCLC3	0.35	40.44	2.94	–	23.31	
		SCLC4	0.40	31.70	2.81		20.30	
		SCLC5	0.35	36.20	2.85		23.20	
		SCLC6	0.40	27.02	2.05		19.23	
OPC + FA + Metakaolin	Lightweight slag aggregate	NSNC	0.40	11.53	1.61	2.22	–	Abouhussien et al., 2015
		HSNC1 / HSSCC1	0.35 / 0.35	39.10 / 48.59	2.62 / 4.17	3.80 / 5.57		
		HSNC2 / HSSCC2	0.35 / 0.35	41.44 / 46.89	2.97 / 4.03	4.82 / 4.85		
		NSSCC1 / HSSCC3	0.40 / 0.35	19.65 / 41.58	2.02 / 3.01	2.85 / 4.70		
		NSSCC2 / HSSCC4	0.40 / 0.35	18.16 / 40.77	1.97 / 2.85	2.76 / 4.42		
		NSSCC3	0.40	18.42	1.90	2.46		
		NSSCC4	0.40	16.78	1.82	2.19		

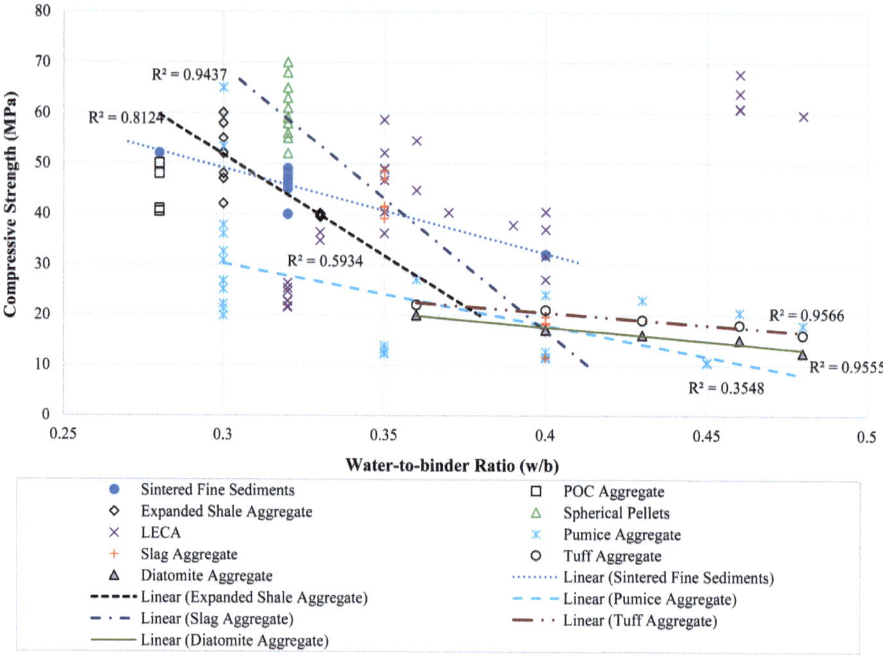

Figure 5.8 *Relationship between compressive strengths and water-to-binder ratios of SCLWCs containing alternative materials as lightweight aggregates: sintered fine sediments aggregates (Hwang and Hung, 2005); palm oil clinkers (POC) (Kanadasan and Razak, 2014a, 2014b, 2015); expanded shale aggregate (Lo et al., 2007; Karahan et al., 2012); spherical pellets (Gesoğlu et al., 2014); lightweight expanded clay aggregate (LECA) as coarse and combinations of both coarse and fine (Mazaheripour et al., 2011; Corinaldesi and Moriconi (2015); Iqbal et al., 2015; Grabois et al., 2016; Karamloo et al., 2016); pumice as coarse and a combination as both coarse and fine (Topçu and Uygunoğlu, 2010; Kurt et al., 2016); and slag aggregate (Abouhussien et al., 2015).*

SCLWC was also produced using LECA as both fine and coarse aggregate by Mazaheripour *et al.* (2011). They incorporated propylene fibers as an additional reinforcement in order to observe the effect of materials on the properties of the concrete. In both sets of mixes, using the same w/b ratio of 0.32, flexural strength values were achieved in the range 4.3–5.4 MPa, about 20% of the respective compressive strengths. It was further observed that, as the volume of propylene fibers was increased, the flexural strength was also improved (see Table 5.5).

During the investigation of the flexural strength of the SCLWCs made of alternative materials, a few new types of material were also found. Yang *et al.* (2015) studied the properties of SCLWC made with expanded clay ceramsite and plastic waste particles. At same w/b ratio of 0.25, the increase in compressive strength was not significant, however, the flexural strength results were found better in the range of 3 to 4.2 MPa, which is about 16% of the 28-day compressive strength. It was observed that, as the plastic content was increased up to 15%, the flexural strength was also improved; however, beyond this replacement level a certain reduction in flexural strength occurred because more free water weakens the interfacial bonding between the cement paste and the plastic particles (Table 5.5).

Karahan *et al.* (2012) reported the flexural strength for expanded shale SCLWC with metakaolin binder in the range of 3.6 to 4.0 MPa which is about 10% of its 28-day compressive strength. Generally, it was observed that the utilization of metakaolin with expanded shale aggregate slightly increased the density of the concrete and even affected the passing and filling ability of the SCLWC but had no positive effect on the flexural strength of the concrete. Furthermore, Abouhussien *et al.* (2015) explored the effect of metakaolin with other binders on the flexural strength of semi-lightweight SCLWC using lightweight slag aggregate (LSA). They prepared eleven mixes with w/b ratios of 0.35 and 0.40 and achieved a 28-day flexural strength in the range 2.2–5.6 MPa (see Table 5.5).

9.4 Splitting tensile strength

As with compressive strength, the splitting tensile strength of the concrete plays an important role in structural design, specifically in the design of roads/highways and airfield slabs, with regard to providing better resistance to cracking (Neville, 2008). Few researchers have investigated the splitting tensile strength of SCLWCs through a consideration of naturally available materials (such as pumice, diatomite and tuff) as an alternative aggregate. Kurt *et al.* (2016) studied the effect of pumice powder on SCLWC made of pumice aggregate with aggregate size in the range of 0–16 mm. With a w/b ratio in the range 0.30 to 0.45, they achieved a 28-day splitting tensile strength for SCLWCs in the range 1.7–4.0 MPa, or about 11% of the compressive strength. Furthermore, Topçu and Uygunoğlu (2010) considered the same material and explored the effect of pumice aggregate on the hardened property splitting tensile strength of the SCLWC. For pumice aggregate SCLWC, the splitting tensile strength values were found to be in the range 1.4–1.8 MPa within a w/b ratio range of 0.36–0.48, about 7% of the compressive strength. They also investigated the splitting tensile strength of SCLWCs made with tuff and diatomite aggregates, using the same w/b ratio as they did with pumice SCLWC. The splitting tensile strength values were found lie in the range 1.3–1.5 MPa and 0.75–1.2 MPa, respectively (see Table 5.5). In all three sets, it was observed that as the w/b ratio increases, so the splitting tensile strength is reduced; however, at the same w/b ratios of between 0.36 to 0.48, the pumice aggregate gives better results compared to SCLWCs made with tuff and diatomite aggregates.

Several researchers studied the use of lightweight expanded clay aggregate (LECA) as an aggregate for the development of SCLWCs. Corinaldesi and Moriconi (2015) developed SCLWC using synthetic fibers; OPC, fly ash and silica fume as the binders. They utilized two types of fibers, the first set contained micro-fibers, had a w/b ratio in the range 0.35–0.36, and gave a 28-day splitting tensile strength in the range 1.66–2.0 MPa, or about 4% of the compressive strength. However, the second set, with macro-fibers and a w/b ratio range of 0.35–0.39, gave values between 1.7 MPa and 2.3 MPa (Table 5.5). It was found that incorporation of micro and macro synthetic fibers into the SCLWC had no significant effect on the splitting tensile strength, although the macro-fibers gave better compressive strength results.

Iqbal *et al.* (2015) prepared high-strength SCLWC using LECA and steel fibers up to 1.25%. They reported splitting tensile strength in the range 4.1–5.6 MPa for w/b ratios in the range 0.46–0.48. Similarly, Karamloo *et al.* (2016) observed the effect of maximum aggregate size on the properties of SCLWC and reported splitting tensile strength results in the range 2–3.4 MPa for w/b ratios ranging between 0.35 and 0.40. Later, LECA as both fine and coarse aggregate with propylene fibers was trialed for SCLWC by Mazaheripour *et al.* (2011). They prepared a total of eight mixes with a varying range of fibers and materials and with a w/b ratio of 0.32. They achieved splitting tensile strengths of between 2.3 MPa and 3.2 MPa, or about 12% of the compressive strength, as can be seen in Table 5.5.

The relationship between the splitting tensile strength and the w/b ratios for the SCLWCs made of naturally available alternative materials as an aggregate is shown in Figure 5.9. It was observed that, as the w/b ratio increases, so the splitting tensile strength decreases for all the materials except SCLWC made with LECA aggregates. This might be due to the composition of LECA materials, as the authors have collectively presented all the available results for the LECA concretes in the given graph. The scatter data for SCLWCs made with LECA as coarse aggregate, as a combination

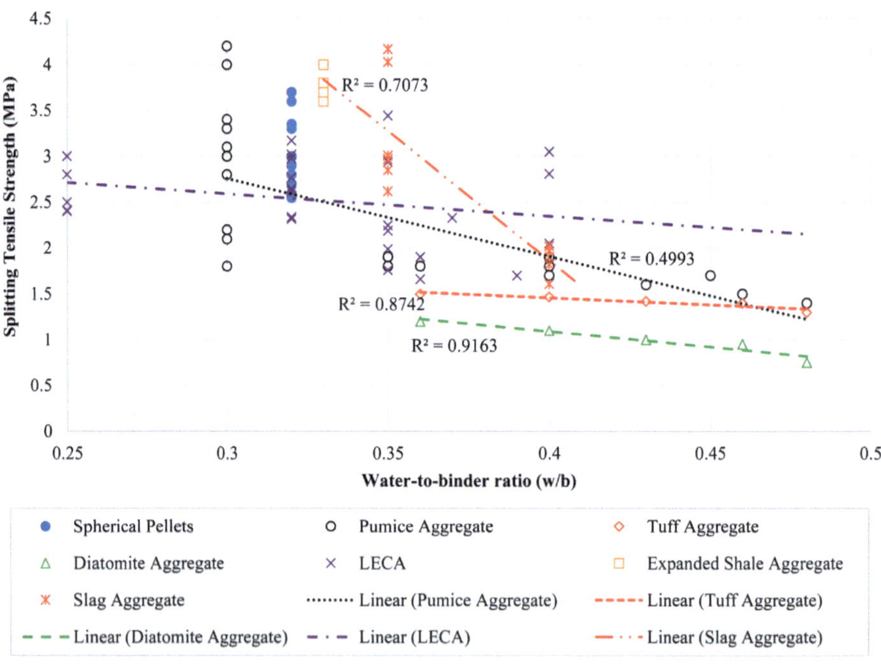

Figure 5.9 *Relationship between splitting tensile strengths and the water-to-binder ratios of SCLWCs containing alternative materials as lightweight aggregates: lightweight expanded clay aggregate (LECA) as coarse and as a combination of both coarse and fine (Mazaheripour et al., 2011; Corinaldesi and Moriconi (2015); Iqbal et al., 2015; Grabois et al., 2016; Karamloo et al., 2016); pumice as coarse and as a combination of both coarse and fine (Topçu and Uygunoğlu, 2010; Kurt et al., 2016); slag aggregate (Abouhussien et al., 2015); tuff aggregate and diatomite aggregate (Topçu and Uygunoğlu, 2010).*

of both fine and coarse aggregates, and with different types of binders, showed a reliability of about 50% between the results. A similar reliability was also found for pumice aggregate SCLWCs. However, SCLWCs made with expanded slag, tuff and diatomite gave more reliable trends for the splitting tensile strength, with reliability values of about 70%, 87%, and 92%, respectively.

9.5 Modulus of elasticity

The modulus of elasticity is one of the most important properties in the design of structural components because it reflects the stiffness of the material. The modulus of elasticity of concrete depends on the moduli of elasticity of its constituents, and also on their quantities by volume: as the volume of conventional crushed stones is reduced or replaced with LWAs in the concrete, so the modulus of elasticity is reduced (Neville, 2008).

The modulus of elasticity of LWAs generally ranges between 5 GPa and 28 GPa. This is lower than for NWAs and even lower than for mortar (CEB/FIP, 1977). Therefore, it is expected that by reducing the volume of NWA in the concrete and replacing it with LWA, the modulus of elasticity will be significantly reduced. Usually, the modulus of elasticity of structural LWCs is found to be in the range 10–24 GPa, whereas for NWC the range varies between 14 GPa and 41 GPa (Shafigh *et al.*, 2014; Aslam *et al.*, 2016b). The modulus of elasticity of SCLWCs prepared by different types of alternative material was investigated by few researchers, as can be seen from Table 5.5. Lo *et al.* (2007) compared the modulus of elasticity of SCC and SCLWCs using conventional and expanded shale (ESA) as LWAs, respectively. With a w/b ratio of 0.30 for SCLWC they achieved E-values in the range 22–27 GPa, or about 20% lower than the conventional SCC. These values were found to be higher than the E-values of LWCs recommended by Shafigh *et al.* (2014). Furthermore, Topçu and Uygunoğlu (2010) performed a detailed study of the effect of various natural aggregate types on the modulus of elasticity of SCLWCs. They prepared three sets of SCLWCs using pumice, tuff and diatomite as lightweight coarse aggregates and considered similar w/b ratios in the range 0.36–0.48. The modulus of elasticity for SCLWCs containing pumice, tuff and diatomite was found to lie in the ranges 5–18 GPa, 6–10.6 GPa, and 3.5–9 GPa, respectively (Table 5.5). It was found that pumice aggregate gave better results for modulus of elasticity of SCLWC compared to concrete made with tuff and diatomite, because pumice aggregate by nature has a rough surface texture which gives stronger bonding with the cement matrix.

Several researchers investigated the modulus of elasticity of the SCLWCs using LECA as an aggregate. Corinaldesi and Moriconi (2015) investigated the properties of SCLWC using the synthetic fiber, classifying the mix proportions based on micro- and macro-fibers. Both sets, containing micro- and macro-fibers with different w/b ratios, showed similar E-values in the range 13–23.3 GPa. In addition, Iqbal *et al.* (2015) investigated the E-values of high-strength SCLWC using LECA as coarse aggregate and also using steel fibers up to 1.25%. With w/b ratios in the range 0.46–0.48 they achieved elasticity values in the range 15–17 GPa. Although these mixes had higher w/b ratios and steel fibers, they showed lower E-values compared to previous studies containing synthetic fibers. The same material was also considered by Karamloo *et al.* (2016) in order to observe the effect of maximum aggregate size on the fracture behavior of SCLWC. Karamloo and colleagues used w/b ratios of 0.35 and 0.40 and achieved values for modulus of elasticity in the range 19–25 GPa.

Few researchers have used LECA as both fine and coarse aggregate for the production of SCLWCs. Grabois *et al.* (2016) investigated the properties of SCLWC made with LECA as both fine and coarse aggregate, and with steel fibers. They chose w/b ratios of 0.33 and 0.40 and achieved a 28-day elastic modulus value in the range 19.4–22 GPa. Similarly, LECA as both fine and coarse aggregate was investigated by Mazaheripour *et al.* (2011) but using propylene fibers as an additional reinforcement instead of steel fibers. They prepared a total of eight mixes with varying fibers and materials, using the same w/b ratio of 0.32, achieving slightly higher E-values, ranging between 21.6 GPa and 25 GPa compared to SCLWCs containing LECA and steel fibers (see Table 5.5).

10. Concluding remarks

This chapter provided a comprehensive review of the potential use of alternative materials as aggregates and binders for the production of sustainable and environmentally friendly self-compacting lightweight concrete (SCLC). The physical, mechanical, and chemical properties of alternative waste aggregate materials were discussed, and their proportions for the mix design were elaborated. In addition, the effect of different types of aggregates on the fresh and hardened properties of SCLC has been discussed.

Lightweight aggregates (LWAs) are natural, manufactured, or industrial by-product. A LWA is the most important ingredient of SCLC. Many types of LWAs were successfully used in SCLC. The specific gravity of these aggregates ranged from 0.8 to 2.3 with a compacted density, for most of the aggregates, less than 1000 kg/m^3, while a few heavers lightweight aggregates, with maximum density up to 1400 kg/m^3, were also used in this type of concrete. In addition, there are several types of alternative pozzolanic binders that increase the filling ability of concrete and improve the interface bonding between aggregates and the binder paste.

Self-compacting concrete (SCC) design methods such as the compressive strength method (CSM), the empirical design method (EDM), the rheology of paste model (RPM), the statistical factorial model (SFM), and the close aggregate packing method (CAPM) were applied to develop SCLC with and without utilization of alternative waste materials as aggregates and binders. It was found that using some of these methods can produce high-strength SCLC with a compressive strength ranging from 60 to 68 MPa, and a density less than 1700 kg/m^3, while the slump flow of the fresh concrete remails about 600 mm. The contribution of fly ash as filler or ternary blend in a SCLC mixture improves the fresh properties and fluidity of concrete and reduces the risk of blockage and segregation. It was also found that the usage of pre-wetted aggregates has no significant effect on self-compaction ability and stability of the SCLC mixtures.

In terms of the mechanical properties of SCLC mixes, several types of aggregates showed higher compressive strength than the minimum strength requirement of 17 MPa as recommended by ACI. The OPBC mixes showed the 28-day compressive strength up to 50 MPa, slag aggregate up to 49 MPa, sintered aggregates up to 52 MPa, expanded shale up to 60 MPa, and LECA aggregates showed a wide range of compressive strengths from 27 to 68 MPa depending on the w/b ratio and addition of fibers. While some natural aggregates showed compressive strength in the range of 18 to 27 MPa for pumice, 16 to 22 MPa for tuff and 13 to 22 MPa for diatomite aggregates.

In general, the SCLC is very versatile for structural applications where both strength and reduced weight are critical. Additionally, this concrete has the advantage of high flowability without segregation and blocking which is ideal to be applied for structural elements with congested reinforcements. However, a proper selection of materials and optimize mix proportions should be considered for the large-scale applications. While, assessing the concrete durability under different environmental conditions and investigating the blended approaches for multiple waste materials are essential. Further, the SCLC still needs detailed investigation on cost-benefit analysis and life-cycle assessment to promote its wide applications and to promote its transition towards sustainable and green construction practices.

References

Abdul Talib, N.R. (2010) Engineering characteristics of Bottom ash from power plants in Malaysia. Unpublished B. Eng. Thesis, University of Technology Malaysia, Johor, Malaysia.

Abdullah, E.S., Mirasa, A.K. and Asrah, H. (2015) Review on the effect of Palm Oil Fuel Ash (POFA) on concrete. *Journal of Industrial Engineering Research* **1**(7 Special), 1–4.

Abouhussien, A.A., Hassan, A.A. and Ismail, M.K. (2015) Properties of semi-lightweight self-consolidating concrete containing lightweight slag aggregate. *Construction and Building Materials* **75**, 63–73.

ACI. (1991) Standard practice for selecting proportions for normal, heavyweight and mass concrete (ACI 211.1-91). Committee Editors. Farmington Hills, MI: American Concrete Institute.

ACI Committee 116R (1990) Cement and Concrete Terminology, (ACI 116R-90), *Manual of Concrete Practice*, Part 2, American Concrete Institute, Detroit.

Asniar, N., Purwana, Y.M. and Surjandari, N.S. (2019, June). Tuff as rock and soil: Review of the literature on tuff geotechnical, chemical and mineralogical properties around the world and in Indonesia. In AIP Conference Proceedings (Vol. 2114, No. 1, p. 050022). AIP Publishing LLC

Ahmad, H., Hilton, M. & Noor, N.M. (2007b) Physical properties of local palm oil clinker and fly ash. In: *Proceedings of 1st Engineering Conference on Energy and Environment (EnCon2007)*. Sarawak, Malaysia: University of Malaysia. UTHM Repository.

Ahmad, S., Sallam, Y.S. and Al-Hashmi, I.A.R. (2013) Optimizing dosage of Lytag used as coarse aggregate in lightweight aggregate concretes. *Journal of the South African Institution of Civil Engineering* **55**(1), 80–84.

Alengaram, U.J., Al Muhit, B.A. and Jumaat, M.Z. (2013) Utilization of oil palm kernel shell as lightweight aggregate in concrete-a review. *Construction and Building Materials* **38**, 161–172.

Altwair, N.M. and Kabir, S.G. (2010) Concrete structures by replacing cement with pozzolanic materials to reduce greenhouse gas emissions for sustainable environment. Paper presented at the American Society of Civil Engineers, 6th International Engineering and Construction Conference (IECC'6), Cairo, Egypt.

Arenas, C., Leiva, C., Vilches, L.F. and Cifuentes, H. (2013) Use of co-combustion bottom ash to design an acoustic absorbing material for highway noise barriers. *Waste Management* **33**(11), 2316–2321.

Arik, H. (2003) Synthesis of Si_3N_4 by the carbo-thermal reduction and nitridation of diatomite. *Journal of the European Ceramic Society* **23**(12), 2005–2014.

Aslam, M., Shafigh, P. and Jumaat, M.Z. (2015) Structural lightweight aggregate concrete by incorporating solid wastes as coarse lightweight aggregate. *Applied Mechanics and Materials* **749**, 337–342.

Aslam, M., Shafigh, P. & Jumaat, M.Z. (2016a) Drying shrinkage behaviour of structural lightweight aggregate concrete containing blended oil palm bio-products. *Journal of Cleaner Production* **127**, 183–194.

Aslam, M., Shafigh, P. and Jumaat, M.Z. (2016b) Oil-palm by-products as lightweight aggregate in concrete mixture: A review. *Journal of Cleaner Production* **126**, 56–73.

Aslam, M., Shafigh, P. and Jumaat, M.Z. (2017) High strength lightweight aggregate concrete using blended coarse lightweight aggregate origin from palm oil industry. *Sains Malaysiana* **46**(4), 667–675.

Aslam, M., Shafigh, P., Jumaat, M Z. and Lachemi, M. (2016) Benefits of using blended waste coarse lightweight aggregates in structural lightweight aggregate concrete. *Journal of Cleaner Production* **119**, 108–117.

Aslam, M., Shafigh, P., Nomeli, M.A. and Jumaat, M.Z. (2017) Manufacturing of high-strength lightweight aggregate concrete using blended coarse lightweight aggregates. *Journal of Building Engineering* **13**, 53–62.

Bellotto, M., Gualtieri, A., Artioli, G. and Clark, S.M. (1995) Kinetic study of the kaolinite-mullite reaction sequence. Part I: Kaolinite dehydroxylation. *Physics and Chemistry of Minerals* **22**(4), 207–217.

Bilim, C. (2011) Properties of cement mortars containing clinoptilolite as a supplementary cementitious material. *Construction and Building Materials* **25**(8), 3175–3180.

Bogas, J.A., Gomes, A. and Pereira, M.F.C. (2012) Self-compacting lightweight concrete produced with expanded clay aggregate. *Construction and Building Materials* **35**, 1013–1022.

Bohan, R. and John R. (2008) "Structural Lightweight Aggregate Concrete." Concrete Technology IS032.06, Skokie, IL: Portland Cement Association (PCA). https://www.escsi. org/wp-content/uploads/2020/09/CT-SLWAC-PCA-2008.pdf.

Bouziani, T. (2013) Assessment of fresh properties and compressive strength of self-compacting concrete made with different sand types by mixture design modelling approach. *Construction and Building Materials* **49**, 308–314.

Bravo, M., Brito, J., Pontes, J. and Evangelista, L. (2015) Mechanical performance of concrete made with aggregates from construction and demolition waste recycling plants. *Journal of Cleaner Production* **99**, 59–74.

Bremner, T.W. and Holm, T.A. (1995) High performance lightweight concrete: A review. *ACI Special Publication* **154**, 1–20.

Bui, V.K., Akkaya, Y. and Shah, S.P. (2002) Rheological model for self-consolidating concrete. *ACI Materials Journal* **99**(6), 549–559.

Cadersa, A.S. and Auckburally, I. (2014) Use of unprocessed coal bottom ash as partial fine aggregate replacement in concrete. *University of Mauritius Research Journal* **20**, 62–84.

Carballosa, P., Calvo, J.G., Revuelta, D., Sanchez, J.J. and Gutierrez, J.P. (2015) Influence of cement and expansive additive types in the performance of self-stressing and self-compacting concretes for structural elements. *Construction and Building Materials* **93**, 223–229.

CEB/FIP manual of design and technology, (1977) Lightweight Aggregate Concrete. First Publication. The Construction Press Ltd., Lancaster, England.

Chandra, S. and Berntsson, L. (2002) *Lightweight Aggregate Concrete: Science, Technology, and Applications*. Norwich, NY: Noyes Publications, William Andrew Publishing.

Chen, H.J., Wu, K.C., Tang, C.W. and Huang, C.H. (2018) Engineering properties of self-consolidating lightweight aggregate concrete and its application in prestressed concrete members. *Sustainability* **10**(1), 142.

Clarke, J.L. (2002) *Structural Lightweight Aggregate Concrete*. London: CRC Press.

Corinaldesi, V. and Moriconi, G. (2004) Durable fiber reinforced self-compacting concrete. *Cement and Concrete Research* **34**(2), 249–254.

Corinaldesi, V. and Moriconi, G. (2011) The role of industrial by-products in self-compacting concrete. *Construction and Building Materials* **25**(8), 3181–3186.

Corinaldesi, V. and Moriconi, G. (2015) Use of synthetic fibers in self-compacting lightweight aggregate concretes. *Journal of Building Engineering*, 4, 247–254.

Dinakar, P., Babu, K.G. and Santhanam, M. (2008) Durability properties of high-volume fly ash self-compacting concretes. *Cement and Concrete Composites* **30**(10), 880–886.

Edamatsu, Y. and Ouchi, M. (2003) A mix-design method for self-compacting concrete based on mortar flow and funnel tests. In O. Wallevik and I. Nielsson (eds) *International RILEM Symposium on Self-compacting Concrete*. Champs-sur-Marne, France: RILEM Publications SARL, pp. 345–354.

Edamatsu, Y., Nishida, N. and Ouchi, M. (1999) A rational mix-design method for self-compacting concrete considering interaction between coarse aggregate and mortar particles. In Å. Skarendahl and Ö. Petersson (eds) *Proceedings of the 1st International RILEM Symposium on Self-compacting Concrete, Stockholm. Sweden (1999)*. Champs-sur-Marne, France: RILEM Publications SARL, pp. 309–320.

Emdadi, Z., Asim, N., Yarmo, M.A. and Shamsudin, R. (2014) Investigation of more environmentally friendly materials for passive cooling application based on geopolymer. *APCBEE Procedia* **10**, 69–73.

Esmaeilkhanian, B., Khayat, K.H., Yahia, A. and Feys, D. (2014) Effects of mix design parameters and rheological properties on dynamic stability of self-consolidating concrete. *Cement and Concrete Composites* **54**, 21–28.

Feng, N. and Peng, G. (2005) Applications of natural zeolite to construction and building materials in China. *Construction and Building Materials* **19**(8), 579–584.

Fidjestol, P. and Dastol, M. (2008) The history of silica fume in concrete from novelty to key ingredient in high performance concrete. In *Proceedings of the Congresso Brasileiro do Concreto, Elkem Materials. 4th–9th September, 2008*.

Gesoğlu, M., Güneyisi, E., Özturan, T., Öz, H.Ö. and Asaad, D.S. (2014) Permeation characteristics of self-compacting concrete made with partially substitution of natural aggregates with rounded lightweight aggregates. *Construction and Building Materials* **59**, 1–9.

Grabois, T.M., Cordeiro, G.C. and Toledo Filho, R.D. (2016) Fresh and hardened-state properties of self-compacting lightweight concrete reinforced with steel fibers. *Construction and Building Materials* **104**, 284–292.

Güneyisi, E., Gesoğlu, M., Booya, E. and Mermerdas, K. (2015) Strength and permeability properties of self-compacting concrete with cold bonded fly ash lightweight aggregate. *Construction and Building Materials* **74**, 17–24.

Han, J., Fang, H. and Wang, K. (2014) Design and control shrinkage behavior of high-strength self-consolidating concrete using shrinkage-reducing admixture and super-absorbent polymer. *Journal of Sustainable Cement-based Materials* **3**(3–4), 182–190.

Hemmings, R.T., Cornelius, B.J., Yuran, P. and Milton, W. (2009) Comparative study of lightweight aggregates. Paper presented at the 2009 World of Coal Ash (WOCA) Conference, May 4–7. 2009, Lexington, KY, USA.

Hwang, C. (2005) The effect of aggregate packing types on engineering properties of self-consolidating cocrete. In Zhiwu Yu, Caijun Shi, Kamal Henri Khayat and Youjun Xie (eds) *1st International Symposium on Design, Performance and Use of Self-consolidating Concrete (SCC'2005-China)*. Champs-sur-Marne, France: RILEM Publications SARL, pp. 337–345.

Hwang, C. L. and Hung, M.F. (2005) Durability design and performance of self-consolidating lightweight concrete. *Construction and Building Materials* **19**(8), 619–626.

Ibrahim, M.W., Hamzah, A.F., Jamaluddin, N., Ramadhansyah, P.J. and Fadzil, A.M. (2015) Split tensile strength on self-compacting concrete containing coal bottom ash. *Procedia-Social and Behavioral Sciences* **195**, 2280–2289.

Iqbal, S., Ali, A., Holschemacher, K. and Bier, T.A. (2015) Mechanical properties of steel fiber reinforced high strength lightweight self-compacting concrete (SHLSCC). *Construction and Building Materials* **98**, 325–333.

Janotka, I., Krajci, L. and Dzivak, M. (2003) Properties and utilization of zeolite-blended Portland cements. *Clays & Clay Minerals* **51**(6), 616–264.

Javed, M.F., Sulong, N.R., Memon, S.A., Rehman, S.K.U. and Khan, N.B. (2018) Flexural behaviour of steel hollow sections filled with concrete that contains OPBC as coarse aggregate. *Journal of Constructional Steel Research* **148**, 287–294.

Kaffetzakis, M.I. and Papanicolaou, C.G. (2016) Bond behavior of reinforcement in lightweight aggregate self-compacting concrete. *Construction and Building Materials* **113**, 641+-652.

Kakali, G., Perraki, T.H., Tsivilis, S. and Badogiannis, E. (2001) Thermal treatment of kaolin: The effect of mineralogy on the pozzolanic activity. *Applied Clay Science* **20**(1–2), 73–80.

Kanadasan, J., Razak, H.A. (2014) *Fresh Properties of Self-Compacting Concrete Incorporating Palm Oil Clinker*. In: Hassan, R., Yusoff, M., Ismail, Z., Amin, N., Fadzil, M. (eds) InCIEC 2013. Springer, Singapore. https://doi.org/10.1007/978-981-4585-02-6_22.

Kanadasan, J. and Razak, H.A. (2014b) Mix design for self-compacting palm oil clinker concrete based on particle packing. *Materials & Design* (1980-2015) **56**, 9–19.

Kanadasan, J. and Razak, H.A. (2015) Engineering and sustainability performance of self-compacting palm oil mill incinerated waste concrete. *Journal of Cleaner Production*, **89**, 78–86.

Karahan, O., Hossain, K.M., Ozbay, E., Lachemi, M. and Sancak, E. (2012) Effect of metakaolin content on the properties self-consolidating lightweight concrete. *Construction and Building Materials* **31**, 320–325.

Karamloo, M., Mazloom, M. and Payganeh, G. (2016) Effects of maximum aggregate size on fracture behaviors of self-compacting lightweight concrete. *Construction and Building Materials* **123**, 508–515.

Khan, M.I., Usman, M., Rizwan, S.A. and Hanif, A. (2019) Self-consolidating lightweight concrete incorporating limestone powder and fly ash as supplementary cementing material. *Materials* **12**(18), 3050.

Khayat, K.H., Ghezal, A. and Hadriche, M.S. (1999) Factorial design model for proportioning self-consolidating concrete. *Materials and Structures* **32**(9), 679–686.

Khayat, K.H., Ghezal, A. and Hadriche, M.S. (2000) Utility of statistical models in proportioning self-consolidating concrete. *Materials and Structures* **33**(5), 338–344.

Kheder, G.R. and Al-Jadiri, R.S. (2010) New method for proportioning self-consolidating concrete based on compressive strength requirements. *ACI Materials Journal*, **107**, 490–497.

Kim, Y.J., Choi, Y.W. and Lachemi, M. (2010) Characteristics of self-consolidating concrete using two types of lightweight coarse aggregates. *Construction and Building Materials* **24**(1), 11–16.

Kotan, T. and Gul, R. (2010) Effect of atmospheric pressure steam curing to mechanical properties of lightweight concrete produced with Erzurum-Pasinler pumice. *Machines, Technologies, Materials - International Virtual Journal*, **4–5**, 66–69.

Kou, S.C. and Poon, C.S. (2009) Properties of self-compacting concrete prepared with recycled glass aggregate. *Cement and Concrete Composites* **31**(2), 107–113.

Kurt, M., Gul, M.S., Gul, R., Aydin, A.C. and Kotan, T. (2016) The effect of pumice powder on the self-compactability of pumice aggregate lightweight concrete. *Construction and Building Materials* **103**, 36–46.

Lee, K.T. and Ofori-Boateng, C. (2013) *Sustainability of Biofuel Production from Oil Palm Biomass*, first ed. Springer, SPS Penang, Malaysia.

Li, J., Chen, Y. and Wan, C. (2017) A mix-design method for lightweight aggregate self-compacting concrete based on packing and mortar film thickness theories. *Construction and Building Materials* **157**, 621–634.

Li, J., Zhao, E., Niu, J. and Wan, C. (2021) Study on mixture design method and mechanical properties of steel fiber reinforced self-compacting lightweight aggregate concrete. *Construction and Building Materials* **267**, 121019.

Li, J., Tan, D., Zhang, X., Wan, C. and Xue, G. (2021) Mixture design method of self-compacting lightweight aggregate concrete based on rheological property and strength of mortar. *Journal of Building Engineering* **43**, 102660.

Lo, T.Y., Tang, P.W.C., Cui, H.Z. and Nadeem, A. (2007) Comparison of workability and mechanical properties of self-compacting lightweight concrete and normal self-compacting concrete. *Materials Research Innovations* **11**(1), 45–50.

Lotfy, A., Hossain, K.M. and Lachemi, M. (2014) Application of statistical models in proportioning lightweight self-consolidating concrete with expanded clay aggregates. *Construction and Building Materials* **65**, 450–469.

Lv, J., Du, Q., Zhou, T., He, Z. and Li, K. (2019) Fresh and mechanical properties of self-compacting rubber lightweight aggregate concrete and corresponding mortar. *Advances in Materials Science and Engineering* **2019**, 1–14.

Lytag® Lightweight Solutions (2017) *Technical Manual – Section 1 Introduction to Lytag® Lightweight Aggregate.* London: Lytag Ltd. www.aggregate.com/sites/aiuk/files/documents/lytag_technical_data_sheet_-_intro_to_lytag_1.pdf.

Madandoust, R., Ranjbar, M.M. and Mousavi, S.Y. (2011) An investigation on the fresh properties of self-compacted lightweight concrete containing expanded polystyrene. *Construction and Building Materials* **25**(9), 3721–3731.

Mazaheripour, H., Ghanbarpour, S., Mirmoradi, S.H. and Hosseinpour, I. (2011) The effect of polypropylene fibers on the properties of fresh and hardened lightweight self-compacting concrete. *Construction and Building Materials* **25**(1), 351–358.

Mehta, P.K. and Monteiro, P.J.M. (2006) *Concrete: Microstructure, Properties and Materials: Third Edition.* New York, NY: McGraw-Hill.

Mo, K.H., Alengaram, U.J., Jumaat, M.Z. and Yap, S.P. (2015) Feasibility study of high-volume slag as cement replacement for sustainable structural lightweight oil palm shell concrete. *Journal of Cleaner Production* **91**, 297–304.

Nagaratnam, B.H., Rahman, M.E., Mirasa, A.K., Mannan, M.A. and Lame, S.O. (2016) Workability and heat of hydration of self-compacting concrete incorporating agro-industrial waste. *Journal of Cleaner Production* **112**, 882–894.

Nepomuceno, M.C.S. (2005) Methodology for the composition of self-compacting concrete. PhD thesis, University of Beira Interior, Covilhã, Portugal.

Neville, A.M. (2008) *Properties of Concrete. Fourteenth Edition.* Kuala Lumpur, Malaysia: CTP-VVP.

Neville, A.M. and Brooks, J.J. (2008) *Concrete Technology.* Singapore: Pearson Education Asia.

Okamura, H. (1997) Self-compacting high-performance concrete. *Concrete International* **19**(7), 50–54.

Okamura, H. and Ozawa, K. (1995) Mix design for self-compacting concrete. *Concrete Library of Japanese Society of Civil Engineers* **25**, 107–120.

Okamura, O. (2003) Self-compacting concrete. *Journal of Advanced Concrete Technology* **1**(1), 5–15.

Okpala, D.C. (1990) Palm kernel shell as a lightweight aggregate in concrete. Building and environment, **25**(4), 291–296.

Ozawa, K. (1989) High-performance concrete based on the durability design of concrete structures. In W. Kanock-Nukulchai et al. (eds) *Structural Engineering and Construction: Achievements, Trends and Challenges. Proceedings of the Second East Asia-Pacific Conference on Structural Engineering and Construction, Chiang Mai, Thailand, 11-13 January 1989*. Bankok, Thailand: EASEC-2 Secretariat, Division of Structural Engineering and Construction, Asian Institute of Technology.

Paschen, S. (1986) Diatomaceous earth extraction, processing and application. *Erzmetall* **39**, 158–161.

Polat, R., Demirboga, R. and Karakoc, M.B. (2010) The influence of lightweight aggregate on the physico-mechanical properties of concrete exposed to freeze thaw cycles. *Cold Regions Science and Technology* **60**(1), 51–56.

Poon, C.S., Lam, L., Kou, S.C. and Lin, Z.S. (1999) A study on the hydration rate of natural zeolite blended cement pastes. *Construction and Building Materials* **1B3**(8), 427–432.

Saak, A.W., Jennings, H.M. and Shah, S.P. (2001) New methodology for designing self-compacting concrete. *ACI Materials Journal* **98**(6), 429–439.

Sadek, M.M. (2020) Properties of self-consolidating concrete containing expanded slate lightweight aggregate. Doctoral dissertation, Memorial University of Newfoundland, St Johns, Canada.

Sadon, S.N., Beddu, S., Naganathan, S., Kamal, N.L.M. and Hassan, H. (2017) Coal bottom ash as sustainable material in concrete-A Review. *Indian Journal of Science and Technology* **10**(36), 1–10.

Sata, V., Jaturapitakkul, C. and Chaiyanunt, R. (2010) Compressive strength and heat evolution of concretes containing palm oil fuel ash. *Journal of Materials in Civil Engineering* **22**(10), 1033–1038.

SCC-EG Self-Compacting Concrete European Project Group). (2005) *The European Guidelines for Self-compacting Concrete: Specification, Production and Use.*

Shafigh, P., Jumaat, M.Z. and Mahmud, H. (2010) Mix design and mechanical properties of oil palm shell lightweight aggregate concrete: A review. *International Journal of the Physical Sciences* **5**(14), 2127–2134.

Shafigh, P., Jumaat, M.Z., Mahmud, H. and Alengaram, U.J. (2013) Oil palm shell lightweight concrete containing high volume ground granulated blast furnace slag. *Construction and Building Materials* **40**, 231–238.

Shafigh, P., Mahmud, H., Jumaat, M. Z., Ahmmad, R. and Bahri, S. (2014) Structural lightweight aggregate concrete using two types of waste from the palm oil industry as aggregate. *Journal of Cleaner Production* **80**, 187–196.

Shafigh, P., Jumaat, M.Z., Mahmud, H.B., and Alengaram, U.J. (2011) A new method of producing high strength oil palm shell lightweight concrete. Materials & Design, **32**(10), 4839–4843.

Shi, C. (2005) Design and application of self-compacting lightweight concrete. In Zhiwu Yu, Caijun Shi, Kamal Henri Khayat and Youjun Xie (eds) *1st International Symposium on Design, Performance and Use of Self-consolidating Concrete (SCC'2005-China)*. Champs-sur-Marne, France: RILEM Publications SARL, pp. 55–64.

Shi, C. and Wu, Y. (2005) Mixture proportioning and properties of self-consolidating lightweight concrete containing glass powder. *ACI Materials Journal* **102**(5), 355.

Shi, C., Wu, Z., Lv, K. and Wu, L. (2015) A review on mixture design methods for self-compacting concrete. *Construction and Building Materials* **84**, 387–398.

Siddique, R., Aggarwal, P. and Aggarwal, Y. (2012) Mechanical and durability properties of self-compacting concrete containing fly ash and bottom ash. *Journal of Sustainable Cement-based Materials* **1**(13), 67–82.

Silva, R.V., Brito, J. and Dhir, R.K. (2016) Establishing a relationship between modulus of elasticity and compressive strength of recycled aggregate concrete. *Journal of Cleaner Production* **112**, 2171–2186.

Soleymani, F. (2012) Pore structure and flexural strength of ZrO_2 nano-powders palm oil clinker aggregate-based binary blended concrete. *Journal of American Science* **8**(6), 187–194.

Sukumar, B., Nagamani, K. and Raghavan, R.S. (2008) Evaluation of strength at early ages of self-compacting concrete with high volume fly ash. *Construction and Building Materials* **22**(7), 1394–1401.

TCEB/FIP manual of design and technology (1977) Lightweight Aggregate Concrete. First Publication. The Construction Press Ltd., Lancaster, England

Topçu, İ.B. and Bilir, T. (2009) Experimental investigation of some fresh and hardened properties of rubberized self-compacting concrete. *Materials & Design* **30**(8), 3056–3065.

Topçu, I.B. and Uygunoğlu, T. (2010) Effect of aggregate type on properties of hardened self-consolidating lightweight concrete (SCLC). *Construction and Building Materials* **24**(7), 1286–1295.

Toraldo, E., Saponaro, S., Careghini, A. and Mariani, E. (2013) Use of stabilized bottom ash for bound layers of road pavements. *Journal of Environmental Management* **121**, 117–123.

Trumble, R. and Santizo, L. (1993) Advantages of using lightweight concrete in a medium rise building and adjoining post-tensioned parking garage. *ACI Special Publication* **136**, 247–254.

Uygunoğlu, T. and Topçu, İ B. (2009) Thermal expansion of self-consolidating normal and lightweight aggregate concrete at elevated temperature. *Construction and Building Materials* **23**(9), 3063–3069.

Wang, R., Ren, M., Gao, X. and Qin, L. (2018) Preparation and properties of fatty acids based thermal energy storage aggregate concrete. *Construction and Building Materials* **165**, 1–10.

Wang, X.H., Wang, K.J., Taylor, P. and Morcous, G. (2014) Assessing particle packing based self-consolidating concrete mix design method. *Construction and Building Materials* **70**, 939–952.

Wu, Z., Zhang, Y., Zheng, J. and Ding, Y. (2009) An experimental study on the workability of self-compacting lightweight concrete. *Construction and Building Materials* **23**(5), 2087–2092.

Yang, S., Yue, X., Liu, X. and Tong, Y. (2015) Properties of self-compacting lightweight concrete containing recycled plastic particles. *Construction and Building Materials* **84**, 444–453.

Yehia, S., Douba, A., Abdullahi, O. and Farrag, S. (2016) Mechanical and durability evaluation of fiber-reinforced self-compacting concrete. *Construction and Building Materials* **121**, 120–133.

Yu, Z., Tang, R., Cao, P., Huang, Q., Xie, X. and Shi, F. (2019) Multi-axial test and failure criterion analysis on self-compacting lightweight aggregate concrete. *Construction and Building Materials* **215**, 786–798.

Zakaria, M.L. (1986). Strength properties of oil palm clinker concrete. UTM *Jurnal. Teknologi* **8**(1), 28–37.

Zhao, H., Sun, W., Wu, X. and Gao, B. (2012) The effect of coarse aggregate gradation on the properties of self-compacting concrete. *Materials & Design* **40**, 109–116.

Chapter 6
Cellular concrete

Lina María Chica Osorio and Albert Leonard Alzate Ramírez

1. Introduction

The need for new infrastructure is increasing. Innovations in construction and infrastructure demand materials that are far stronger and lightweight. At the same time, environmental issues require people to save raw resources and to enhance the energy efficiency of new and rehabilitated buildings (Feneuil, 2019). A new construction material has to meet many requirements, such as low production cost, low environmental impact, high thermal insulation, high sound insulation, adequate mechanical strength, high durability, high fire resistance, and manufacturing/placement efficiency. Cementitious materials satisfy many of these requirements. In particular, cellular concrete is a promising material that offers the advantages of concrete while improving on its performance, containing as it does a larger number of air voids that help to save resources, reduce the material density and improve its thermal resistance (Feneuil, 2019).

Cellular concrete is a type of lightweight concrete. The addition of foam or gas generation within the cementitious mixture gives it a cell structure. The pore structure reduces the cellular concrete weight but may compromise its strength and durability, changing its practical applications (Wang *et al.*, 2019). The materials used to make lightweight concrete are the same as those used for normal concrete except that the chemical agents produce the air cells. Cellular concrete is relatively homogeneous compared to conventional concrete, as it does not contain coarse aggregate. The properties of cellular concrete depend on its microstructure and composition, methods of pore-formation and curing (Fouad, 2006). Density can be varied between 300 kg/m³ and 1800 kg/m³. The production of a stable cellular concrete mix depends on many factors: the selection of the foaming agent or gas forming reagents, the method of preparation, the material selection, the mixture design strategies and type of production. The main advantages of cellular concrete compared to conventional Portland cement concrete are weight reduction (up to 80%); excellent acoustic and thermal isolation; high resistance to fire; lower costs in raw materials, easier transporting and application; and finally, it does not need compacting, vibration or leveling (Chica and Alzate, 2019).

Foamed concrete is now widely used internationally, but growing pressures for more sustainable construction technologies necessitates the development of ultra-low density foamed concrete that current technology is unable to achieve (Jones *et al.*, 2016).

The purpose of this chapter is to provide an overview of the key aspects in the production and application of cellular concrete.

2. Fundamental terms

Binder agent: any material that holds other materials together, through a chemical reaction. Portland cement and geopolymers are binders.

Bleeding: In a suspension, when particles are settling and the water displaces and rises to the top.

Consistency: the degree to which a freshly mixed concrete, mortar, grout, or cement paste resists deformation.

Durability: the ability to last a long time without significant deterioration.

Foaming agent: a chemical reactive with surfactant properties that facilitates the generation of foam.

Geopolymer: an inorganic material that forms long-range, covalently bonded, non-crystalline network. It is a binding system that hardens at low temperature.

OPC: ordinary Portland cement.

Permeability: defined as a measure of the water flow under pressure in a saturated porous medium.

Plastic viscosity: defined as the proportional coefficient between shear stress and shear rate under a state of steady shear.

Pozzolan: siliceous or aluminosilicate materials which possess cementitious value. In water, pozzolans react chemically with calcium hydroxide at ordinary temperature to form binder compounds.

Shrinkage: defined as the volume changes of a hardened concrete mixture due to the loss of moisture.

Stability: the ability of foam to withstand spontaneous bubble collapse or breakdown from external causes.

Thermal conductivity: the rate at which heat passes through a material, expressed as the amount of heat that flows per unit time through a unit area with a temperature gradient.

Workability: a property of freshly mixed concrete which determines the ease and homogeneity with which it can be mixed, placed, consolidated and finished.

3. Definition and background

Cellular concrete is a lightweight concrete. It is created by a uniform distribution of air bubbles throughout the paste mix. In CT-13: ACI Concrete Terminology – An ACI Standard, the American Concrete Institute (ACI) defines cellular lightweight concrete as a low-density product consisting of Portland cement, cement–silica, cement–pozzolan, lime–pozzolan, or lime–silica pastes, or pastes containing blends of these ingredients and having a homogeneous void or cell structure, attained with gas-forming chemicals or foaming agents. Also, ACI in 523 defines foamed cellular lightweight concrete as "a mixture of cement, water and preformed foam". The purpose of the foam is to supply a production mechanism of a high ratio of air cells that, when they are mixed with cement, produce a porous solid (Gomez, 2015).

The history of cellular concrete begins with a series of developments related to autoclaved aerated concrete. The first attempts go back to Zernikov (1877), who 'boiled' calcium sand-based cement in high-temperature steam but achieved little stability and firmness. Wilhelm Michaelis followed a similar line but used a reduced water content. The result was a hard and waterproof calcium hydro silicate, for which he was granted a patent in 1880. This is the basis of all steam-cured building materials. In 1889 E. Hoffman successfully tested and patented the method of 'aerating' concrete. The aeration was produced by carbon dioxide generated in the reaction between hydrochloric acid and limestone. Americans Aylsworth and Dyer used aluminium powder and calcium hydroxide to attain a porous cementitious mixture for which they also received a patent in 1914. Later patents involved the reaction between zinc dust and the alkalis in the cement mixture, hydrogen peroxide, sodium or calcium hypochlorite and air foaming (E-VERDE Building Solutions, no date).

The first cement-based cellular concrete was patented in 1923 by Johan Axel Eriksson. The discovery was almost accidental. Eriksson, Assistant Professor for Building Techniques at the Royal Institute of Technology in Stockholm, was working on a variety of aerated concrete samples. Running short of time, he decided to speed up the curing process on a porous mass of burnt shale limestone, water and aluminum powder by placing the sample in the laboratory autoclave. The porous mass survived the overnight autoclaving and the resulting cured brick possessed greatly increased strength and a new, stronger crystalline composition. In the heat and pressure of the steam curing, the silica and lime components had become fused into a form of calcium silicate hydrate crystal, similar to the volcanic rock, known in nature as Tobermorite (E-VERDE Building Solutions, no date). This concrete is known as Ytong (Amran *et al.*, 2015).

In the early 1930s, the demand for the aerated concrete products led to the building of further plants. In 1934 Siporex was patented in Switzerland, manufactured through a vapor-curing process invented by Eklund. Foamed concretes have been used in the Soviet Union since 1938, where the manufacturing processes introduced by Kudriashoff were used even though their usage was restricted to non-structural elements. In 1950 aerated concrete using the coal slag from thermoelectric plants was introduced into the United Kingdom for load-bearing elements.

The first comprehensive review of cellular concrete was presented by Valore in 1954. A detailed treatment by Rudnai, Short and Kinniburgh, summarizing the composition, properties and uses of cellular concrete, appeared in 1963 (Abd Saloum *et al.*, 2015).

By 1970, cellular concrete was successfully applied as a cementing agent in oil wells and as a filler in excavation (Panesar, 2013). The first large-scale project involving cellular concrete was in 1980 in Scotland in the Falkirk railway tunnel, using approximately 4500 m^3 of concrete with a density of 1100 kg/m^3 (Nandi *et al.*, 2016). In the late 1950s foaming agents based on hydrolyzed proteins came onto the market. These improved the stability of the air cells to maintain an acceptable control of the density.

Around 1990 synthetic foaming agents were created, which gave highly stable air cells an extended life in the concrete plastic state, and gave the concretes increased durability (Gomez, 2015). In the last twenty years, the improvement in the production of superplasticizers and the creation of hybrid foaming agents (a mixture of protein and synthetic agents), has allowed the use of foamed concrete on a larger scale and extensive efforts have been made to study the characteristics and mechanical behavior with the idea of using cellular concrete in mainstream construction applications.

Another improvement of significance has been the development of equipment for foam production. Siporex, Ytong, Duros, Hebel and H+H are the most important companies worldwide that making cellular concrete. A timeline of cellular concrete development is shown in Figure 6.1. The desire to increase performance has led to a growing interest to understand the behavior of cellular concrete, focusing on a range of aspects such as the assessment of many kinds of additives, determination of the rheological properties and microstructure, evaluation of mechanical properties, numerical modeling, and the development of new applications based on density (Chica and Alzate, 2019).

4. Advantages and disadvantages of cellular concrete

(Sallal, 2018) listed the advantages of using cellular concrete in construction. A complete list of these advantages is as follows:

- Overall weight reduction due to the reduction in concrete density, which leads to lower loads, resulting in lower amounts of steel reinforcement.
- Seismic force reduction because the structure has less weight. Also, foam concrete forms a rigid, well-bonded solid after hydrating. It is effectively a free-standing (monolithic) structure and, once hardened, imposes lower lateral loads on any adjacent structures (Jalal et al., 2017).
- Damage by hurricanes, cyclones and floods is reduced compared to conventional concrete-based structures.
- Cost reduction in raw materials, transportation and workforce. Also, the capital investment for foam concrete is low.
- The application of foam concrete can save time over conventional ground treatment methods for settlement reduced construction. There is a lower waiting time for the consolidation of subsoils, eliminating the need for surcharging. The need for the removal and replacement of borrow soils is lower. It can reduce or eliminate the need for piling, sand drains, or grade beams. Increased construction speed. Also, it is easier to cut channels and holes for electrical wiring, sockets and pipes (Jalal et al., 2017). More rapid installation of services reduces installed costs.
- Foam concrete has less maintenance costs due to its durability.
- Improvement in thermal insulation and fire protection.
- Improved behavior in the 'fluid' state because it is naturally self-leveling and self-compacting. Foam concrete has a high flowing capability, which fills hollow spaces. Also, cellular concrete pumps use relatively low pressures via hoses over longer distances (Jalal et al., 2017).
- Improved freeze/thaw resistance because it has low water absorption.

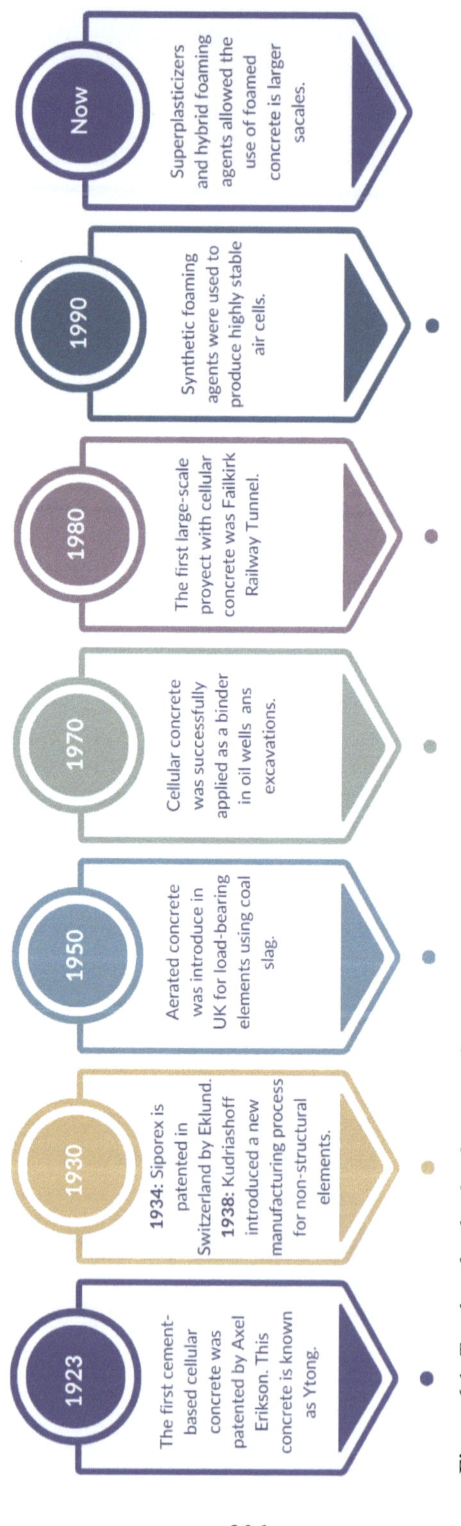

Figure 6.1 *Timeline for the development of cellular concrete.*

- Foam concrete can control air humidity in a room by absorbing and distributing moisture during the day and night. Also, in foam concrete houses more favorable average temperatures during the day are maintained (Jalal *et al.*, 2017).
- Eco-compatibility: Foam concrete does not produce toxic substances, as it is ecologically clean (Jalal *et al.*, 2017).

Concerning the disadvantages, foam concrete is very sensitive to water content in the mixture. Foam concrete can be difficult to place and finish because of its porosity, and the mixing time can be longer than for normal concrete to ensure proper mixing (Sallal, 2018).

5. Classification

The primary classification of cellular concretes is based on the mechanism through which bubbles are generated. Whichever method is used, the issue is how to trap air and distribute it uniformly through the mass of the cementitious mix. Using this criterion, cellular concrete can be divided into gas concretes and foamed concretes (Narayanan and Ramamurthy, 2000; Fouad, 2006). The difference between the two groups is defined by their porous system and the average size of the pores.

In gas concretes, gas-forming chemicals are mixed into lime or cement mortar during the liquid or plastic stage, resulting in an increased volume. When, later, the gas escapes, it leaves a porous structure. The gas-forming process takes place as a consequence of the chemical reactions. Aluminum powder, hydrogen peroxide/ bleaching powder and calcium carbide liberate hydrogen, oxygen and acetylene, respectively. Gas concrete is only manufactured at large plants and it is delivered to the consumer as pre-fabricated elements (Strommasina-Corp, no date).

Foam concrete is reported to be the most economical and controllable pore-forming process as there are no chemical reactions involved. The introduction of pores is achieved through mechanical means, either by pre-formed foaming (foaming agent mixed in as a part of the mixing water) or by mix foaming (foaming agent mixed with the mortar) (Narayanan and Ramamurthy, 2000). The most commonly used foaming agents are synthetic and protein based, but detergents, glue resins and saponins may also be used (Chica and Alzate, 2019). The technology of foam concrete manufacture allows production in small batches in the immediate vicinity of the construction site. So, foam concrete is produced by small enterprises and the production output is dozens of times as low as the gas concrete lock manufacturing plants (Strommasina-Corp, no date). With the in-mixing method, foaming agent is added, foaming agent is added to a mixer, which creates bubbles due to its high speed of rotation. The additive must trap air (encapsulate) and ensure it is distributed uniformly. This method is easy to perform, standardized and widely used. However, it can produce a large volume of imperfect bubbles, which compromises the amount of included air. The method of intensive mixing has many advantages: for example, it provides a homogenous mix, promotes accelerated hydration and the effective use of cement, and keeps together fine aggregate and agglomerated cement. The intensity of mixing depends upon the speed of the mixing elements.

The pre-formed foam method requires the use of compressed air to create the bubbles that are later going to be added to the mortar to create the cell structure. The foam must be very stable so that it does not dissolve. The pre-formed foam might be dry or wet. Dry foam is quite stable and generates bubbles with sizes below 1 mm, which facilitates uniform mixing and pumping. Wet foam produces bubbles between 2 mm and 5 mm but is less stable compared to dry foam. Pre-formed foam is an expensive operation compared to the mixing method, but it creates a more efficient foam of higher quality. The pre-formed foam method is the more extensively used method due to some comparative advantages with the in-mixing method: the low consumption of foaming agent and a direct relation between the amount of agent used and the included air content in the mixture (Chica and Alzate, 2019).

In both processes, the mixture presents a homogenous structure that contains small air bubbles that are independent of each other, and which in size, in already-cured elements, have a diameter of between 0.1 mm and 1.0 mm and are almost spherical in shape. The mixing technology affects the density, strength and geometry of the air cells (Namsone *et al.*, 2017).

A further way in which to classify cellular concrete depends on the curing process. Using this criterion, cellular concrete may be classified as either autoclave or non-autoclave. Autoclave cellular concrete is formed at high temperature, high pressure (12 atmospheres) and is steam cured (autoclave cured) (Strommasina-Corp, no date). Under these conditions the reaction is accelerated and product output is increased. Usually, gas concretes are cured in autoclaves and are called autoclaved aerated concrete (ACC). Non-autoclave cellular concretes are the usual pored cement and sand slurry cured at standard temperature and not processed.

In a similar manner, cellular concrete may be classified according to the production methods, i.e. pre-casting and on-site casting. Precast is used for walls panels, slabs and construction bricks. To produce high-quality elements with pre-casting fabrication, it is necessary to maintain a constant ambient temperature and curing system. The on-site casting process is used for structural and secondary elements. These cast-on-site elements can be cured with air and water spraying or vapor.

6. Materials used in cellular concrete manufacture

6.1 Binders

The most important binders in cellular concretes manufacture are Portland cement, calcium and sulphoaluminate cements. Geopolymers (alkali-activated aluminosilicate binders) are a recent innovation and are more sustainable than Portland cement in terms of reduced energy consumption and lower CO_2 emissions (Raj *et al.*, 2019). New developments under consideration include the replacement of some proportion of the cement, and assessing materials for their pozzolanic activity, which helps the hydration process. The materials that may be used to partly replace the cement are generally residues from other industries and include micro silica, blast furnace slag, fly ash, lime, sugarcane filter cake, and demolition and construction waste. The purpose of each of these materials can be different: it may be to improve consistency and long-term resistance, or simply to lower production costs (Chica and Alzate, 2019).

A residue from the manufacture of phosphoric acid, known as phosphogypsum, was used together with a small portion of Portland cement to produce cellular concretes. It was shown that only a portion of this material acts as a cementing material, stimulating the production of calcium silicate and ettringite and improving the compressive strength (Tian *et al.*, 2016). Cellular concrete was also manufactured using zeolites as the binder material (Poznyak and Melnyk, 2014). Waste-paper sludge ash can be used as an alternative binder and OPC replacement of between 5% to 30% in weight can change and improve the compressive strength (Sharipudin *et al.*, 2016).

The mineral mixtures used as either alternative binders or as a replacement for the Portland cement can be as follows:

- Blast furnace slag + fly ash + microsilica (23%, 15% and 12% of cement weight) improves strength and workability, which makes these suitable to be cast on site.
- Fly ashes + microsilica improves the interlinks paste-aggregate and allows enhancements in workability, compression strength and thermal isolation.
- Lime + microsilica improves the width and uniformity to the pores, which results in greater thermal isolation and better strength.
- Fly ash (80%) + blast furnace slag (20%) (with hydrogen peroxide as foaming agent) improves the production of cellular concretes of density 1270 kg/m^3

Alkali-activated slag cements have the characteristics of high viscosity, quick hydration, setting, hardening, and high strength, which makes them suitable for the preparation of foamed concrete (He *et al.*, 2019). Geopolymer systems made from alkali-activated fluid cracking catalyst residue aerated by recycled aluminum foil powders have been proposed (Font *et al.*, 2017). This material exhibits interesting properties, such as pore pressure distribution in the matrix, low natural density (600–700 kg/m^3), relatively high compressive strength (2.5–3.5 MPa), a low open/closed porosity ratio (1.15) and the lowest thermal conductivity. This opens up an interesting means of using alternative binders in foamed concrete fabrication.

6.2 Foaming agents and stabilizers

The properties of foaming agents have become a much more important concern for foamed concrete (Hou *et al.*, 2019). Foam is a major component of foamed concrete, which can be manufactured either by the use of expansive agents and/or foaming agents. Which foaming agent is selected has an impact on the properties of the foam as it affects the surface tension and gas–liquid interfacial properties. The nature of the agent also modifies the properties of the thin liquid film that separates the bubbles (Ranjani and Ramamurthy, 2010). Furthermore, the type of agent has a noticeable effect on the thermal resistance and sorptivity coefficient but less of an effect on mechanical properties. This is significant because the type of foaming agent used will directly influence the applications for which the cellular concrete can be used (Panesar, 2013). It is observed that foamed slurry collapse might frequently occur due to the deterioration of the prefabricated foams in cement pastes before hardening. This being so, more stable prefabricated foams without bubble growth and bleeding are required to be encapsulated and maintained in fresh and hardened cement pastes in the formation process of foamed concrete (Hou *et al.*, 2019). Aluminum powder, CaH_2, H_2O_2, TiH_2, and MgH_2 are the expansive agents most used in the fabrication of autoclaved aerated concrete.

The most commonly used foaming agents are synthetic and protein based, but detergents and glue resins may also be used. The action of a protein agent is by its own degradation, which generates the bubbles. This process reduces surface tension and creates interfaces for the air bubbles. Foam produces no reaction on the concrete but serves as a layer that traps the air. The effectiveness of the foam depends on temperature and pH. Agents based on protein agents allow close pores, which is reflected in more stronger structure, providing a stable web of empty spaces. Standard protein-based foaming agents are made with protein hydrolysate from animal proteins out of horn, blood, bones of cows, pigs and other remainders of animal carcasses. Vegetal protein, such as saponins, are also used in the fabrication of protein foaming agents (Simo Research Institute, no date).

Synthetic foaming agents reduce the surface tension of the dilution, which are amphiprotic substances and strongly hydrophilic. They create a complex chemical environment and therefore the compatibility of the surfactant and the cement is critical in allowing the desired entry of air and the development of the cell microstructure. Synthetic agents are easier to handle, cheaper and require less energy for long-term storage. Synthetic foaming agents lead to more stable foam concrete specimens than those obtained with protein for a fixed water/cement ratio (Falliano *et al.*, 2018). Juan He and colleagues used three foaming agents: sodium α-olefin sulfonate (AOS), sodium dodecyl sulfate (K12) and sodium alcohol ether sulphate (AES), with the same foam stabilizer of silicone resin polyether emulsion FM-500 (MPS) to prepare foam alkali-activated slag cement (He *et al.*, 2019). Results demonstrated that the foam stability, physical and mechanical properties were best for the AOS foaming agent.

Protein-based foaming agents generate smaller isolated spherical air bubbles compared to air voids produced by synthetic foaming agents. Oxides and silica nanoparticles have also been used as coadjuvants in the formation of triphasic foams in which the boundaries of the bubbles are much stronger.

It is found that the presence of solid particles can increase the rigidity of the inter-faces and decrease the permeability of the foam. Micro- or nanosized solid particles have been used to increase the stability of prefabricated foams. Nanosized materials like nanosilica or nanotubes, for example, are introduced and enable self-cleaning, self-healing, and improved mechanical performance and durability of foamed con-crete (Krämer *et al.*, 2017).

High-volume particle-stabilized foams showed neither bubble growth nor bleed-ing over more than 4 days. Some studies have used nanoparticles to improve foam quality. Results shows that the addition of nanosilica in foaming agents improves the stability of the prefabricated foam due to the increased viscosity and smaller change in surface tension (Hou *et al.*, 2019). However, improvements for other properties of foamed concrete, such as thermal conductivity, water absorption and shrinkage, were not obvious when foaming agent with silica nanoparticles was used. Graphene nanoparticles demonstrated no improvement on foamed performance.

Many researchers have utilized different stabilizers to enhance the properties of foams and foamed concretes. Xanthan gum has been utilized as a foam stabilizer for the aggregate, the liquid film and to improve controllability with geopolymers. This stabilizing method significantly enhances the pore size distribution of foam concretes (Hajimohammadi *et al.*, 2018). While the majority of pores in the stabilized samples with xanthan gum are in the 0–0.005 mm range, the pore size in the control sample (without stabilization) is mainly in the range 0.005–0.2 mm and 0.4–1.47 mm regions. By increasing the concentration of xanthan gum, the pore size distribution becomes narrower. This is evident by noting the gradual shift in pore size to a smaller diameter (Hajimohammadi *et al.*, 2018). Another stabilizer commonly used is hydroxypropyl methylcellulose (She *et al.*, 2018).

6.3 Water

Many factors control water demand in cellular concrete: the composition of the binder materials, the type of filler and the required workability. The water/cement (w/c) ratio varies from 0.4 to 1.25, the latter value being the case when a superplasticizer is not used (Sallal, 2018). A low w/c ratio results in a stiff mix and the breakage of bubbles during the mixing. Also, it brings a higher proportion of small pores with a larger surface area, finally resulting in thinner pore walls and more connected pores. A higher w/c ratio is too weak to trap the bubbles and leads to segregation (Nambiar and Ramamurthy, 2006b).

An appropriate combination of superplasticizers and mineral mixtures can reduce the water consumption for certain fluid properties (Ramamurthy *et al.*, 2009). It is recommended that the water used for the foamed concrete be fresh and of drinking water quality. Organic elements such as oils, acids, alkalis and salts may have a negative effect on foaming agents, specifically on protein-based agents.

6.4 Aggregates and fillers

Cellular concretes are achieved by carefully selecting the most suitable raw material. For cellular concretes with non-structural applications, artificial aggregates are not used: instead, fillers are used that reduce the cement consumption without increasing its weight. For structural cellular concrete (densities greater than 1200 kg/m^3) many types of aggregate are used. The most usual aggregates are natural sands. The filler materials most employed are industrial residues with pozzolanic activity. The type of filler material used influences the various properties of the foamed concrete. As the filler materials become finer, they produce more uniform and narrower air-void distribution thereby improving the mechanical behavior.

Sand: For foamed concretes, river sand is preferred, which is washed and should have a minimum 20% fines. Dust in the sand increases the demand for water and cement, without adding to the mechanical properties; it also increases shrinkage. A small number of fines contributes towards strength. Crushed sand, due to its sharp edges, may destroy the foam mechanically (Thakrele, 2014). It was found that 0.60 mm sand exhibits the highest compressive and flexural strengths as well as flexural toughness compared to coarser sand gradations (Lim *et al.*, 2014).

Silica fume: In their early stages, foamed concretes with low density are very weak and require a longer period to harden completely. Incorporation of silica fume can significantly improve early stage compressive strengths. In addition, silica fume can reduce bleeding in the fresh state and improve consistency as well as cohesion between the aggregate and cement paste (Chung *et al.*, 2019).

Construction civil waste (CCW): A way to minimize the environmental impact and to achieve sustainability is to increase the use of construction civil waste (CCW) in the development of materials for civil engineering. This has not yet been widely applied in cellular concrete production. Using recycled aggregates becomes more attractive in cases where high mechanical resistance and low water absorption are not required. CCW is a highly heterogeneous material group, physically, compositionally, and there are within-source variabilities (Jones *et al.*, 2012). As a result, the uptake of such materials has been limited. However, some studies using recycled aggregates in foamed concrete fabrication found it to be a good partial or full replacement for natural sand (Jones *et al.*, 2012; Lermen *et al.*, 2019). Fine recycled concrete aggregate (<1.18 mm) is used as a sand replacement for between 5% to 30% in weight. The addition of this fine waste significantly increases the compressive strength (Sharipudin *et al.*, 2016). Other waste used is crushed brick powder, which may enhance the compressive strength especially at a 50% sand replacement ratio (Aliabdo, Abd-Elmoaty and Hassan, 2014).

Soil: In recent years, soil has been used as an eco-friendly building material (Cong and Bing, 2015). Soil-based foamed concrete is a novel lightweight construction material. The application of soil in foamed concrete as a sand replacement is another alternative to reduce density (Yang and Chen, 2016). For comparable dry densities, soil-based foamed concrete has better thermal insulation than sand-based foamed concrete. Although the foamed concrete is not used as a load-bearing material, it is necessary to improve the strength of this soil-based foamed concrete without degradation of the thermal insulation performance. Generally, fly ash (30–70 wt%), silica fume (up to 10 wt% by weight of cement) and fibers are added to the mixture to customize the mechanical properties of foamed concrete (Cong and Bing, 2015). Up to 20% of laterite soils can be used in foamed concrete production to improve workability (Falade and Ukponu, 2013). Using soil as a filler can produce foamed concrete with a density of 800 kg/m^3, a compressive strength of 7.5 MPa, a thermal conductivity of 0.16 W/m.k and it can be used as a water-resistant lightweight concrete (Yang and Chen, 2016), (Cong and Bing, 2015).

Recycled tyre crumbs: The use of recycled tyre crumb as a component of construction materials has emerged as a potentially sustainable solution to the environmental problem associated with final disposal. There are very limited studies focusing on the insulation properties of recycled tyres in lightweight cellular concrete. However, the introduction of tyre crumb into concrete mixtures poses various challenges, most notably the lower Young's modulus, stiffness and compressive strength relative to conventional aggregates. These lower values can be attributed to the poor adhesion between the rubber particles and the cement paste in the interfacial zone. A homogeneous distribution of tyre crumbs within the foamed concrete structure has been achieved (Kashani *et al.*, 2017). Both the sound and thermal insulation are improved as the rubber content is increased, compared to a sample of similar density but without tyres. There is also a significant reduction in the rate of water permeability compared to a sample without rubber but having the same total porosity. However, mechanical properties are reduced due to the rubber.

Expanded polystyrene (EPS): EPS is another alternative for replacing natural aggregates with lightweight aggregates in concretes, despite its extremely low density and hydrophobic nature. EPS beads tend to float, and this may cause segregation and poor distribution in the matrix. However, EPS used as aggregate in foamed concrete results in a super lightweight material, whose bulk density is less than 500 kg/m^3 (Wu, Chen and Liu, 2013). For an EPS contribution of between 45% and 82% by volume, a significant improvement is observed in terms of resistance to chemical attack and corrosion, but thermal conductivity and fire resistance are lower (Sayadi *et al.*, 2016). For EPS foamed concrete, the compressive strength ranges from 0.7 MPa to 2.5 MPa (Wu, Chen and Liu, 2013). Furthermore, a low elastic modulus and a high residual to ultimate strength ratio ensures EPS's excellent deformation and energy absorption capacity (Wu, Chen and Liu, 2013).

EPS foamed concrete has the potential to be used for non-structural elements (cladding panels, curtain walls, composite flooring systems), protective layers (good energy absorption) and insulated concrete.

Plastic waste: Increased production and consumption of polymeric material has resulted in an increasing waste output as it reaches the end of its life. There exists a substantial literature on the issue of recycling plastic waste. The addition of polymeric waste materials to the cementitious matrix has been investigated since the late 1990s. The addition of polymer waste to concrete in various forms (fibers or aggregates), has a number of beneficial effects on the resulting material properties: weight and density reductions, improved mechanical energy absorption, better toughness, enhanced ductility and impact resistance, plus an improvement in insulation capacity (Ruiz-Herrero *et al.*, 2016). Alongside the above improvements, there may be – depending on the type of additions – undesired effects, such as a reduction in the compression strength or durability of the material. The addition of polyethylene (PE) and polyvinyl chloride (PVC) to foamed concrete and mortar was studied for non-structural applications by Ruiz-Herrero *et al.* (2016). The resulting materials have low mechanical properties but can be easily enhanced by improving the matrix properties – in particular, and importantly, decreasing thermal conductivity. Recycled thermoplastic powder has been used, derived from polypropylene-based molded furniture (after a damaged or useful life) with a particle size finer than 1.18 mm and a specific gravity of 0.9 (Chandni and Anand, 2018). Based on density, this plastic filler foamed concrete yielded strengths suited either for insulation or load-bearing.

6.5 Fibers and reinforcements

Much work has been carried out to improve the cement matrix in foamed concrete by optimizing the formulations of cement paste, e.g. by the addition of fibers, epoxy resin, silicon fume, fly ash and other fillers (Hou *et al.*, 2019). Additions of fiber reinforcement enhance performance in flexural strength, ductility and on the toughness index. In reinforced concretes, soon after cracking, large numbers of fibers bridge the crack and provide resistance to crack opening (Rasheed and Prakash, 2018). Some of the fibers utilized are alkali-resistant fiber glass, kenaf, steel, palm fiber and polypropylene fibers. The use of fiber can change the typical behavior of a cellular concrete, as it introduces a ductile elastic–plastic region. The volumetric fraction of the fiber reinforcement varies between 0.25 and 0.4 of the mixture (Chica and Alzate, 2019).

Some experimental results for cellular concrete with fiber reinforcement are:

Glass fibers: dispersed glass fibers can reinforce the structure in cellular concrete. However, reinforcement with fiber glass has been demonstrated as being effective in part due to the capability of the fibers to transmit the strength and thus prevent the progressive collapse of the cellular structure (Akthar and Evans, 2010). The compressive, splitting and flexural strengths of glass-fiber-reinforced foamed concrete show significant increases as the percentage of glass fibers is increased (Hamad, 2014)

Steel fibers: high-performance steel-fiber-reinforced cellular concrete (SFRCC) has recently been suggested as an attractive material in structural engineering because of its desirable engineering characteristics. However, the influence of the steel fibers on the mechanical properties of SFRCC are still very limited (Wang *et al.*, 2019). In general, the steel fiber shows a great reinforcement effect on the mechanical behavior of SFRCC, mainly on the splitting tensile strength and ductility. Fiber characteristics (fiber shape, volume fraction and aspect ratio) play an important role in affecting the relationship between compressive strength and the splitting tensile strength of SFRCC.

Synthetic fibers: synthetic fibers can be used as microfiber or fibrillated. The addition of even a small amount of microfiber significantly improved the post-cracking behavior when compared to specimens with macrofibers. Polyolefin-based fibers are commonly used to reinforce cellular concrete. These kinds of fiber prevent the premature fracture of cellular concrete and lead to improved post-cracking stiffness and ductility. The addition of even a small amount of microfiber significantly improves the post-cracking behavior when compared to specimens with only macrofibers (Rasheed and Prakash, 2018). The use of polyvinyl alcohol fibers for ultra-lightweight cellular concrete reinforcement has also been reported as avoiding the fragile failure and augmenting tensile strength (Yan and Chen, 2016).

7. Mix proportion

Practically, there are no specific mix proportion methods that yield targeted properties in foamed concrete (Amran *et al.*, 2015). Often, a trial-and-error process is adopted to achieve foamed concrete with particular desired properties (Ramamurthy *et al.*, 2009). There is a limited standard specification for the method of proportioning of conventional and non-conventional foamed concrete (Kumar *et al.*, 2018). Some efforts have been made by deriving a few formulas for mix design. The main target is the amount of air that should be entrained into the mortar base mix (Kumar *et al.*, 2018).

Design aid ACI 523-1975 relates plastic density and compressive strength: by using the cement content and the w/c ratio, a given strength and density can be chosen. This standard is based on McCormick's work (Ramamurthy *et al.*, 2009). ASTM C 796-97 also, provides a method of calculating the foam volume required to make cement slurry with a known w/c ratio and target density. Kearsley and

Wainwright proposed equations to calculate the mix proportions based on cement and foam contents. Target density can be obtained by solving Equations (1) and (2) (Amran *et al.*, 2015):

$$p_m = x + x\left(\frac{w}{c}\right) + x\left(\frac{a}{c}\right) + x\left(\frac{s}{c}\right) + x\left(\frac{a}{c}\right)\left(\frac{w}{a}\right) + x\left(\frac{s}{c}\right)\left(\frac{w}{s}\right) + RD_f \times V_f \tag{1}$$

$$1000 = \left(\frac{x}{RD_c}\right) + x\left(\frac{w}{c}\right) + x\left(\frac{\frac{a}{c}}{RD_a}\right) + x\left(\frac{\frac{s}{c}}{RD_s}\right) + x\left(\frac{a}{c}\right)\left(\frac{w}{a}\right) + x\left(\frac{s}{c}\right)\left(\frac{w}{s}\right) + V_f \tag{2}$$

where

p_m = target casting density, kg/m^3

$\dfrac{s}{c}$ = sand/cement ratio

x = cement content, kg/m^3

$\dfrac{w}{a}$ = water/ash ratio

$\dfrac{w}{c}$ = water/cement ratio

$\dfrac{w}{s}$ = water/sand ratio

$\dfrac{a}{c}$ = ash/cement ratio

V_f = volume of foam

RD_f = relative density of foam

RD_a = relative density of ash

RD_c = relative density of cement

RD_s = relative density of sand

Kumar *et al.* (2018) have developed a novel design method for cellular concrete when foam density is fixed at 50 kg/m^3 and the weight of solid matter (lime + sand) is in the ratio of 1:1.35. The procedure includes:

Step 1: Total water content calculation by w/c design ratio
Step 2: A target plastic/green density X (kg/m^3) is calculated by adding total weight of solid matter and water for mix proportion.
Step 3: Volume of air V_a required is derivate from Equation (3):

$$V_a = C_W \times W_{TW} + C_C \times W_C + C_L \times W_L + C_S \times W_S \tag{3}$$

where

$$X = \text{targeted green density} \left[\frac{kg}{m^3}\right]$$

$$C_W = \frac{1}{X} \frac{1}{\text{water specific gravity} \times 1000}$$

$$W_{TW} = \text{Mass of water in batch} \, [kg]$$

$$C_C = \frac{1}{X} \frac{1}{\text{cement specific gravity} \times 1000}$$

$$W_C = \text{Cement content} \, [kg]$$

$$C_L = \frac{1}{X} \frac{1}{\text{limestone specific gravity} \times 1000}$$

$$W_L = \text{limestone content} \, [kg]$$

$$C_S = \frac{1}{X} \frac{1}{\text{sand specific gravity} \times 1000}$$

$$W_S = \text{sand content} \, [kg]$$

Step 4: The volume of foam V_f and weight of foam W_f required are calculated using Equations (4) and (5), as follows:

$$W_f = \rho_f V_f \tag{4}$$

$$V_f = V_a + \frac{W_f}{1000} \tag{5}$$

Step 5: Amount of water in mortar base mix is calculated as Total water content minus Weight of foam.

8. Properties

8.1 Fresh state

8.1.1 Stability

The characteristics of the foam before the lightweight structure sets and maintains its shape has a great impact on the properties of foamed concretes. The tendency of the foams to coalesce and collapse during the preparation process brings challenges in controlling the properties of cellular structures. Consequently, it is critical to improve the stability of fresh foams in order to produce high-quality cellular structures using a predictable and reliable approach (Hajimohammadi *et al.*, 2018).

The fundamental physical mechanisms causing foam instability are: coarsening caused by interbubble gas transport; gravitational drainage from the films; and coalescence of adjacent bubbles due to the rupture of interbubble lamellae. Drainage rate is often used to characterize the water retention ability of the foam. A stable mix in foamed concrete does not suffer any of the above phenomena.

Experimentally, the stability of cellular concrete is measured by observing the height reduction over a period of 24 hours of mixtures contained in polycarbonate cylinders 500 mm deep and 75 mm in diameter (coated with polyethylene film) (Jones *et al.*, 2016). Using this method, and by considering the internal forces affecting the bubbles and the surrounding paste fraction of foamed concrete, a mechanism to explain instability is available (Jones *et al.*, 2016). The underlying cause of instability is considered to be the buoyancy of the bubbles, which allows them to float out of a fresh mix and ultimately cause complete separation of the gas and solid phases. The degree of buoyancy is directly related to bubble size, and this becomes significantly larger at lower densities; larger bubbles are consequently much more buoyant and hence lower-density mixes are more prone to instability.

Minor concentrations of bubbling agent have a positive effect on the stability of foamed concrete (Kuzielová *et al.*, 2016). The type of foaming agent also has an influence on mix stability. For a lower density, a lower stability of foam will be expected. The density ratio (the ratio of the density in the fresh state to that in the hardened state) should be close to 1:1 to achieve stability (Ramamurthy *et al.*, 2009).

The use of Portland cement replacement materials such as calcium sulfoaluminate (enhances the foam stability), silica fume, and quicklime (degrades the foam stability) and nanosilica (attributes a longer lifetime and increases viscosity) has a great impact on foam stability (Hou *et al.*, 2019). Hydroxypropyl methylcellulose can slow the coalescence and disproportionation of the bubbles by adsorbing at the air bubble surface and increasing the viscosity of cell-wall paste, thus preventing gas transfer and hindering physical drainage between the gaseous and liquid phases (She *et al.*, 2018). The addition of superplasticizers also enhanced the stability by 43%.

8.1.2 Consistency

ACI 238.1R-1 defines consistency as the degree to which a freshly mixed concrete, mortar, grout or cement paste resists deformation. The consistency and stability of freshly mixed foamed concrete is essential in preventing a separation of bubbles and cement mortar as well as breaking the bubbles, which would affect the hardened properties. These properties are also related to the fresh-mix rheological properties (Lim *et al.*, 2014). Usually, a flow table test is used to determine the consistency of the freshly mixed mortar, as described in ASTM C 1437. A higher flow table spread value for the mixture indicates a higher fluidity of the mortar mix (Lim *et al.*, 2014). An inverted slump flow cone described by ASTM C 1611 is used to determine slump and optimal w/c ratio as well (Lim *et al.*, 2014). An equation developed to relate slump flow and spread (in mm) (Jones *et al.*, 2003) is given in Equation (6):

$$slump\ flow = 273.67e^{0.003\,spread} \tag{6}$$

Paste consistency must be well chosen: if it is too fluid, bubbles tend to rise and escape from the mix, which can result in segregation; if it is insufficiently fluid, bubbles tend to break or collapse (Panesar, 2013; Feneuil, 2019). Cellular concrete is designed to achieve a desired flowability and compactability, both of which are partly influenced by the water content and the amount of foam used. The addition of foam considerably reduces the mix consistency and increases foam volume and necessitates a higher water-to-solids ratio (Chandni and Anand, 2018).

8.1.3 Rheology

The ease which concrete is processed is an extremely important aspect of its successful use in building economical structures and gives it excellent long-term performance and durability (Yahia, Mantellato and Flatt, 2015). The rheological properties of fresh concrete are very important because of their effects on the quality of casting and the forming process, and properties of hardened concrete (Zhang *et al.*, 2019). The rheology of concrete also affects mixing, handling, transportation, pumping, consolidation, finishing and surface quality after hardening. Rheology is an effective tool for characterizing the workability, predicting flow behavior, stability, and even the compactability of cement paste, mortar and concrete (Jiao *et al.*, 2017). Foamed concrete is a very complex fluid, which not only contains particulate materials but also contains entrapped air. What is more, the rheological behavior of concrete evolves over time because of cement hydration, which is a complicated subject (Jiao *et al.*, 2017). The variation of shear stress with shear rate in foamed concrete is shown in Figure 6.2.

Initially, the shear stress increases gradually with time but there is no flow. When the stress reaches the static yield stress, the concrete begins to flow (Jiao *et al.*, 2017). Foamed concrete often shows non-Newtonian behavior such as shear thinning with yield stress. As a result, non-Newtonian constitutive equations such as provided by the Bingham, Power Law, and Herschel–Bulkley models are applied. The fundamental physical parameters for describing rheological properties in these models mainly include yield stress and plastic viscosity. A plot of apparent viscosity versus shear rate on a log–log scale gives a straight line for power-law fluids. However, a similar plot for Herschel–Bulkley fluids forms a curve on a log–log scale (Ahmed *et al.*, 2009).

Experimental results from Ahmed *et al.* (2009) indicate that cement foams at low shear rates show a good fit with the Herschel–Bulkley model. Also, for high-slump concrete it has been found that the flow is non-Newtonian and is best described using the Herschel–Bulkley model (Daoud, 2008). For normal-slump concrete, the Bingham model provides a good description of the flow (Daoud, 2008).

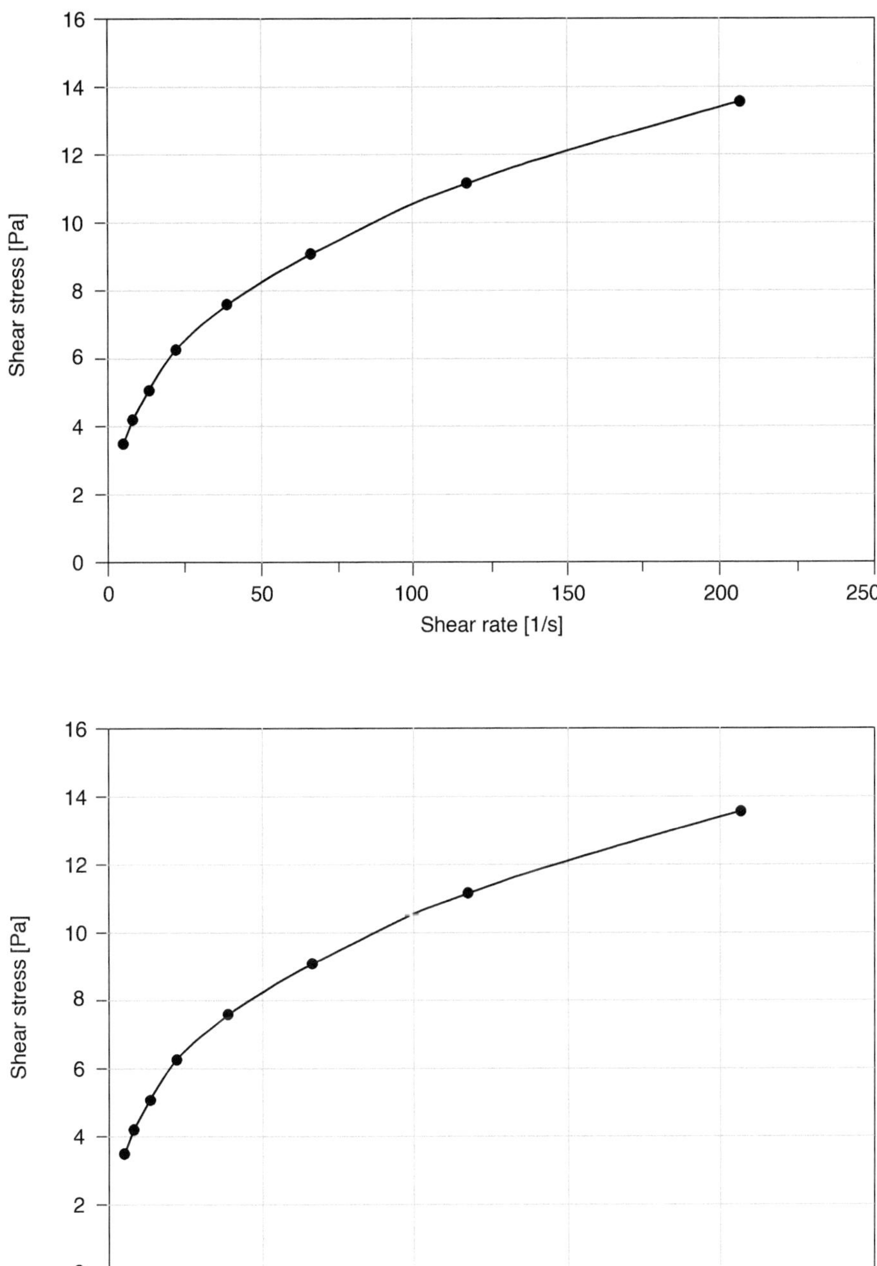

Figure 6.2 *Typical rheograms for foamed concrete.*

Yield stress and plastic viscosity vary in a complex way with the composition and this makes rheology measurement a versatile way of controlling the quality of fresh-concrete production (Banfill, 1991).

The plastic viscosity depends on the ratio of volume fraction to the packing density of particles in the mixture and is a good indicator for the compactability, machinability and segregation resistance. Its measure is important because it is a workability indicator.

Yield stress is caused by the force required to break down the network structure formed by the colloidal interaction and rigid links between cement particles, as well as the adhesion and friction between aggregate particles. The magnitude of the yield stress depends on cement proportioning and, specifically, on the effect of a superplasticizer. Yield stress can be used to characterize the filling ability and stability of fresh concrete (Jiao et al. 2017). For a short time the processes such as pumping and stirring, the yield stress are not relevant, but for longer term process such as those impacted by gravity or sedimentation, establishing the presence of a true yield stress can be important (Malvern Instruments 2012). It was found that when the cement paste yield stress value is high (above 10 Pa), cement foams evolve significantly before setting, leading to uncontrolled final morphology (Feneuil *et al.*, 2019). For lower reference cement paste yield stress values, remarkable morphological control can be achieved. All the reference cement paste yield stress values in this study were below 100 Pa. In practice, stable cement foams can be produced when cement paste yield stress is low (Feneuil *et al.*, 2019).

8.1.4 Workability

The term 'workability' is broadly defined and no single test method measures all aspects of workability (ACI Committee 238, 2008). The American Concrete Institute (ACI) Standard 116R-90 (ACI 1990b) defines workability as: 'that property of freshly mixed concrete which determines the ease and homogeneity with which it can be mixed, placed, consolidated, and finished' (Wong *et al.*, 2001) (ACI Committee 238, 2008). Neville (1996) succinctly defines workability as: 'the amount of useful internal work necessary to produce full compaction'.

There are many methods used to measure workability. Tattersall (1991) divided these workability test methods into three classes (ACI Committee 238, 2008):

- Class I Qualitative: Tests to be used only to describe flow qualitatively
- Class II Quantitative empirical: Tests to be used as a simple quantitative statement of behavior in a scenario: slump and flow table spread, for example. Single-point tests belong to this category.
- Class III Quantitative fundamental (rheology): Tests to use in conformance with standard definitions: viscosity, yield stress, fluidity. Multi-point tests fall into this class.

Workability is affected by every component of the concrete and essentially by every condition under which concrete is made. A list of factors includes:

properties and amount of cement; grading, shape, angularity and surface texture of aggregates; proportion of aggregates; amount of air entrained; types and amounts of chemical admixtures; temperature of the concrete; mixing time and method; and time since water and cement made contact. These factors interact, so that changing the proportion of one component to produce a specific characteristic requires that other factors be adjusted to maintain workability (Wong *et al.*, 2001). Workability depends not just on the properties of the concrete but also on the nature of the application. For example, a very dry concrete mixture may seem to have very low workability when it is, in fact, appropriate for the given application (ACI Committee 238, 2008).

The benefits of entrained air or bubbles on workability include: an increase in the volume of the paste; a fluidification of the cement paste; and an improvement in concrete rheology (Aïtcin, 2015). Air entrainment typically improves the consistency of the concrete while reducing bleeding and segregation. In foamed concretes, smaller and more numerous bubbles can give full play to the ball-and-lubrication effect, then improve the rheological properties to a high degree. Bubbles in the range 10–600 μm have a larger correlation with concrete slump and yield stress (Zhang *et al.*, 2019). Water-reducing admixtures disperse cement particles and improve workability, increasing the consistency and reducing segregation. A higher w/c ratio would result in a lower relative viscosity and a weaker bubble-maintaining capacity in the cement paste (Liu *et al.*, 2016).

A summary (from the literature) of the properties of cellular concrete in the plastic state is presented in the Table 6.1.

8.2 Hardened state

8.2.1 Physical properties: Density, porosity, drying shrinkage and permeability

The key property of cellular concrete is its low density, usually between 300 kg/m^3 and 1800 kg/m^3. Thus, a wide range of applications is available. In the hardened state, properties of concrete such as density, permeability, shrinkage and thermoacoustic isolation are intimately related to porosity and pore size distribution (Narayanan and Ramamurthy, 2000b). The porosity of the foamed concrete is totally dependent on the pore structure of the matrix and it has been established that the value varies between 28% to 90%.

The pore structure of foamed concrete consists of gel pores, capillary pores, and air-voids (air entrained and entrapped pores). The morphology of the hardened cement depends on the bubble size distribution and on the evolution of the bubbles during mixing until cement hardening (Feneuil, 2019). The difference in pore structure is largely due to the foam type and foam stability. Different levels of foam stability in the hardened state can be defined according to Feneuil (2019): fully stable; largely stable; small stable area(s); unstable; and collapse. Figure 6.3 provides an example illustrating the different levels of stability. The better the foam stability, the smaller the pore and narrower the pore size distribution will be formed (He *et al.*, 2019),

Table 6.1 *Properties for foamed concrete in the fresh state.*

Author	Foaming agent/filler	Density (kg/m³)	Air content (%)	Workability/Slump (mm)	Spreadability (% or mm)	Rheological properties	
Panesar (2013)	OPC/Protein/sand	1648–2362	32–6	230–15			
	OPC/Synthetic/sand	1581–2188	35–12.5	225–1			
Kunhanandan et al. (2008)	OPC/organic foaming agent/sand	945–1731			120–70% (by foam volume)		
	Class F fly ash/ organic/sand	814–1499			50–10% (by foam volume)		
Liu et al. (2016)	OPC/vegetable protein foaming agent	400–800				W/C	η [Pa.s]
						0.40	0.4075
						0.45	0.2737
						0.50	0.0594
						0.55	0.0255
						0.60	0.0159
Lim et al. (2014)	OPC/synthetic foaming agent/sand Sand < 2.36 0.46 < w/c < 0.54			413–623	180–326 mm		
	Sand < 1.18 0.48 < w/c < 0.56	1300		398–587	174–267 mm		
	Sand < 0.90 0.50 < w/c 0.58			445–503	208–270 mm		
	Sand < 0.60 0.52 < w/c < 0.60			333–496	194–245 mm		

(Continued)

Table 6.1 *(Continued)*

Author	Foaming agent/filler	Density (kg/m³)	Air content (%)	Workability/Slump (mm)	Spreadability (% or mm)	Rheological properties
Nambiar and Ramamurthy (2006a)	OPC/synthetic/fly ash class F/sand	1000–1500			25–75%	
Hilal *et al.* (2015)	OPC/protein EABASSOC/fly ash class S/silica fume/sand	1300–1900			140–180 mm (conventional foam mix) 290–350 (mix with additives)	
Jones *et al.* (2003)	OPC/synthetic/fly ash class F/sand	1000 OPC OPC+fly ash		430–570 result not available	175–245 mm 155–250	
Harith (2018)	OPC/polyurethane/fly ash class F/sand	1650 Without fly ash 1640 With fly ash		200 230	452 478	

(Continued)

Table 6.1 *(Continued)*

Author	Foaming agent/filler	Density (kg/m³)	Air content (%)	Workability/Slump (mm)	Spreadability (% or mm)	Rheological properties
Ahmed *et al.* (2009)	Cement class G/ foaming agent B237/ plasticizer B165	Not reported				0.2-2 Pa.s Herschel–Bulkley model parameters (by foam quality) in pressure test $0 < \tau_y \text{[Pa]} < 8.30$ $0.59 < m < 1.02$ $0.18 < k \text{[Pa.s}^m\text{]} < 3.48$
Xie *et al.* (2018)	OPC/foaming agent FP-180/Lithium bentonite	300 w/c = 0.6 w/c = 0.4 600 w/c = 0.6 w/c = 0.4		200–180 125–150 250–210 150–160		$\tau = ky^n$

as shown in Figure 6.4. Fine and close pores result in a compact texture with high strength and low permeability (Yu *et al.*, 2010). The pore connectivity is related to the formation methods and the type of foaming agents used, and it changes considerably with the mix composition and method of curing (Chica and Alzate, 2019). Narayanan

Fully stable   Largely stable area   Small stable area   Unstable   Collapse

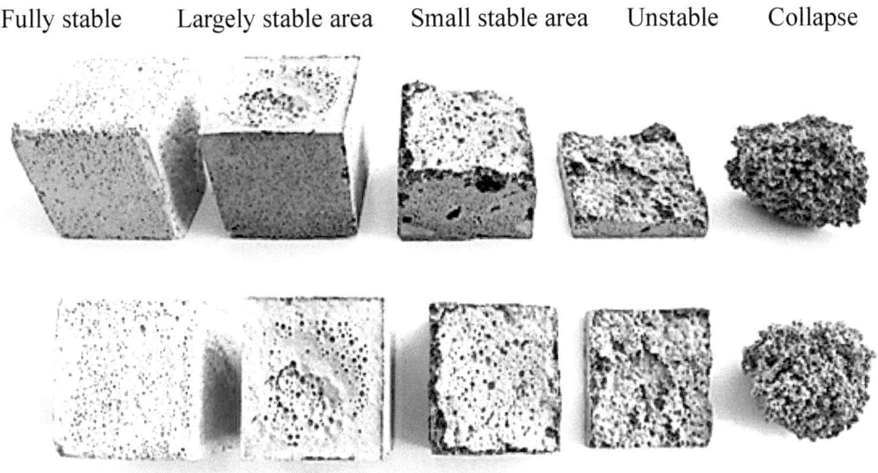

Figure 6.3 *Samples illustrating the different levels of foam stability.*

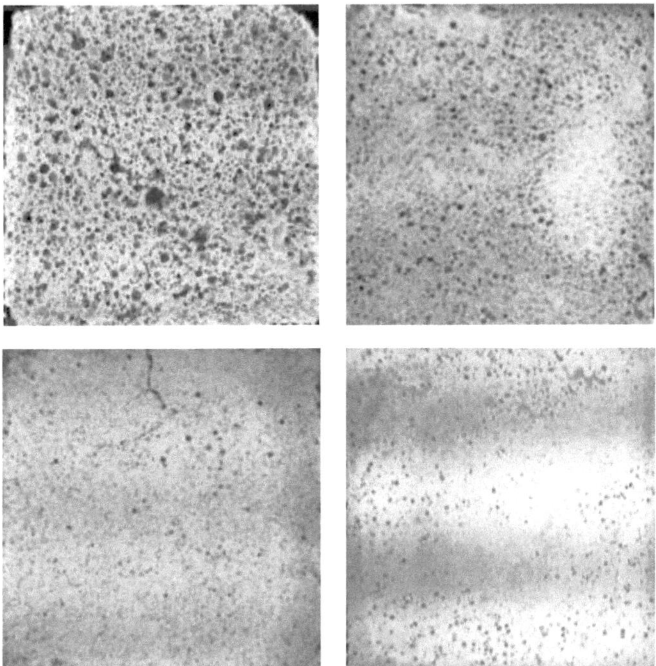

Figure 6.4 *Porosity structure due to changes in foam stability.*

inferred that a transition zone exists at the void–paste interface of aerated concrete analogous to the one in the aggregate–paste interface of normal concrete (Narayanan and Ramamurthy, 2000a). The larger pores in aerated concrete can be treated as aggregates of zero density and can be thought of as acting as inclusions in the matrix. The transition zone in aerated concrete is less porous than the one in normal concrete. This zone controls water absorption and permeability and, therefore, durability (Cui and Cahyadi, 2001).

In general, the water absorption of foamed concrete is almost twice that of the normal concrete at similar water-to-binder ratio (Amran *et al.*, 2015). The water absorption capacity of concrete depends on a combination of pore size, uniformity of pore distribution, the continuity of the pore system, and other miscellaneous factors (Shon *et al.*, 2018; Lermen *et al.*, 2019). Others have developed a relationship between porosity and dry density from experimental data (Kearsley, EP. Wainwright, 2002) and given by Equation (7):

$$p = 18700\gamma_d^{-0.85}$$

$$(7)$$

where p is porosity (%) and γ_d dry density (kg/m^3). Equation (7) shows the significant influence that porosity has on density.

Drying shrinkage is considered to be one of the drawbacks of foamed concrete and usually occurs during the first 20 days of casting time. The typical range for drying shrinkage of foamed concrete is between 0.1% and 0.35% of the total volume of the hardened concrete matrix, which is 4–10 times greater than for normal concrete (Amran *et al.*, 2015). Table 6.2. gives a summary with the physical and mechanical characterization by density intervals.

Table 6.2 *Cellular concrete properties with respect to density. Data adapted from Narayanan and Ramamurthy (2000b), Aini et al. (2017) and own results.*

Dry density (kg/m³)	Compressive strength (MPa)	Elastic modulus (GPa)	Thermal conductivity (W/m.K)	Volumetric contraction (%)	Porosity (%)
400	0.5–1.0	0.8–1.0	0.07–0.11	0.30–0.35	Close to 90%
500	1.0	1.24–1.84	0.08–0.13	–	
600	1.0–1.5	2.0–2.5	0.11–0.17	0.22–0.25	50–80%
800	1.5–2.0	2.0.–2.5	0.17–0.23	0.20–0.22	30–50%
1000	2.5–3.0	2.5–3.0	0.23–0.30	0.15–0.18	
1200	4.5–5.5	3.5–4.0	0.38–0.42	0.009–0.11	
1400	6.0–8.0	5.0–6.0	0.50–0.55	0.07–0.009	<30%
1600	7.5–10.0	10.0–12.0	0.62–0.66	0.006–0.07	

8.2.2 Mechanical properties: Compressive strength, modulus of elasticity, tensile and flexural strength

Concrete foams are characterized by their low density, which implies low resistance. The w/c and sand-to-cement ratios, curing regime, type of foaming agent used, foam production method and voids distribution are the key parameters affecting the mechanical properties of foamed concrete (Ramamurthy *et al.*, 2009). Nevertheless, it should be highlighted that mechanical behavior also depends on many factors: element form and size, load path and age (Khaw, 2010; Amran *et al.*, 2015).

The compressive strength f_c' is more influenced by porosity and pore size distribution than by density (Chung *et al.*, 2019). Kearsley determined that the compressive strength decreases exponentially with the density reduction (Kearsley 2002). The relationship between porosity and compressive strength is modeled by some researchers as in Equations (8)–(12):

$$f_c' = f_{c,0}(1-p)^n \qquad \text{Balshin} \qquad (8)$$

$$f_c' = f_{c,0}e^{-kp} \qquad \text{Ryshkevitch} \qquad (9)$$

$$f_c' = k_s \ln\left(\frac{p_0}{p}\right) \qquad \text{Schiller} \qquad (10)$$

$$f_c' = f_{c,0} - k_h p \qquad \text{Hasselmann} \qquad (11)$$

$$f_c' = 39.6(\ln t)^{1.174}(1-p')^{3.6} \qquad \text{Kearsley and Wainwright} \qquad (12)$$

where:

f_c' is the compressive strength of concrete with porosity p;
$f_{c,0}'$ the compressive strength at zero porosity;
p the porosity (volume of voids expressed as a fraction of the total concrete volume);
n is a coefficient, which need not be constant;
p_0 the porosity at zero strength;
k, k_s, k_h are empirical constants;
t the time since casting (days);
p' the mature porosity measured after 365 days.

Robler and Odler determined that the relations between compressive strength and porosity can best be expressed in the form of a linear plot (Kearsley, EP. Wainwright, 2002). The parameters from Robler and Odler are listed in Table 6.3.

With respect to the tensile and flexural strengths, they are not normally determined experimentally but are obtained from correlations from the compressive strength. The flexural strength of foamed concrete is lower than that of the equivalent normal-weight and lightweight aggregate concretes. The inclusion of fibers in foamed concrete transformed the basic material character from a brittle to a ductile elastoplastic material with enhanced flexural strength. The tensile strength f_t decreases as density decreases and is strongly related to the composition of the matrix (Feneuil, 2019). A typical expression to determine tensile strength is given by Equation (13):

$$f_t = k(f_c')^n \qquad (13)$$

Table 6.3 *Robler and Odler equations for the compressive strength. Adapted from Kearsley and Wainwright (2002).*

Equation	Roble and Odler fit
Balshin	$f_c = 540(1-p)^{14.47}$
Ryshkevitch	$f_c = 636e^{-17.04p}$
Schiller	$f_c = 81.5\ln\left(\dfrac{0.31}{p}\right)$
Hasselmann	$f_c = 158 - 601p$

where k and n are parameters obtained experimentally. Table 6.4 gives some model constants for Equation (13).

An equation (Equation (14)) of the same form as Equation (13) is used to model the elastic modulus, E:

$$E = k(f_c')^n \tag{14}$$

Table 6.5 gives some initial values for the elastic modulus.

The modulus of elasticity of foamed concrete varies between 1 kN/m² and 12 kN/m² for dry density between 500 kg/m³ and 1600 kg/m³, which was found to be 25% lower than for normal concrete (Ra *et al.*, 2019).

Feneuil (2019) found that there exist three methods for improving mechanical properties:

- Increasing the strength of the concrete matrix is the role of autoclaving during the fabrication. The use of reinforcing fibers and carbon nanotubes is an alternative.

Table 6.4 *Initial values for k and n for empirical modulus of elasticity approximation. Adapted from Amran et al. (2015).*

k	n	Reference
0.20	0.70	Oluokun (1991)
0.23	0.67	CEB-FIP MC90 (1993)
1.03	0.50	Byun *et al.* (1998)
0.23	0.67	CFM Model code (1990)

Table 6.5 *Initial values for k and n for tensile strength. Adapted from Amran et al. (2015).*

k	n	Reference
0.99	0.67	McCormick (1967). Fly ash
0.42	1.18	McCormick (1967). Sand
9.10	0.33	Rowe (1987)

- Using a lighter concrete matrix. This can be obtained, for example, by the use of lightweight aggregates and the replacement of sand with fly ash.
- Controlling of the bubble morphology. There is no consensus of opinion on the effect of bubble morphology, in the proposed strength–density model.

The assessment of all kinds of pozzolanic additives allows the improvement of the mechanical properties (Chica and Alzate, 2019). Also, there are other ways to increase the compressive strength: by using less foam, finer sand and air curing (Sallal, 2018).

8.2.3 Functional properties: Acoustic and thermal isolation, fire resistance, freezing and thawing resistance

Foamed concrete is one of the most commonly used inorganic insulation materials. However, compared to organic insulation material, foamed concrete has higher thermal conductivity, which restricts its application (Li *et al.*, 2020). Concrete is inert and fireproof and does not easily conduct sound, which suggests it would be a good material for insulation (Averyanov, 2018). Cellular concrete has been shown to have good freeze-thaw resistance, fire resistance and sound absorption, and to have superior thermal insulating properties which improve with lower plastic densities (Averyanov, 2018).

Foamed concrete has relatively high acoustic absorption compared to normal concrete. The pore structure is responsible for this: the closed, disconnected pore network blocks sound propagation. With regard to thermal conductivity, density is the key factor. The number of pores and their arrangement are essential for thermal insulation. Smaller pores give better insulation (Averyanov, 2018). Thermal conductivity tends to increase with density (Chung et al., 2019). Some authors report a linear relation between density and thermal conductivity (Feneuil, 2019). Thermal conductivity values by density are shown in Table 6.2. In foamed concretes, lower thermal conductivity was achieved with carbon dioxide and sulfoaluminate cement (Li *et al.*, 2020). In geopolymers, it was found that the thermal conductivity varied between 0.2 W/m.K and 0.23 W/m.K. (Hajimohammadi *et al.*, 2018).

Foamed concrete, with its porous structure, gives a significant improvement in fire resistance, as at a high temperatures the heat transfer through porous materials is affected by radiation (radiation is an inverse function of the number of air–solid interfaces traversed) (Sayadi *et al.*, 2016). The fire resistance is associated with changes in the mechanical properties of foamed concrete when it is exposed to high temperatures (Tan *et al.*, 2017). Some fillers, such as EPS, produce a reduction in fire resistance. It has been reported that cellular concretes with densities of 950 kg/m^3 are able to withstand fire for up to 3.5 hours and that concretes with a density of 1200 kg/m^3 are able to withstand fires for 2 hours (Amran *et al.*, 2015).

Concrete faced with cycles of freezing and thawing is evaluated through the very severe ASTM C666 Standard Test Method (Aïtcin, 2015). Cellular concrete has good freeze–thaw (F–T) resistance compared to non-aerated concrete but a cellular concrete mixture with high porosity did not necessarily result in higher resistance to F–T. The size of the air voids affects the F–T resistance: it is very important to produce millions of evenly distributed, uniformly small sized air bubbles or cells smaller than 300 μm in order to reduce F–T damage (Shon *et al.*, 2018). When concrete is exposed to F–T cycles, its water absorption becomes very important because highly porous concrete containing many capillary pores and open-cell pores tends to absorb much more water than normal concrete. Eventually, free water in the larger pores of the concrete is easily frozen, making such concrete is vulnerable to frost damage (Shon *et al.*, 2018).

8.2.4 Durability

A durable material helps the environment by conserving resources and reducing waste (Namsone *et al.*, 2017). In accordance with the American Concrete Institute's definition, 'durability is the ability to last a long time without significant deterioration'. Durability depends upon the effect of external (environmental) factors, such as changes of temperature, water, humidity, and upon internal factors, such as shrinkage (Namsone *et al.*, 2017). Durability is to be considered an important problem, especially in conditions of wet and cold weather.

An accelerated chloride ingress test developed from Jones and McCarthy shows that the mechanical performance of foamed concrete is equivalent to conventional concrete, with enhanced corrosion resistance at lower density (Ramamurthy *et al.*, 2009). The main properties affecting durability are mechanical strength, water absorption, shrinkage and frost resistance. Strength and frost resistance have been discussed in previous sections.

High water resistance is a necessary condition for increased durability (Namsone *et al.*, 2017). Water absorption is related to porosity. The permeability coefficient of lightweight foamed concrete is proportional to the unit weight and inversely proportional to the pore ratio (Ramamurthy *et al.*, 2009). Aerated concretes, especially ultra-low density and autoclaved gas silicates, are characterized by high open porosity and increased water absorption, which reduce durability.

Shrinkage is a serious problem because it is the main source of cracking. There are three forms of shrinkage: autogenous shrinkage; drying shrinkage; and shrinkage due to carbonation of portlandite, which is long-term process and depends on permeability. Shrinkage causes a reduction in strength, increases heat conductivity and increases the risk of damage during F–T cycles.

Namsone *et al.* (2017) report two methods by which the durability properties of foamed concrete can be increased: modifying the mix by active micro/nano components; and by applying advantageous mixing techniques, such as turbulent mixing (Namsone *et al.*, 2017). Also, adding fibers increases the tensile strength, which reduces the risk of shrinkage and stabilizes the fresh mix. The effectiveness of the fibers depends upon modulus of elasticity, the tensile resistance and ultimate deformation. Glass fibers are a good option for improving the durability of cellular concretes.

9. Applications

The use to which cellular concrete can be put depends fundamentally on its strength, which is directly related to its density (see Table 6.6). By controlling the ratios of

Table 6.6 *Common uses of cellular concrete, according to its final density.*

Density (kg/m³)	Application	Compressive strength to 28 days (MPa)
300–600	• Acoustic and thermal isolation • Fire protection • Covering rooftops	0.48–3.1
700–800	• Leveling mortars • Landscaping work • Grouting • Pipelines • Prefabricated building blocks	1.0–4.0
900–1100	• Precast elements: partitions, balcony railings and fence walls	1.00–6.00
1200–1800	• Precast walls or in-situ cast walls • Screen floors and on floors for noise reduction, insulation and weight reduction. • Reinforced loading walls • Seismic energy absorption systems	2.8–16.0

cement, sand, water and foaming agent, a wide range of densities can be achieved, to suit different applications (Namsone *et al.*, 2017).

Today, lightweight cellular concrete is gaining popularity in many construction applications, such as reducing, minimizing dynamic forces, mitigating settlement, and absorbing the forces arising in subsurface structures due to earthquakes (Tiwari *et al.*, 2017).

Construction and precast elements are common uses for cellular concrete. Sallal (2018) reports the use of foamed concrete in 800 houses of brick and slab construction in the Maysan province of Iraq. The use of bricks made of foamed concrete may contribute to reducing the use of cement in construction by up to 40% (Sallal, 2018). Autoclaved aeriated concrete masonry walls are used to build structural walls with steel truss type reinforcement (Jasiński and Drobiec, 2016).

Foamed concrete has a good energy absorption capability and can be used as *seismic isolation material for tunnels*. The mechanical properties and associated seismic isolation effects of foamed concrete layers in a rock tunnel was studied in the Galongla tunnel project (Ma *et al.*, 2019). The foamed concrete is strain-rate dependent and highly influenced by its density, and also has a high volumetric compressibility. The seismic isolation effect is mainly determined by the tunnel layer properties. Isolation effects would be better at lower densities, greater thickness and a smaller residual friction coefficient (Ma *et al.*, 2019).

For other *geotechnical applications*, Tiwari studied total and effective shear strength parameters, consolidation characteristics, hydraulic conductivity and Poisson ratio values. Using this information, recommendations were made for the use of foamed concrete as backfill for mechanically stabilized earth walls (Tiwari *et al.*, 2017). In excavations of poor soil, foamed concrete forms 100% of stable lightweight foundations and can be designed using the principle of equilibrium (Jalal *et al.*, 2017).

For *road construction*, due to the dense cell structure of foamed concrete, as the material is compressed during an impact the resistance of the foamed concrete increases, absorbing kinetic energy, which is especially useful in the design of aircraft arrestor systems at airports. In the case of soft compressible soils, foamed concrete can be used to minimize settlement (Jalal *et al.*, 2017). To reduce the subgrade weight and additional stress, the use of lightweight foamed concrete as a subgrade filler in roads is proposed (Huang *et al.*, 2017). Foamed concrete with a target density of 500–800 kg/m^3 can meet the requirements of both the static and dynamic conditions of ballast-less track subgrade. Lightweight foamed concrete with a target density of 650 kg/m^3 can be used to cast the bottom layer of a subgrade bed after considering safety factors (Huang *et al.*, 2017).

A stable, long-lasting solution can be obtained as a result of the excellent freeze/thaw resistance and low-density properties of foamed concrete (Jalal *et al.*, 2017).

Foams are being used in a *number of petroleum industry applications* that exploit their high viscosity and low density. Foamed cement slurries can have superior displacement properties relative to non-foamed cement slurries (Ahmed *et al.*, 2009). Foamed cement slurries are considered to have better drilling-fluid displacement properties than non-foamed cement slurries. This is because foamed cement is expected to have a higher shear viscosity than the base slurry. Foamed cements lower the overall slurry fluid loss and thereby help to control gas and fluid influx into the setting cement.

10. Future trends

Increased strength-to-weight ratio is the main goal. Although cellular concrete has been extensively investigated since the 1950s, some aspects need more intensive research (Chica and Alzate, 2019). These include:

- Applications involving extreme loading, extreme temperatures and aggressive environments
- Stability
- Durability
- Elastic modulus, Poisson ratio, and creep
- Rheological behavior and modeling

Also, to achieve extensive industrial use, it will be necessary to develop advanced foaming agents, chemical admixtures and reinforcements for ultra-high performance.

References

Abd Saloum, Q., Zaid Abdullah, M. and Adnan Hashim, A. (2015) The preparation of foam cement and determining some of its properties. *Engineering and Technology Journal* 33(1), 61–69.

ACI Committee 238 (2008) *Report on Measurements of Workability and Rheology of Fresh Concrete.*

Ahmed, R.M. *et al.* (2009) Rheology of foamed cement. *Cement and Concrete Research*, **39**, 353–361. doi: 10.1016/j.cemconres.2008.12.004.

Aïtcin, P.C. (2015) Entrained air in concrete: Rheology and freezing resistance, *Science and Technology of Concrete Admixtures*. Elsevier Ltd. Available at: https://doi.org/10.1016/B978-0-08-100693-1.00006-0.

Akthar, F.K. and Evan, J.R.G. (2010) High porosity (>90%) cementitious foams, *Cement and Concrete Research* 40(2), 352–358. https://doi.org/10.1016/j.cemconres.2009.10.012.

Aliabdo, A.A., Abd-Elmoaty, M. and Hassan, H.H. (2014) Utilization of crushed clay brick in cellular concrete production. *Alexandria Engineering Journal* **53**, 119–130. http://dx.doi.org/10.1016/j.aej.2013.11.005.

Amran, Y.H.M., Farzadnia, N. and Abang Ali, A. (2015) Properties and applications of foamed concrete: A review. *Construction and Building Materials* **101**, 990–1005. doi: 10.1016/j.conbuildmat.2015.10.112.

Averyanov, S. (2018) *Analysis of construction experience of using lightweight cellular concrete as a subbase material by A thesis presented to the University Of Waterloo in fulfilment of the thesis requirement for the degree of Master of Applied Science in.* University Of Waterloo.

Banfill, P.F.G. (1991) *Rheology of Fresh Cement and Concrete*, Rheology Review. Available at: https://doi.org/10.4324/9780203473290.

Beltrán, J.M. and Chica, L. (2023) 'On fresh state behavior of foamed cement pastes and its influence on hardened performance', *Construction and Building Materials*, **368**, p. 130518. Available at: https://doi.org/10.1016/j.conbuildmat.2023.130518.

Chandni, T.J. and Anand, K.B. (2018) Utilization of recycled waste as filler in foam concrete. *Journal of Building Engineering* **19**, 154–160. doi: 10.1016/J.JOBE.2018.04.032.

Chica, L. and Alzate, A. (2019) Cellular concrete review: New trends for application in construction. *Construction and Building Materials* **200**, 637–647. doi: 10.1016/j.conbuildmat.2018.12.136.

Chung, S.-Y. *et al.* (2017) Pore characteristics and their effects on the material properties of foamed concrete evaluated using micro-CT images and numerical approaches. *Applied Sciences* **7**(6), 550. doi: 10.3390/app7060550.

Chung, S.Y. *et al.* (2019) Comparison of lightweight aggregate and foamed concrete with the same density level using image-based characterizations. *Construction and Building Materials* **211**, 988–999. doi: 10.1016/j.conbuildmat.2019.03.270.

Cong, M. and Bing, C. (2015) Properties of a foamed concrete with soil as filler. *Construction and Building Materials* **76**, 61–69. http://dx.doi.org/10.1016/j.conbuildmat.2014.11.066.

Cui, L. and Cahyadi, J.H. (2001) 'Permeability and pore structure of OPC paste', *Cement and Concrete Research* **31**(2), pp. 277–282. Available at: https://doi.org/10.1016/S0008-8846(00)00474-9.

Daoud, O. (2008) *Correlating Concrete Mix Design to Rheological Properties of Fresh Concrete.* An-Najah National University, Palestine.

Jiao, D., Shi, C., Yuan, Q., An, X., Yu, L. and Li, H. (2017) Effect of constituents on rheological properties of fresh concrete-A review, *Cement and Concrete Composites* **83**, 146–159, https://doi.org/10.1016/j.cemconcomp.2017.07.016.

E-VERDE Building Solutions (no date) *History of Autoclaved Aerated Concrete.* Available at: http://e-verde.com/history_aac.htm (Accessed: 23 September 2019).

Falliano, D. *et al.* (2018) Experimental investigation on the compressive strength of foamed concrete: Effect of curing conditions, cement type, foaming agent and dry density. *Construction and Building Materials* **165**, 735–749. doi: 10.1016/J.CONBUILDMAT.2017.12.241.

Feneuil, B. (2019) Cement foam stability: Link with cement paste rheological properties. Doctoral thesis, Université Paris-Est, Créteil, France.

Feneuil, B., Roussel, N. and Pitois, O. (2019) Optimal cement paste yield stress for the production of stable cement foams. *Cement and Concrete Research* **120**(November 2018), 142–151. doi: 10.1016/j.cemconres.2019.03.002.

Font, A. *et al.* (2017) Geopolymer eco-cellular concrete (GECC) based on fluid catalytic cracking catalyst residue (FCC) with addition of recycled aluminium foil powder. *Journal of Cleaner Production* **168**, 1120–1131. doi: 10.1016/j.jclepro.2017.09.110.

Fouad, F.H. (2006) Cellular concrete. In J.F. Lamond and J. Pielert (eds) *Significance of Tests and Properties of Concrete and Concrete-making Materials.* West Conshohocken, PA: ASTM International, pp. 561–569.

Falade, F. and Ukponu, B. (2013). The Potential of Laterite as Fine Aggregate in Foamed Concrete Production. *Civil and Environmental Research,*Vol.3, No.10.

Gomez, M. (2015) An introduction to cellular concrete and advanced engineered foam technology.

Hajimohammadi, A., Ngo, T. and Mendis, P. (2018) Enhancing the strength of pre-made foams for foam concrete applications. *Cement and Concrete Composites* **87**, 164–171. doi: 10.1016/j.cemconcomp.2017.12.014.

Hamad, A.J. (2014) Materials, Production, Properties and Application of Aerated Lightweight Concrete: Review. *International Journal of Materials Science and Engineering* Vol. 2, No. 2. doi: 10.12720/ijmse.2.2.152-157.

Harith, I.K. (2018) Study on polyurethane foamed concrete for use in structural applications, *Case Studies in Construction Materials* **8**, 79–86. doi: 10.1016/J.CSCM.2017.11.005.

He, J., Gao, Q, Song, X., Bu, X. and He, J. (2019) Effect of foaming agent on physical and mechanical properties of alkali-activated slag foamed concrete. *Construction and Building Materials* **226**, 280–287. doi: 10.1016/j.conbuildmat.2019.07.302.

Hilal, A.A., Thom, N.H. and R. Dawson, A. (2015) The use of additives to enhance properties of pre-formed foamed concrete. *International Journal of Engineering and Technology* 7(4), 286–293. doi: 10.7763/IJET.2015.V7.806.

Hou, L. *et al.* (2019) Effect of nanoparticles on foaming agent and the foamed concrete. *Construction and Building Materials* **227,** Article 116698. doi: 10.1016/j.conbuildmat.2019.116698.

Huang, J.-j. *et al.* (2017) Experimental study on use of lightweight foam concrete as subgrade bed filler of ballastless track. *Construction and Building Materials* **149**, 911–920. doi: 10.1016/j.conbuildmat.2017.04.122.

Jalal, M. *et al.* (2017) Foam concrete. *International Journal of Civil Engineering Research* 8(1), 1–14. Available at: http://www.ripublication.com (Accessed: 19 February 2018).

Jasiński, R. and Drobiec, Ł. (2016) Study of autoclaved aerated concrete masonry walls with horizontal reinforcement under compression and shear. *Procedia Engineering* **161**, 918–924. doi: 10.1016/j.proeng.2016.08.758.

Jones, M.R., Mccarthy, M.J. and Mccarthy, A. (2003) Moving fly ash utilisation in concrete forward : A UK perspective. In T. Robl and T. Adams (eds) *2003 WOCA Peoceedings Papers.* Lexington, KY: University of Kentucky Press, pp. 1–24. (International Ash Utilization Symposium, Center for Applied Energy Research, University of Kentucky, Paper #113, January.)

Jones, M.R., Ozlutas, K. and Zheng, L. (2016) Stability and instability of foamed concrete. *Magazine of Concrete Research* **68**(11), 542–549. doi: 10.1680/macr.15.00097.

Jones, Zheng, Yerramala and Rao (2012) Use of recycled and secondary aggregates in foamed concretes. *Magazine of Concrete Research* **64**(6), 513–525. https://doi.org/10.1680/macr.11.00026.

Kashani, A. *et al.* (2017) A sustainable application of recycled tyre crumbs as insulator in lightweight cellular concrete. *Journal of Cleaner Production* **149**, 925–935. doi: 10.1016/j.jclepro.2017.02.154.

Kearsley, EP. and Wainwright, P. (2002) 'Effect of porosity on the strength of concrete', *Cement and Concrete Research,* **32**, pp. 233–239.

Khaw, Y.H. (2010) Performance of lightweight foamed concrete using laterite as sand replacement. Bachelor's thesis. Faculty of CIvil Engineering & Earth Resources, Universiti Malaysia Pahang. doi: 10.1017/CBO9781107415324.004.

Krämer, C. *et al.* (2017) Application of reinforced three-phase-foams in UHPC foam concrete. *Construction and Building Materials* **131**, 746–757. doi: 10.1016/J. CONBUILDMAT.2016.11.027.

Kumar, R., Lakhani, R. and Tomar, P. (2018) A simple novel mix design method and properties assessment of foamed concretes with limestone slurry waste. *Journal of Cleaner Production* **171**, 1650–1663. doi: 10.1016/j.jclepro.2017.10.073.

Kunhanandan Nambiar, E.K. and Ramamurthy, K. (2008) Fresh state characteristics of foam concrete. *Journal of Materials in Civil Engineering* **20**(2), 111–117. doi: 10.1061/ (ASCE)0899-1561(2008)20:2(111).

Kuzielová, E., Pach, L. and Palou, M. (2016) Effect of activated foaming agent on the foam concrete properties. *Construction and Building Materials* **125**, 998–1004. doi: 10.1016/j. conbuildmat.2016.08.122.

Lermen, R.T. *et al.* (2019) Effect of additives, cement type, and foam amount on the properties of foamed concrete developed with civil construction waste. *Applied Sciences* **9**(15), 2998. doi: 10.3390/app9152998.

Li, T. *et al.* (2020) Effect of foaming gas and cement type on the thermal conductivity of foamed concrete. *Construction and Building Materials* **231**. doi: 10.1016/j. conbuildmat.2019.117197.

Lim, S.K. *et al.* (2014) Strength and toughness of lightweight foamed concrete with different sand grading. *KSCE Journal of Civil Engineering* **19**(7), 2191–2197. doi: 10.1007/ s12205-014-0097-y.

Liu, Z. *et al.* (2016) Effect of water-cement ratio on pore structure and strength of foam concrete. *Advances in Materials Science and Engineering* **2016**(11), 1–9. doi: 10.1155/2016/9520294.

Ma, S., Chen, W. and Zhao, W. (2019) Mechanical properties and associated seismic isolation effects of foamed concrete layer in rock tunnel. *Journal of Rock Mechanics and Geotechnical Engineering* **11**(1), 159–171. doi: 10.1016/j.jrmge.2018.06.006.

Nambiar, E.K.K. and Ramamurthy, K. (2006a) Influence of filler type on the properties of foam concrete. *Cement and Concrete Composites* **28**(5), 475–480. doi: 10.1016/j. cemconcomp.2005.12.001.

Nambiar, E.K.K. and Ramamurthy, K. (2006b) Models relating mixture composition to the density and strength of foam concrete using response surface methodology. *Cement and Concrete Composites* **28**(9), 752–760. doi: 10.1016/j.cemconcomp.2006.06.001.

Nambiar, E.K.K. and Ramamurthy, K. (2007) Air-void characterisation of foam concrete. *Cement and Concrete Research* **37**(2), 221–230. doi: 10.1016/j.cemconres.2006.10.009.

Namsone, E., Šahmenko, G. and Korjakins, A. (2017) Durability properties of high- performance foamed concrete. *Procedia Engineering* **172**, 760–767. doi: 10.1016/j.proeng.2017.02.120.

Nandi, S. *et al.* (2016) Cellular concrete and its facets of application in civil engineering. *International Journal of Engineering Research* **5** (Special Issue 1), 2347–5013. doi: 10.17950/ijer/v5i1/009.

Narayanan, N. and Ramamurthy, K. (2000) 'Structure and properties of aerated concrete: A review. *Cement and Concrete Composites* **22**(5), 321–329. doi: 10.1016/S0958-9465(00)00016-0.

Narayanan, N. and Ramamurthy, K. (2000a) 'Microstructural investigations on aerated concrete', *Cement and Concrete Research* **30**(3), pp. 457–464. Available at: https://doi. org/10.1016/S0008-8846(00)00199-X.

Narayanan, N. and Ramamurthy, K. (2000b) 'Prediction models based on gel-pore parameters for compressive strength of aerated concrete', *Concrete Science and Engineering* **2**(12), pp. 206–212.

Narayanan, N. and Ramamurthy, K. (2000c) 'Structure and properties of aerated concrete: A review', *Cement and Concrete Composites,* **22**(5), pp. 321–329. Available at: https://doi. org/10.1016/S0958-9465(00)00016-0.

Neville, A.M. (1996) *Properties of Concrete*. 4th Edtion, Pearson Higher Education, Prentice Hall, Englewood Cliffs.

Panesar, D.K. (2013) Cellular concrete properties and the effect of synthetic and protein foaming agents. *Construction and Building Materials* **44**, 575–584. doi: 10.1016/j.conbuildmat.2013.03.024.

Poznyak, O. and Melnyk, A. (2014) Non-autoclaved aerated concrete made of modified binding composition containing supplementary cementitious materials. *Budownictwo i Architektura* **13**(2), 127–134. Available at: http://wbia.pollub.pl/files/85/content/files/1961_127-134.pdf (Accessed: 22 November 2017).

Raj, A., Sathyan, D. and Mini, K.M. (2019) Physical and functional characteristics of foam concrete: A review. *Construction and Building Materials* **221**, 787–799. doi: 10.1016/j.conbuildmat.2019.06.052.

Ramamurthy, K., Kunhanandan Nambiar, E.K. and Indu Siva Ranjani, G. (2009) A classification of studies on properties of foam concrete. *Cement and Concrete Composites* **31**(6), 388–396. doi: 10.1016/j.cemconcomp.2009.04.006.

Ranjani, I.S. and Ramamurthy, K. (2010) Relative assessment of density and stability of foam produced with four synthetic surfactants. *Materials and Structures/Materiaux et Constructions* **43**(10), 1317–1325. doi: 10.1617/s11527-010-9582-z.

Rasheed, M.A. and Prakash, S.S. (2018) Behavior of hybrid-synthetic fiber reinforced cellular lightweight concrete under uniaxial tension: Experimental and analytical studies. *Construction and Building Materials* **162**, 857–870. doi: 10.1016/j.conbuildmat.2017.12.095.

Ruiz-Herrero, J.L. *et al.* (2016) Mechanical and thermal performance of concrete and mortar cellular materials containing plastic waste. *Construction and Building Materials* **104**, 298–310. doi: 10.1016/j.conbuildmat.2015.12.005.

Sallal, A.K. (2018) Use foam concrete in construction works. *International Journal of Research in Advanced Engineering and Technology* **4**(2), 15–20.

Sayadi, A.A. *et al.* (2016) Effects of expanded polystyrene (EPS) particles on fire resistance, thermal conductivity and compressive strength of foamed concrete. *Construction and Building Materials* **112**, 716–724. doi: 10.1016/j.conbuildmat.2016.02.218.

Sharipudin, S.S. *et al.* (2016) Strength properties of lightweight foamed concrete incorporating waste paper sludge ash and recycled concrete aggregate. In N.A. Yacob, M. Mohamed and M.A.K.M. Hanafiah (eds) *Regional Conference on Science, Technology and Social Sciences (RCSTSS 2014)*. Singapore: Springer Nature, pp. 3–15. doi: 10.1007/978-981-10-0534-3_1.

She, W. *et al.* (2018) Application of organic- and nanoparticle-modified foams in foamed concrete: Reinforcement and stabilization mechanisms. *Cement and Concrete Research* **106**, 12–22. doi: 10.1016/j.cemconres.2018.01.020.

Shon, C.S. *et al.* (2018) Freezing and thawing resistance of cellular concrete containing binary and ternary cementitious mixtures. *Construction and Building Materials* **168**, 73–81. doi: 10.1016/j.conbuildmat.2018.02.117.

Simo Research Institute (no date) Types of foaming agents. SIMO Research Institute of Inorganic Chemistry. Available at: http://m.simo-chem.com/news/types-of-foaming-agents-14983507.html (Accessed: 28 October 2019).

Strommasina-Corp (no date) The most important facts about gas concrete - Strommashina. Available at: http://strommashina.com/about/the-most-important-facts-about-gas-concrete (Accessed: 7 October 2019).

Tan, X. *et al.* (2017) 'Influence of high temperature on the residual physical and mechanical properties of foamed concrete', *Construction and Building Materials* **135**, 203–211. Available at: https://doi.org/10.1016/j.conbuildmat.2016.12.223.

Tian, T. *et al.* (2016) Utilization of original phosphogypsum for the preparation of foam concrete. *Construction and Building Materials* **115**, 143–152. doi: 10.1016/j.conbuildmat.2016.04.028.

Tiwari, B. *et al.* (2017) Mechanical properties of lightweight cellular concrete for geotechnical applications. *Journal of Materials in Civil Engineering* **29**(7), 06017007. doi: 10.1061/(ASCE)MT.1943-5533.0001885.

Wang, X. hua *et al.* (2019) Effect of steel fibers on the compressive and splitting-tensile behaviors of cellular concrete with millimeter-size pores. *Construction and Building Materials* **221**, 60–73 doi: 10.1016/j.conbuildmat.2019.06.069.

Wong, G.S. *et al.* (2001) *Portland-cement concrete rheology and workability: Final Report. FHWA-RD-00-025.*

Xie, Y. *et al.* (2018) Effects of bentonite slurry on air-void structure and properties of foamed concrete. *Construction and Building Materials* **179**, 207–219. doi: 10.1016/j.conbuildmat.2018.05.226.

Yahia, A., Mantellato, S. and Flatt, R.J. (2016) 7 - Concrete rheology: A basis for understanding chemical admixtures,Editor(s): Pierre-Claude Aïtcin, Robert J Flatt, *Science and Technology of Concrete Admixtures*, Woodhead Publishing, Pages 97–127, ISBN 9780081006931, https://doi.org/10.1016/B978-0-08-100693-1.00007-2.

Yang, Y. and Chen, B. (2016) Potential use of soil in lightweight foamed concrete. *KSCE Journal of Civil Engineering* [Korean Society of Civil Engineers], **20**(6), 2420–2427. doi: 10.1007/s12205-016-0140-2.

Yu, X.G. *et al.* (2010) 'Pore Structure and Microstructure of Foam Concrete', Advanced Materials Research, **177**, pp. 530–532. Available at: https://doi.org/10.4028/www.scientific.net/AMR.177.530.

Zhang, X. *et al.* (2019) Effect of bubble feature parameters on rheological properties of fresh concrete. *Construction and Building Materials* **196**, 245–255. doi: 10.1016/j.conbuildmat.2018.11.088.

Zhen, W.U., Bing, C. and Ning, L. (2013) Fabrication and Compressive Properties of Expanded Polystyrene Foamed Concrete: Experimental Research and Modeling. *Journal of Shanghai Jiaotong University*, 61–69.

Index

G

H